LEGAL ASPECTS OF THE FIRE SERVICE

By Lawrence J. Hogan. A.B., M.A., J.D.

Edited by Ilona Modly Hogan, A.B., J.D.

About the Cover:

The Rescue, painted on canvas in 1855 by Sir John Everett Millais, hangs in the National Gallery of Victoria, Melbourne, Australia. It was inspired by the artist seeing firemen bravely fighting a fire. The artist said, "Soldiers and sailors have been praised on canvas a thousand times. My next picture shall be of the fireman." The writer, Charles Dickens, suggested to Millais that the following excerpt from Gay's poem, "Trivia," would be an appropriate quotation for his painting:

> *"The fire-man sweats beneath his crook'd arms,*
> *A leather casque his vent'rous head depends,*
> *Boldly he climbs where thickest smoak ascends;*
> *Mov'd by the mother's streaming eyes and pray'rs*
> *The helpless infant through the flames he bears,*
> *With no less virtue, than through hostile fire*
> *The darden hero [Aeneas] bore his aged sire."*

John Everett Millais, 1929 – 1896 (British)
The Rescue 1855
Oil on canvas, 121.5 x 83.6 cm
Felton Bequest 1924

Amlex, Inc.
P.O. Box 3495
Frederick, Maryland 21705-3495
Telephone: 301-694-8821
Fax: 301-694-0412
E-Mail: amlex@radix.net

Printed in the United States of America

THIRD EDITION

Library of Congress Catalogue No: 95-94446

Hogan, Lawrence J.
Legal Aspects of The Fire Service
    1. Laws affecting fire service — United States

Dedicated to my children,
Terry, Larry, Matt, Mike, Pat and Tim
and to the dedicated men and women of the fire service.

ACKNOWLEDGEMENTS

I wish to express my appreciation to all the men and women from my classes at the National Fire Academy and in our "Legal Aspects of the Fire Service Seminars" who urged me to write this book.

I also wish to acknowledge the outstanding help and encouragement I received from Ilona Modly Hogan, my wife/team teacher/partner/editor who shares my interest in the subject matter.

A special note of thanks is due to Amy Wokasien of Publications Concepts of Frederick, Maryland, who transformed my manuscript into the book you now hold in your hand. Her competence and intelligence were tremendous assets in completing the project. She was unswervingly pleasant and cooperative throughout a grueling, time-consuming ordeal for both of us.

LH

ABOUT THE AUTHOR

LAWRENCE J. HOGAN, an attorney who has been admitted to practice before the Supreme Court of the United States, has been developing and teaching courses for FEMA'S Emergency Management Institute and the National Fire Academy since 1982. He also teaches the legislative process to federal executives for the U.S. Office of Personnel Management's Public Affairs Institute.

For many years he has taught law courses for the University of Maryland and has lectured on legal subjects at Frostburg State University, Brandeis University and the Maryland Fire and Rescue Institute.

His articles on legal and business subjects have been published by numerous magazines and journals. He has delivered thousands of speeches and has conducted hundreds of training programs for state, local and federal government executives, business groups and candidates for elective office.

For six years he served as a member of the U.S. Congress where he was the leading architect of the District of Columbia Court Reorganization and Criminal Procedures Act, and served on the Subcommittee which revised the Federal Rules of Evidence. For four years he was the elected County Executive of Prince Georges County, Maryland's largest county.

Prior to his political career he served for ten years with the FBI.

Mr. Hogan received Bachelor of Arts (dual majors in history and philosophy) and Juris Doctorate degrees from Georgetown University. He received a Master of Arts degree in communications from the American University.

In addition to teaching legal modules in National Fire Academy courses, Larry Hogan and his wife, Ilona, conduct seminars throughout the country on "Legal Aspects of the Fire Service."

"The study of law should be introduced as part of a liberal education, to train and enrich the mind...I am convinced that, like history, economics, and metaphysics — and even to a greater degree than these — the law could be advantageously studied with a view to the general development of the mind."

Justice Louis D. Brandeis

AUTHOR'S NOTE

CAUTION: This book has been written for laymen, not for lawyers. It is not intended as an exhaustive legal treatise on any of the subjects covered. Its objective is to give fire service personnel a general familiarity with some legal matters affecting the fire service. We hope it will help them to avoid legal problems in making management decisions.

However, THE READER IS CAUTIONED NOT TO TAKE LEGAL ACTION RELYING ON THE INFORMATION HEREIN. The law differs from state to state and from time to time. Legislatures change and repeal statutes, court decisions are overturned, and regulations are revised by administrative agencies. Before taking ANY legal action, it is ALWAYS imperative that you consult a competent attorney who is familiar with the law, regulations and court decisions in your state as well as on the federal level.

Suggestions from readers are always welcome, especially for recommendations regarding material which should be included in subsequent editions of "Legal Aspects of the Fire Service" or in the legal seminars which the author conducts throughout the country.

CONTENTS

CHAPTER SEVEN 77

TORTS

CHAPTER ELEVEN   129

FAIR LABOR STANDARDS ACT

CHAPTER THIRTEEN   171

DRUG TESTING

CHAPTER FOURTEEN 181

AGE DISCRIMINATION

CHAPTER FIFTEEN 185

SEXUAL HARASSMENT

CHAPTER TWENTY   **271**

HEALTH AND SAFETY

AIDS AND THE FIRE SERVICE

CHAPTER TWENTY-TWO 291

WOMEN AND THE FIRE SERVICE

PREFACE

Legal problems are increasing, not only for fire service personnel, but for everyone in society. Our society is becoming more and more complex and people are learning more and more about the law and their rights. As a consequence, our very litigious society is getting more so.

There is also a continuing, steady trend toward finding government liable and limiting immunities.

The first jury award against a city for over $1 million came in 1962. In 1975 there were 20 per year. Now there are over 400 per year. In 1989 there were 18 million suits filed in the United States, triple the number thirty years ago.

Many Americans seem to view the courts as a lottery where you might, with luck, hit the jackpot. Many people feel that life can be made risk-free and when an accident happens, someone is to blame and should be made to pay.

A woman sued McDonald's Corp. because she was scalded when SHE spilled a cup of hot coffee between her legs. (Shouldn't she have assumed the coffee was hot?) An Albuquerque jury awarded her $2.9 million.

McDonald's lost another unbelievable case. A jury awarded $375,000 to a man who sued the fast-food chain because one of its employees, who had worked long hours, caused an auto crash in which the plaintiff was injured. The Oregon Court of Appeals affirmed the lower court's decision. Under traditional tort law the employer was only liable for the torts of an employee which were committed within the scope of the employee's employment. Does this mean that the fire department will be liable if a fire fighter, who has worked long hours on a major incident, injures someone in an automobile accident in his/her private car?

A woman was visiting a friend in the hospital. The chair she was sitting on broke and she hit her head on a shelf. All she required was medication and some physical therapy. She sued the hospital for $75,000 and the jury gave her $45,500.

A six year-old girl, playing on a neighbor's backyard swing, hit her hip. She sued Sears, manufacturer of the seat, on the grounds that it should have been made out of plastic or rubber instead of metal. Sears settled out of court for $535,000.

A woman who fell while boarding a bus broke her arm and hurt her shoulder. The jury gave her $270,000.

A woman, leaving a K-Mart store, noticed streams of water flowing across the sidewalk coming from the garden shop. In trying to step over the first stream, she fell, fracturing her knee cap. She sued for $150,000. The jury found her 25% liable and gave her only $75,000. K-Mart argued unsuccessfully that she could have avoided the water by walking around it and that she had padded her medical bills unconscionably.

A New York man, trying to commit suicide, jumped in front of a subway train. The motorman managed to stop the train without killing him, but he was injured. He sued the New York Transit Authority for negligence in not stopping the train soon enough to avoid hitting him! The transit authority settled out of court for $650,000.

On the night of June 28, 1984, Bernard McCummings and an accomplice attacked a 72-year-old man in a New York City subway station, striking him, choking him and pinning him to the ground while rifling his pockets. Two plainclothes transit police heard the man's screams for help and ran to his aid. As McCummings was fleeing, one of the officers ordered him to stop. When he failed to do so, the officer shot him twice, severing his spinal cord. The 72-year-old victim had to be hospitalized because of the injuries he received from the muggers. McCummings was convicted of the crime and spent more than two years in prison. Astoundingly, he successfully sued the transit authority and a state court jury awarded him $4.3

million in damages. The U.S. Supreme Court refused to review the case, leaving the $4.3 million judgment to stand. Who says crime doesn't pay?

BACKLASH AGAINST TREND BEGINS

A federal court awarded the police chief and the city $22,258 in legal fees and costs against an ex-officer who filed a frivolous lawsuit challenging his termination. [1]

A Houston constable was awarded $3 million in his counterclaim against a woman who falsely accused him of sexual harassment. The plaintiff's lawyer also received a $70,000 sanction for bringing the suit without verifying the evidence. The plaintiff's courtroom testimony was so incredible that the presiding judge read her the "Miranda" warnings in open court. She dropped the suit midway through the trial.[2]

The new Republican majority which took control of Congress in January, 1995, promised to reform this runaway tort system. The legislation would, among other things, authorize judges to require losing plaintiffs to pay the other party's legal fees and expenses as well as their own. It would also put a monetary cap on punitive damages.

MANAGERS NEED A WORKING KNOWLEDGE OF THE LAW

It is essential that managers — regardless of their field — have a working knowledge of the law as it relates to their activities. This will help them to avoid problems which might get them and their employers into legal trouble.

Liability is merely another matter with which the modern fire service manager must be concerned. The better manager you are, the less legal risk you will have. By learning about the law and instituting those programs, forms and procedures which reduce the chances of getting into legal trouble, and training those under your supervision regarding their legal responsibilities, you will minimize your liability risk.

However, there are no guarantees. Even the most enlightened and well-prepared manager may get sued and may lose the suit. This should not cause paranoia, but should be accepted as part of the territory as a manager. A manager faces many complicated challenges. The law is only one of them.

Our goal is not to make you firehouse lawyers, but to give you a general familiarization with the law and help you to recognize some of the legal problems confronting you so you can conduct yourself and your operations in such a way that you will reduce the risk.

Our focus is to help you understand the law and how it affects you, your department, your government and your community. Our goal is to give you a better understanding of the legal issues which affect the fire service and provide some insights as to how you should respond to them when they confront you.

As a good manager, you should know your duties and responsibilities and what authority you have.

The law impacts on every decision you will make during an emergency. The entire matrix of laws — criminal and civil — affect fire service operations in numerous ways. So, understandably, most fire service personnel worry about being sued for actions that they take, or fail to take. Don't be worried; be prepared. Don't be intimidated by the law. Prepare yourself by learning as much about how the law affects your work as you can and be guided accordingly.

Study the most cost effective, reasonable means to reduce liability. Your rules, regulations and procedures should reflect the most effective means for running your department, based on state-of-the-art standards to maximize safety and efficiency as well as to protect the rights of everyone with whom you deal.

Make sure your personnel know what is expected of them. Make clear to them what their duties are, the limits of their authority and the standards against which their performance will be measured. Tell them what acts are prohibited and the consequence of violations.

HAVE COMPETENT LEGAL COUNSEL

No matter how much you learn about the law, when you have

legal problems or potential legal problems, you should consult a competent attorney. It is very important to have access to a good attorney whom you can consult regularly, even BEFORE you are in legal trouble. Just as you regularly have your automobile serviced, just as you have regular routine medical examinations, you should follow this same type of preventive maintenance with regard to legal matters. Invite a lawyer to lecture about various legal matters in your training programs. All firefighters need this training. For example, they should know how to recognize and preserve evidence of arson which they might find at a suppression scene and they should know where they can properly search during a cause-and-origin investigation and the legality of evidence they find during their search.

GET TO KNOW YOUR GOVERNMENT'S ATTORNEYS

Become personally acquainted with your jurisdiction's office of prosecution. Ask the attorneys there to explain criminal procedures in your state. Find out what their requirements are when the fire department presents violations for possible prosecution. Ask them the requirements and procedures for getting search and arrest warrants. Solicit their suggestions regarding revision of your forms, such as code violation citations, and procedures to make the process more efficient and to minimize mistakes. When you are scheduled to be called as a witness in a court case, ask for advice on how to testify.

It would also be fruitful to have a similar meeting with the attorneys who handle civil matters for the fire department.

YOUR GOVERNMENT SHOULD PROVIDE YOU LEGAL COUNSEL

The most important thing you will read in this book is this: Make sure your government protects you legally by providing an attorney for you if you are sued individually for something which happens within the scope of your employment and, if you lose the case, make sure your government pays the judgment.

If not, it can have a very chilling effect on a conscientious public servant's enthusiasm for doing the job. It is unfair to you who must make important decisions instantaneously under great pressure. It is also not fair to your community because if you are worried about losing all your assets through personal liability, you may be somewhat tentative in how you perform your duties. If your government does not give you this support, you and others similarly situated should initiate an effort to have the law or policy changed so you will have this kind of protection.

The object of this book is to give you a general overview of the law and to whet your appetite to learn more.

CHAPTER ONE

INTRODUCTION TO LAW AND THE LEGAL SYSTEM

WHAT IS "THE LAW"?

The law is a system of rules and regulations governing conduct in society promulgated by courts, legislatures and administrative bodies at all levels of government. It is an orderly, peaceful way of resolving disputes.

The law serves various functions:

- Gives powers and duties to government;
- Creates a mechanism through which government operates;
- Sets forth the rules for our relations with each other as members of society;
- Creates a system through which disputes can be resolved.

WHY DO WE NEED LAWS?

Without laws, civilized society would be completely unworkable. There would be chaos. Part of the price we pay for living in our society with all of its benefits is to accept and abide by the consensus rules which have been developed. In our democracy, power and authority are derived from the will of the people, as communicated through elections to the legislative, executive and judicial branches of government. Those branches of government make, enforce and interpret those rules. When those among us fail to abide by these rules, which we call laws, they are called to answer for the breach. Authoritative sources of law:

- Statutes
- Cases
- Federal and state constitutions
- Regulations

OUR LEGAL SYSTEM IS DIVERSE AND COMPLEX

In studying the law, it is essential to understand the tremendous diversity of our legal system. We have fifty separate, independent states as well as the federal government. Each has its own constitution, its own laws and regulations, its own court system, and its own procedures.

Within the states there are counties and a variety of other local sub-entities numbering hundreds of thousands. They all make and enforce laws. The states enact laws through their legislatures which govern behavior within the state and the local jurisdictions enact laws governing behavior within their geographic boundaries. On the national level the U.S. Congress enacts laws affecting everyone within all fifty states.

It is important to remember that the Federal government derived its authority from the states, not vice versa, even though over the past several decades the power of the central government has grown greatly at the expense of state autonomy. The Republican majority which took over control of Congress in January, 1995 pledged to reverse this trend.

The American legal system is overwhelming when viewed in its entirety, but by calmly focusing on the segment of the law which directly affects you and your operations, it becomes manageable.

The law is not static, it is dynamic and perpetually evolving. It grows and it changes, albeit slowly, to reflect the consensus mores of our society and it evolves to address new problems which face society from developing technology. The laws affecting the operation of horses and buggies evolved into today's laws affecting automobiles.

Congress, state legislatures and the legislative bodies of county and local governments are continually enacting new laws or modifying old ones. Interpretations of the law by the courts at federal and state levels change the law. Administrative agencies revise their regulations.

When a non-lawyer asks, "What is the law on that point?" The lawyer must answer, "It depends on many factors. What state are you talking about? How have the legislatures and the courts in that state addressed the question? Has that matter been addressed by appellate courts in that state?" Even after answering those questions, you still cannot be sure what the law is. Even after extensive research, the lawyer can only make an educated guess as to what the law is. It is not possible to predict with certainty what the outcome of a legal matter will be because there are often no clear answers to legal questions. Good lawyers disagree with each other. In each case which goes to court — and most do not — there are two sides and two interpretations as to what the law is. Judges themselves disagree. There may be similar cases adjudicated within the same state where judges came to diametrically opposite conclusions. In any given courthouse, the judges presiding there in individual courts may disagree with each other as to what the law is. Appellate courts often reverse the decisions of lower courts and higher appellate courts often reverse the decisions of lower appellate courts.

When appellate courts overturn the decisions of lower courts, it is often by a divided opinion. In other words, even outstanding jurists disagree with each other as to what the law is. The case may then go to the Supreme Court of the United States where the decision of the State's highest court may be reversed. Very often decisions of the U.S. Supreme Court are decided by 5 to 4 votes.

Although the majority opinion prevails, a well-reasoned dissenting opinion may later become the majority opinion when the personnel of the Supreme Court changes. When a president appoints replacement Supreme Court justices, their legal philosophy may be different from the previous majority's consensus about that issue and, as a result, many of those 5 to 4 decisions will be overturned.

As you approach the study of law, you might well ask, "If the judges themselves, even justices of the U.S. Supreme Court, don't agree as to what the law is, how am I supposed to know? That is a good question without a satisfactory answer.

When they come, legal problems ensnare us with a tangled web of complexity. They don't lend themselves to quick and easy answers. Why is it difficult to predict the outcome of a legal matter?

- Diversity of the system
- Difficulty of proof
    - physical
    - witnesses
    - legality
- Conflicting court decisions
- Changing legislation
- Unclear statutes and regulations
- Immunities
- Statute of limitations
- Defendant may have an adequate defense
- Unpredictability of juries
- Fallibility of judges
- Procedural problems
- Jurisdictional problems, etc.

An attorney must pursue his/her case on the basis of the facts and the law as they are, not as the lawyer and the client wish they were. The attorney cannot fabricate evidence or mislead a judge as to what the law is. There are criminal and ethical penalties for doing so. In every lawsuit, there are lawyers arguing on each side, disagreeing as to what the law is.

Each state and some local governments have their own judicial system which interprets the law and adjudicates disputes between litigants within that jurisdiction.

The federal judicial system interprets the U.S. Constitution and federal laws. The federal courts also adjudicate legal disputes between litigants related to federal matters, and cases related to state matters when the dispute is between citizens of different states.

In both the state and federal judicial systems there are courts of original jurisdiction where trials take place, mid-level appellate courts and a high court — usually called the Supreme Court. Appellate courts serve as checks on the lower courts. The state's highest court serves as a check on the state's lower appellate courts. The Supreme Court of the United States serves as a check on the states' supreme courts.

The highest court in the country is the Supreme Court of the United States which consists of nine justices appointed by the President and confirmed by the U.S. Senate. The Supreme Court of the United States not only hears appeals from lower federal courts, but also from the highest courts in the states. It is important to note, however, that when the U.S. Supreme Court hears a case involving state law, it applies substantive state law, but federal procedural law.

FOUNDATIONS OF OUR LEGAL SYSTEM

The foundations of our legal system are:
- Common law (derived from custom, usage and cases from early British law to the present time, evolving through court decisions);
- Statutory law (laws passed by legislative bodies);
- Administrative law (regulations for the interpretation and enforcement of statutes);
- Constitutions (state and federal).

COMMON LAW

The Common Law System on which our American legal system is founded is a distinctive system of laws originated in England hundreds of years ago in which the king's judges made the law by deciding the cases which were brought before them. Earlier cases became precedents for deciding subsequent cases with similar issues and facts. These accumulated cases became a vast body of case law. Now statutory law has become as important as case law.

STATUTORY LAW

Statutory law relates to those laws enacted by the legislative bodies of federal, state or local governments. Laws passed by local governments are usually called ordinances. A compendium of statutes or ordinances is called a code, such as the Fire Code.

Statutes enacted by a state legislature must conform to that state's constitution and the U.S. Constitution.

ADMINISTRATIVE LAW

Administrative law refers to the rules and regulations promulgated by administrative agencies, boards and commissions to carry out the powers, authority and obligations delegated to them by legislative bodies. These entities are quasi-judicial and quasi-administrative. Because it is not possible for statutes to anticipate every detail with respect to matters within the purview of those laws, the legislative bodies create these administrative agencies to interpret, enforce and carry out the legislative intent in enacting the statute.

It is important to remember that these agencies have only those powers specifically delegated to them by the legislatures. If they exceed that authority, their acts are "ultra vires" — beyond their authority — and invalid.

Some of these administrative, regulatory bodies are the U. S. Environmental Protection Agency, the Occupational Safety and Health Administration, the local planning and zoning board, the board of appeals, the fire board etc. At the Federal level, administrative rules and regulations are embodied in the <u>Code of Federal Regulations</u> which assists us in interpreting the <u>United States Code</u>.

CONSTITUTIONAL LAW

The U.S. Constitution is the fundamental foundation of our entire legal system. Neither the U.S. Congress nor any state or local government can do anything which violates the U.S. Constitution. The Supreme Court of the United States is the final arbitrator of what contravenes the Constitution.

COURT DECISIONS

Throughout this book we will be referring to "cases". A case is a judicial or administrative proceeding in which facts or a controversy are presented in technical legal form for decision making. The objective of such a proceeding is to enforce rights and remedy wrongs. For example — seeking money for breach of contract or tort, or prosecuting a defendant for a crime.

The term "case" also refers to the opinion and decision of an appellate court deciding appeals.

A case has two functions:

1. It authoritatively decides the particular controversy.
2. It establishes a precedent (or a possible precedent) for the resolution of future controversies with similar facts and issues.

TRIAL

A trial is a legal forum where differences are resolved.

A person who feels his/her rights have been infringed — that another person has breached a duty — can file a legal suit. The person bringing the suit — the plaintiff — must then offer admissible evidence to prove the legal elements of the cause of action. The one against whom the suit is brought — the defendant — then is given the opportunity to offer a defense to disprove the plaintiff's case or to justify his/her own actions. This process is called a trial and is governed by strict rules of procedure and rules of evidence.

TRIAL COURT

A trial court consists of a judge with or without a jury and it performs four basic functions in the administration of legal justice:

*   Fact-finding: It decides the facts which are in conflict in particular controversies that are brought before it in technical legal form. It also decides which witnesses are believable.
*   Interprets laws: It determines which rules of law are applicable and how they are to be applied.
*   Applies laws: It applies these rules of law to the facts in that particular case.
*   Follows procedures: It does all of the above while following strict, prescribed procedures.

Under our system of justice, legal principles are applied to the facts of a particular case. Courts interpret the applicable law. Henceforth, this decision — especially when rendered by an appellate court — is then the law in that jurisdiction and must be obeyed. It is important to remember that the courts only resolve disputes between litigants. They do not render advisory opinions. Litigants can seek a declaratory judgement, asking the court to rule on a question of law, but it must be related to a dispute between contending litigants or the court won't act. It is not an advisory opinion in the usual sense.

As we have seen, the decisions of lower courts are subject to review by higher courts. The facts are established at the trial level, either by the jury or, if there is no jury, by the judge acting alone. The judge always decides what the law is, or more accurately, what s/he THINKS the law is.

CRIMINAL CASE BALANCES INTERESTS

In criminal cases, the legal system tries to balance the interests of society against the rights of the accused. The U.S. Constitution provides several protections to the accused and these have been substantially broadened through decisions of the U.S. Supreme Court. There is a strong underlying current in criminal law that a person is innocent until proven guilty. The question which the law addresses is not: "Did the person do the act which gave rise to the criminal case?" but rather: "Does the government have enough admissible evidence to establish all the elements of the crime and prove beyond a reasonable doubt that the person is guilty of the criminal offense."

These are two distinctly different concepts and the distinction is important. In criminal law it is important to keep in mind this burden which the government always has in prosecuting crimes.

CRIMINAL LAW VS. CIVIL LAW

The same act can give rise to an action in both criminal law and civil law. The criminal case is brought by the state, the objective of which is to punish the wrongdoer for an offense against society and to deter others from committing the same crime. A civil action is brought by an individual, the objective of which is to compensate the plaintiff for the loss sustained through the defendant's actions. The rules of procedure and the rules of evidence are different for the two types of action. In a criminal case, the standard of proof is "beyond a reasonable doubt." The usual standard in a civil case is "by a preponderance of the evidence."

APPEALS

Appellate courts hear appeals from trial courts within their jurisdiction. They decide if the trial court made errors in interpreting the law. Except in rare cases, appellate courts do not consider the facts in a case. They rely on the fact-finding which was done by the trial court. Appellate courts hear no witnesses and look at no evidence. Attorneys for the appellant and appellee file briefs and make counter oral arguments regarding legal issues, following which, appellate judges decide the law, often by a divided vote.

Examples of trial court errors viewed by appellate courts are:

— A prejudicial ruling in admitting or excluding evidence;
— A mistake in selecting and applying the rules;
— Insufficient evidence to prove the case;
— A violation of the constitution.

The legal decision by an appellate court becomes a precedent. Future cases involving similar facts and circumstances must be decided the same way in that state (or federal circuit) until that decision is overturned. Sometimes courts in future cases will narrowly distinguish the new case from the precedent and arrive at a different conclusion.

GOVERNMENTAL TAKING

For government to take private property, it must have a public purpose and must provide just compensation. The due process clause places a substantive as well as a procedural limitation on the exercise of governmental power. It requires that government be fair in dealing with individuals.

At the core of the due process clause is the right to notice and a hearing at a meaningful time and in a meaningful manner. Ordinarily, due process requires that a person be given a hearing before a property interest is taken away. This includes a person's intangible rights, personal or real property, and his/her liberty. Only in extraordinary circumstances involving the necessity for quick action by the government or the impracticability of providing any meaningful pre-deprivation process may the government dispense with the requirement of a hearing prior to the taking.

GOVERNMENT POWER, A DOUBLE-EDGED SWORD

A government entity's police and eminent domain powers go far beyond crisis management. The city or state certainly has the power to protect its citizens by avoiding a condition which it considers potentially, though perhaps not imminently, dangerous.

EMERGENCY ACTION IS AN EXCEPTION

Fire department personnel should know that the U.S. Supreme Court has repeatedly held that summary governmental action taken in emergencies and designed to protect the public health, safety and general welfare does not violate the due process clause of the Constitution. States, in carrying out their police power responsibilities, are given great leeway in adopting summary procedures to protect public health and safety, even in the absence of an emergency in the usual sense. Mere errors of judgment, or actions mistaken or misguided do not violate due process, but malicious, irrational and plainly arbitrary actions do. The latter are not within the legitimate purview of the state's power. In other words, the government's power is not unlimited.

The U.S. Supreme Court has said, "The exercise of emergency powers is particularly subject to abuse. Emergency decision-making is, by its nature, abbreviated. It normally does not admit participation by or input from those affected. Judicial review is often greatly curtailed or non-existent. Exigent circumstances often prompt actions that severely undermine the rights of citizens, actions that might be eschewed after more careful reflection or with the benefit of safeguards that normally constrain governmental actions. Whether government officials invoked emergency powers when they knew, or well should have known, that no exigency justified use of such draconian measures, is therefore highly relevant."[3]

CHAPTER TWO

ANATOMY OF A LAWSUIT

C omplex and highly technical rules of procedure govern the filing or pleadings and the conduct of a trial. The purpose of these rules is to make the process as fair as possible. Criminal procedure differs somewhat from civil procedure.

Here's how a lawsuit develops. Two parties have a dispute. They attempt to resolve the conflict among themselves, but are not successful. They consult lawyers. The lawyers research the law and the facts and negotiate with each other in an attempt to resolve the dispute, to find a compromise and avoid going to court. They are unsuccessful.

One of the parties files suit. That person is called the "plaintiff" while the person sued is called the "defendant." The defendant might countersue. If it is a jury trial, the jury decides what the facts are and the judge decides what the law is. If it is not a jury trial, the judge decides both the facts and the law. One side wins; one side loses. The losing party (or the side against whom the judge rules on a question of law) has the right to appeal to various higher courts, including eventually to the U.S. Supreme Court, the highest court in the nation.

A HYPOTHETICAL CASE

That's a quick summary of how it works, but let's cover the process in more detail, using a hypothetical case.

Joe Firefighter, while suppressing a fire, is overcome by smoke because, according to Joe, the breathing apparatus he was wearing, manufactured by the XYZ Company, malfunctioned. Joe was hospitalized for two weeks, had high medical expenses and sustained permanent damage to his lungs which impairs his breathing.

He contacts the XYZ Company about his problem. The company refers Joe's claim to its insurance company. The matter is negotiated back and forth. The company finally denies liability and refuses to pay Joe anything (or offers him a monetary settlement which he rejects as being too low.) The insurance company believes Joe's claim is without merit.

Joe consults an attorney, I.M. ("Ima") Sharpe. Ima researches the law and the facts and, in due course, she contacts the attorney for the XYZ Company (or its insurance carrier) and negotiates on behalf of Joe, making arguments about what a good case Joe has, etc. In spite of lengthy negotiations over several weeks, the lawyers are unable to reach a satisfactory settlement agreement.

Ima tells Joe the Company will not settle the dispute. She asks, "Do you want to sue them?"

"What's involved in that?" Joe asks. "How good are my chances?"

"It's always difficult to predict the outcome of a case," I. M. Sharpe explains. "Juries are unpredictable. Judges make mistakes. We might have trouble proving our case. They might have a good defense. They'll try to show you didn't have the mask on properly and your injury was your own fault, not theirs."

"It was NOT my fault," Joe says, "Their breathing apparatus was defective. How much will it cost me to sue them?"

"Well, I charge $150 per hour for my time, $100 for my associates and $50 for my legal assistants who would do much of the research," Ima Sharpe tells Joe.

"About how much will that cost me?" Joe asks.

Ima asks, "How long is a fire hose?"

"It depends on a lot of factors," Joe says.

"That's my answer to you about how much it will cost. It's impossible to tell. At this point, I have no way of telling how many hours my employees and I will have to work on your case. We'll have to interview witnesses, research the law, conduct depositions and interrogatories, prepare our strategy, meet with witnesses, and so forth as well as handle the trial. I don't know how long the judge will keep us waiting before our case is called when our trial date comes up and I don't know how long the trial itself will last."

"Aren't your rates pretty high?"

"No. The lawyer who defended President Clinton in 1994 in the sexual harassment case charged $435 per hour. We lawyers don't sell a product. The only way we earn money is by selling our time. There are only 24 hours in a day and we have to sleep for a few of those. So, for every hour or fraction of an hour we work on your case, you will be billed."

"We're talking thousands of dollars, aren't we?"

"Yes, we are."

"I don't have that kind of money," Joe says. "I guess I'm out of luck."

CONTINGENT FEES

"There is another arrangement which we could make," Ms. Sharpe says. "It's called a contingent fee arrangement. You will not have to pay me any hourly legal fees. If we lose the case, you owe me nothing. If we win, you owe me one-third of whatever we recover." (This percentage varies and is subject to negotiation.)

Joe decides to retain Ima Sharpe on the contingent fee arrangement.

(The contingent fee arrangement enables individuals such as Joe, who otherwise could not afford the legal fees, to have competent legal counsel. Lawyers are required to perform some legal services for the poor on a pro bona basis — Latin meaning "for the good." In other words, for free. There are also public and private organizations which provide legal services for the poor. In a criminal case, of course, the defendant is entitled to have the government appoint and pay a lawyer to represent him/her.

(Whether or not to file suit is always an agonizing decision for the plaintiff. In a criminal case it is also difficult for the government's prosecutor who must decide whether or not to prosecute an offender. The prosecutor has a limited staff and a very heavy workload. S/he must decide after a cost-benefit analysis whether or not to prosecute. S/he will ask, "How good is our case? Does the defendant have a good defense? How much time and money will it take to prosecute? How does this case rate from a priority viewpoint compared with all of our other cases? Can we plea bargain with the defendant's lawyer and get him to plead guilty to a lesser offense and save us the trouble and expense of going through a long trial?"

(It is frustrating for the fire department's investigators who may have worked for months to make a case against an arsonist only to have the prosecutor decline to prosecute or accept a guilty plea for a lesser offense.

(It is similarly frustrating and highly irritating to a fire service officer when someone files a civil suit against him/her charging negligence or some other civil wrong and the government's attorney, or the attorney for the department's insurance company, decides to settle the matter out of court with a cash payment to the plaintiff rather than defend the suit. The government's lawyer might conclude that it would cost far more to defend the case than to settle it. Often it is the insurance company making this decision and, if their lawyers think it will be a costly trial and they can settle the case for a reasonable price, the case will be settled, usually with little regard for the merits of the litigation. The fire department officer is upset when the case is settled, however, because s/he has been denied the opportunity to clear his/her good name. The government's lawyer, however, has to decide whether or not defending it is a good expenditure of tax payers' money when it could be settled for a reasonable expenditure. As with the prosecutor, the government's civil lawyer must assess how high a priority this case should have compared with all the other cases pending in the office.)

Let's return to Joe Firefighter and his attorney, I.M. Sharpe.

Ima will now research the law, conduct interviews and investigate the case to gain a thorough knowledge of both the law and the facts related to Joe's case.

Armed with this information, she will then file the first pleading in the case.

THE SUMMONS

The first step is the serving of a summons on the defendant, the XYZ Company. The summons, which is prepared by the plaintiff's attorney, but actually comes from the Clerk of the Court, advises who the plaintiff is and states the time within which the plaintiff is obliged to enter its appearance or a default judgment will be entered against it. In most states the summons is accompanied by the complaint.

THE COMPLAINT

The complaint will tell the complete story of how Joe was injured, claim that the XYZ Company is responsible and allege facts to show why the defendant should be held liable. The complaint will ask for a remedy, usually compensatory and perhaps punitive money damages in certain amounts. The facts on which plaintiff's claim are based are set forth in numbered paragraphs. Plaintiff will not be allowed to present any evidence at trial which is not contained in the complaint, but the complaint can later be amended by the plaintiff to include such information. Ima will file this complaint with the Clerk of the Court and serve a copy on the Company's registered agent.

RETURN OF PROCESS

When the summons/complaint have been served, the process server (who may be the sheriff or a private person) makes a "return," showing that the defendant has received the papers.

The defendant will make his/her appearance by filing an answer to the complaint or by filing a motion, challenging the jurisdiction of the court or asking for more time, etc.

A MOTION

Any time either lawyer wants the judge to do something, a "motion" is filed. It is a formal pleading filed with the court, explaining why the court should do a certain thing and asking the court to do it. A motion can be filed at any time by either party.

THE ANSWER

The company will turn the complaint over to its lawyer (or its insurance company who will, in turn, give it to its lawyer) who will study it and research the law and the facts, and assess the complaint from both perspectives. The lawyer might conclude that Joe has a very good case and recommend that negotiations be opened with Joe's lawyer in an effort to reach a mutually acceptable settlement.

Let's assume the company rejects this advice or the attorney doesn't think Joe has a good case.

The defendant's attorney will prepare the answer and file it with the Court and serve a copy on the plaintiff. The defendant in its answer will respond, paragraph by paragraph, to the items in plaintiff's complaint, denying or admitting each fact, or stating that defendant does not know whether something is true or not, thereby requiring plaintiff to later prove it. The answer might also:

- point out mistakes or inaccuracies
- challenge plaintiff's service of process
- deny some or all of plaintiff's allegations about defendant
- make technical objections unrelated to the merits of the case
- dispute the legal rules underlying plaintiff's complaint
- claim the damages requested are excessive
- state that the statute of limitations has run
- state that the action was brought in the wrong court
- claim that the plaintiff's own negligence caused his injuries because he was not wearing the mask properly or because his beard prevented a tight seal
- make a counterclaim against plaintiff, (unlikely in this case), and
- make a motion to dismiss plaintiff's complaint.

The defendant might say that, even if all plaintiff's facts are true, they don't show that defendant has done anything for which the company is legally accountable. This motion is called a "demurrer" or motion "to dismiss for failure to state a claim on which relief can be granted."

If the defendant does not file an answer by the prescribed date (which is established by that jurisdiction's rules of procedure), Ima will file a motion for, and receive, a default judgment against the company. In other words, Joe would win at that point if the motion is granted.

THE ISSUES

In its answer, the defendant might have admitted some of the plaintiff's facts in the complaint and denied others. The result is that the facts which are in dispute usually become clear when the complaint and answer are compared. The disputed items which they highlight become the issues which will form the basis for the trial. The trial will focus exclusively on those disputed facts.

The function of a trial is to resolve disputed facts.

THE TRIAL

There may be a delay of several months or even years before the case can be scheduled because the court's docket is so crowded.

If plaintiff and defendant agree as to what the facts are, the judge will hold a hearing on the demurrer and the attorney for each side will argue why it should or should not be granted. If the judge grants it, the case is over, except, of course, that the losing party can appeal the judge's ruling to a higher court.

Following the complaint and answer, but before the trial, a number of motions might be filed by each party. Either party could make a motion for "judgment on the pleadings." The party making the motion claims on the basis of the pleadings that no facts are in dispute, so there is no reason for a trial and the judge should decide the case now. Let's assume the judge denies this motion.

DISCOVERY

In the interim between the complaint and the answer and the trial, both parties will engage in discovery by which each side learns about the other party's case. Each side must furnish the other side its theory of the case, the points which each side will try to prove, the physical evidence and a list of witnesses to be called, etc. Each side will use interrogatories (written questions which the other side must answer) and depositions (a session with attorneys for both parties present during which a witness is questioned and cross-examined under oath and a stenographer records the proceedings.)

Be aware that in litigation involving your government, lawyers will have access to all of your relevant files and records under the discovery process. Each side is obliged to exchange information with the other side regarding all elements of its case before the case ever gets to court. This is to encourage negotiation and settlement of the dispute before it gets to court and to maximize the court's time when it does get to court.

On the basis of what is learned during discovery, a motion for summary judgment may be made as a way to end the case without a trial. The moving party says, in effect, "While there is a dispute about certain facts, the dispute is illusory and there is really no genuine issue of fact to justify holding a trial." Affidavits would be filed to support the motion. The judge's decision on the motion can be appealed.

There will be a number of pre-trial conferences, usually held in the judge's chambers with both attorneys present. The judge will try to narrow the issues to be tried and will try to nudge the parties to negotiate a settlement.

The case might be tried before the judge without a jury. In this case the judge decides both the facts and the law. If one of the parties requests a jury trial, the jury, after hearing the evidence, will decide what the facts are and the judge will decide what the law is.

JURY SELECTION

If a jury trial is requested, the jury must be selected.

A panel of potential jurors is ordered to appear in the courtroom. The judge will listen to explanations from some jurors as to why it would be a burden for them to serve. S/He may or may not excuse them. S/He will question jurors to find out if any of them have a bias in favor of one of the parties or that type of case. If so, s/he will dismiss those persons. Each lawyer, after questioning the potential jurors, has an opportunity to challenge jurors "for cause." If the judge agrees, those jurors are dismissed. Each lawyer also has a number of "peremptory challenges" with which s/he can discharge a certain number of potential jurors without having a reason. At the conclusion of this process, the jurors are selected. Usually the number is twelve, but in some jurisdictions fewer are allowed, especially in civil cases. Alternates will also be selected so a full complement will be available to deliberate in case one or more of the jurors must later be excused for some reason such as illness.

After the jury has been selected and takes its oath, the trial can begin.

THE TRIAL BEGINS

The sequence in which the facts in dispute will be argued generally parallels the pleadings. The plaintiff has the burden of going first and proving his/her case. If s/he does not successfully prove the case, s/he loses.

We have seen what happens during the pre-trial period:

- Complaint is filed by plaintiff
- Answer is filed by defendant
- Motions are made by both parties
- Both parties engage in discovery
- Pre-trial conferences are held
- The jury is selected

In brief, the following are the steps in a trial:

- Plaintiff makes opening statement, describing the nature of the case and summarizing the major facts which s/he intends to prove.
- Defendant makes opening statement.
  [Note: If it is a jury trial, these opening statements are usually more dramatic and elaborate because juries are more responsive to emotionalism than judges are.]
- Plaintiff offers case. Calls witnesses. All evidence is presented through witnesses.
- Plaintiff conducts direct examination of his/her witness.
- Defendant cross-examines plaintiff's witness. The purpose of cross-examination is to raise doubts about the credibility of the witness and the truthfulness of his/her testimony.
- Plaintiff conducts re-direct examination of witness. During

re-direct examination, plaintiff's attorney will try to undo any harm which defendant's cross-examination might have done and to clarify and reiterate the facts favorable to plaintiff.

- Defendant conducts re-cross-examination of plaintiff's witness. The same process is repeated for each of plaintiff's witnesses.
- Motion for a directed verdict by defendant. The judge will only grant this motion if plaintiff has offered insufficient evidence to support his/her claim. If the judge grants it, the trial is over. If it is denied, the trial continues. Either party can appeal from the judge's ruling.

Defendant offers defense:
- Defendant conducts direct examination of his/her first witness.
- Plaintiff cross-examines defendant's witness.
- Defendant conducts re-direct examination of his/her witness.
- Plaintiff conducts re-cross-examination of defendant's witness.
The same process is repeated for each of the defendant's witnesses.
- Plaintiff presents rebuttal evidence.
  Note: After both parties have presented their case, either or both attorneys may make a motion for a directed verdict. If it is granted, with the judge deciding in favor of one of the parties, the jury is discharged. The judge only grants this motion when s/he believes that a jury would not be justified in finding for the other party on the basis of the evidence presented during the trial.
- Both parties draft instructions to the jury which they would like the judge to make. Each attorney will slant the drafted instructions to present the facts in the light most favorable to his/her case.
- Plaintiff makes closing argument to the jury.
- Defendant makes closing argument to the jury.
- Plaintiff makes closing remarks.
- The judge reads instructions to the jury. The judge might take the instructions drafted by the plaintiff or those drafted by the defendant, or a combination of both, or s/he might write his/her own independently. The purpose of the instructions is to tell the jury what its decision should be, depending on what they believe

the facts to be after they assess the evidence they heard during the trial.

- The jury retires to deliberate. The jurors discuss the case and take votes until a verdict is agreed upon. It must be unanimous (or a majority if allowed by state law).
  Note: If the jury can't reach agreement, that is called a hung jury. In such a case the plaintiff (or the prosecutor in a criminal case) must decide whether or not to bring the case to trial again.
- The jury returns and announces its verdict.
- Judgment is entered.
- Losing party may make motions:
  — for judgment notwithstanding the verdict. The party making this motion says, in effect, "On the basis of the evidence presented during the trial, no reasonable jury could possibly come to that verdict." (In a criminal case, the prosecutor is not permitted to make this motion if a defendant is acquitted by the jury.)
  — for a new trial
  — for an appeal.

Joe may win or he may lose. If he loses, he has the right to appeal, but only from the judge's rulings on the law, not on the facts. If he wins, he then has the problem of collecting. From a reputable company that usually is not a problem, but with an individual or an unscrupulous or insolvent company, bankruptcy might be filed to avoid paying the judgment or the unsuccessful defendant simply might not have any money. In which case, Joe and his contingent-fee lawyer have won only a Pyrrhic victory and no money.

BEING A WITNESS IN COURT

Being a witness in a courtroom is similar to being a witness at a legislative hearing except the rules are more precise and rigid.

Someone from the fire department might appear in court as a complaining witness in a code enforcement case or a criminal arson trial or as a witness or a defendant in a negligence or other liability suit.

The most critical element in courtroom testimony is preparation. Having a firm understanding of the facts of the case gives a witness credibility before the judge and jury.

Remember, when conflicting evidence is introduced at a trial (as it usually is) the jury must decide whom to believe. What the jury believes becomes the facts of the case.

Generally, a witness must have personal knowledge of the facts about which s/he is testifying. For example, an inspector could not testify about code violations based upon another inspector's report. This would be hearsay. The inspector testifying must have "seen" the violation personally. The inspector could inspect the property before his/her court appearance, thus enabling him/her to testify from personal knowledge that the violations cited remained outstanding at the time of his/her inspection.

The major exception to this rule of first-hand knowledge in order to testify is for expert witnesses. There are some other exceptions to the hearsay rule which are beyond our treatment here, but generally lay witnesses can only testify about facts within their personal knowledge.

Expert witnesses give opinions, not facts. This would allow an inspector, as an expert witness, to go beyond the facts to describe why the conditions found did not comply with the code and the fire hazards which this usually presents, or that the physical evidence found at a fire scene was the type of material usually found at the site of fires started by arson.

The lawyer will "qualify" the expert witness by asking questions to demonstrate formal education, training in the area of expertise, practical experience and other professional credentials. In other words, the lawyer calling the expert witness tries to demonstrate that the witness truly is an expert.

A difficult but essential part of effective courtroom testimony is to remain calm, especially when being cross-examined by the attorney for the other side. The witness' credentials may be attacked as well as the substance of his/her testimony. The jury will be looking very closely to see how the witness and his/her testimony hold up under the pressure of cross-examination.

TIPS FOR WITNESSES

Some other tips for witnesses to keep in mind:

- Don't argue.
- Never lie
- Answer only the question asked. Stick to the facts. Don't volunteer information.
- Keep your testimony simple. Don't use technical jargon that the jury might not understand.
- Don't try to outsmart the attorney. You may be playing into his trap.
- Speak clearly, slowly and with enough volume to be comfortably heard. Try to project honesty, sincerity and professionalism.
- Don't try to be funny. Trials are serious business.
- Dress appropriately and neatly. If you are in uniform, make sure it is clean and presentable.
- Use diagrams and photos where possible.
- If the lawyer cross-examining you is nasty, don't respond in kind. By refraining from being nasty yourself, you will favorably impress the jury.
- When testifying about dead bodies, be respectful, not cold and callous.

CHAPTER THREE

THE U.S. CONSTITUTION, CHARTER FOR OUR GOVERNMENT

Constitutional law is an essential element in your study of law. A constitution is a fundamental political and legal charter for the people of a particular jurisdiction. It defines the character of the government by specifying the nature and extent of the sovereign power, by allocating this power (separation of powers) and by setting forth basic principles and limitations for the exercise of this power by the branches of government and by listing the rights of the people.

The Declaration of Independence did not create our nation, nor did the American Revolution. The true birth of the United States of America came with the adoption of the U.S. Constitution.

The U.S. Constitution divides governmental powers into legislative, executive and judicial branches. Each branch of government is designed to balance and check the powers of the other two branches. The Constitution, in the first ten amendments, lists powers that the U.S. Government may not bring to bear on its citizens. The U.S. Constitution is designed to curb the power of the government and to protect citizens from governmental abuse.

The U. S. Constitution is our highest legal authority, the supreme law of the land. But, by itself, our Constitution is merely a piece of paper. It has no intrinsic importance. It is our attitude and the attitude of all generations of Americans since 1789 toward that document which gives it such great value. It is not the constitution which is powerful, but rather the system of government which it created.

Our Nation has a government of laws, not of men. In other words, those who govern us may not act arbitrarily. They must operate within the strictures of the Constitution and our laws. Its beautiful, power-curbing system of checks and balances has insured protection of our cherished liberty. It puts leashes on the President, the Congress and the courts.

COURTS ARE GUARDIANS OF THE CONSTITUTION

The courts are the guardians of the Constitution. They address unconstitutional actions and declare them null and void.

The Constitution can be amended. Under Article V of the Constitution, Congress, with a vote of two-thirds in both houses, can propose amendments, "or on the application of the legislatures of two-thirds of the several states, shall call a convention for the proposing of amendments." In both cases, they only become part of the Constitution when ratified by three-fourths of the states, or by conventions in three-fourths of them. The Constitution has never been amended through the constitutional convention method. It will be recalled that, on two occasions, Congress passed the Equal Rights Amendment, but in both cases it failed to win approval by the legislatures of three-fourths of the states.

The Constitution is actually a bare-bones document. It sets forth the basic framework of our federal government, lists basic rights and principles and leaves the rest to the courts.

In addition to the U.S. Constitution, each of the fifty states has its own constitution and these are important documents in their own right as the highest law within the states.

State constitutional provisions in conflict with the federal constitution are unenforceable.

Local ordinances in conflict with State law or the State constitution are also invalid. It is important for those in local government to remember that local government has only such powers as are delegated to it by the State government.

COURTS CURB GOVERNMENT'S ACTIONS TO PROTECT PRIVACY

The court's function is to oversee and test government's warrant requests to protect the privacy rights of citizens. Understanding the court's perspective can help to ensure that investigators do not exceed their authority and have properly prepared their warrant requests. In this area, close cooperation between the fire department and the prosecutors can be invaluable. Again, proper training of personnel and having effective procedures and forms will reduce exposure to liability.

POLICE POWER IS LIMITED

Although the police power (which is the basis for all fire safety regulations) is very broad, it is not without limits. The state and federal constitutions, statutes and the common law all restrict how this power can be used.

The broadest constitutional protection is the concept of due process. Due process affects stop-work orders, licenses, permits, code violations, and variance requests. As is true with many legal concepts, it is difficult to define due process precisely. The U. S. Supreme Court has described due process as "whatever process is due" which is not particularly instructive. Fairness is the key element. Due process means providing people with a fair opportunity to know the charges against them and a fair opportunity to defend their rights and interests. The need of government to function effectively is balanced against the need to protect citizens against intrusive, arbitrary abuse by government.

DUE PROCESS CLAUSE

Syndicated newspaper columnist Nat Hentoff wrote on January 14, 1995:

Several years ago, I was in Jerusalem for a conference on the ethics of journalism — a boundless subject. Also present was a prominent philosopher,

*David Hartman, who has his own institute in Jerusalem. We had never met.
During a break, Hartman stopped me in the corridor. Without any pleasant-
ries, he said, "What has been mankind's greatest achievement?"*

*For once I did not falter at being asked so grand a question. "Due
process," I answered.*

*"Correct," said Hartman. He strode on. And we never
spoke again.*

Due process — fundamental fairness — is the foundation of our
system of justice. "Nor shall any state," the 14th Amendment guarantees, "deprive any person of life, liberty or property without due
process of law."

The due process clause of the U. S. Constitution places a substantive, as well as a procedural, limitation on the exercise of governmental power. Due process means not only fair procedures, but fair trials
in the courts and fair treatment by administrative bodies.

To establish a violation of substantive due process, the plaintiff
must prove that the government's action was clearly arbitrary and
unreasonable, having no substantial relation to the public health,
safety, morals or general welfare.

DISCRIMINATION

Governmental power, if wielded in an abusive, irrational or
malicious fashion, can cause grave harm.

One of the elements of a civil rights action is that the defendant is
either a government agency or an individual acting "under color or
authority." Section 1983 makes civil rights claims under the major
civil rights statutes, such as the Civil Rights Acts of 1866, 1964 and
1991, applicable to state and local governments. Because civil rights
actions are designed to protect citizens against the actions of government, not to protect citizens from other citizens, the civil rights
statutes waive governmental immunity.

To succeed in a Section 1983 civil rights suit for damages for a
substantive due process violation, a plaintiff must at least show that
state officials are guilty of grave unfairness in the discharge of their
legal responsibilities.

SCOPE OF DUE PROCESS IS BROAD

Based upon deprivation of constitutional "rights, privileges or immunities," civil rights actions can take many forms.

In *Poe v. Ullman* (1961), U.S. Supreme Court Justice Harlan wrote, "The full scope of the liberty guaranteed by the Due Process Clause cannot be found in or limited by the precise terms of the specific guarantees elsewhere provided in the Constitution. This 'liberty' is not a series of isolated points pricked out in terms of the taking of property, the freedom of speech, press and religion, the right to keep and bear arms, the freedom from unreasonable searches and seizures and so on. It is a rational continuum which, broadly speaking, includes a freedom from all substantial arbitrary impositions and purposeless restraints." [4]

Civil rights actions can take many forms. Among the more common are the following:

- taking of property without due process
- false imprisonment
- right to reasonable medical care
- right to travel/right of access
- right to be secure from unreasonable force
- right to equal treatment

Government officials are justified in using force, even deadly force, in carrying out legitimate governmental functions. But when the force is excessive, or used without justification or for malicious reasons, there is a violation of substantive due process.

In determining whether due process has been violated, the court will consider such factors as the need for the application of force, the relationship between the need and the amount of force that was used, the extent of injury inflicted, and whether force was applied in good faith or maliciously for the very purpose of causing harm.

The 14th amendment's due process clause protects property as well as life and liberty. To the extent that arbitrary or malicious use of physical force violated substantive due process, there is no sound basis for exempting the arbitrary or malicious use of other governmental powers from similar constitutional constraints.

FOURTH AMENDMENT

The Fourth Amendment to the U.S. Constitution protects us from "unreasonable searches and seizures" and sets forth restrictions on government's ability to get search and arrest warrants. It reads as follows:

"The right of the people to be secure in their persons, house, papers and effects, against unreasonable searches and seizures, shall not be violated, and no warrants shall issue, but upon probable cause, supported by oath or affirmation, and particularly describing the place to be searched and the person or things to be seized.

Only unreasonable searches and seizures are prohibited, but unreasonableness cannot be defined in rigid and absolute terms.

In the *Murray v. United States* case, the Supreme Court of the United States held that, if officers illegally enter a building in violation of the Fourth Amendment and see evidence in plain view, that evidence need not be excluded if they again enter under a valid search warrant that is wholly independent of the initial illegal entry. The test of the independent source is: (a) would they have sought the warrant if they had not earlier entered the building; not (b) would they have sought the warrant if they had seen no evidence during the illegal entry.[5]

The Fourth Amendment is a restriction against government action only, not the actions of private individuals such as private detectives. There was no violation of the Fourth Amendment when the accused's wife produced her husband's belongings since there was no governmental action.[6]

The Fourth Amendment protects property as well as privacy and applies in civil as well as criminal cases. Seizures are covered even if there is no search. It protects people as well as places.

To be protected by the Fourth Amendment, the person must have a reasonable expectation of privacy. The test for legitimate expectation of privacy is: a) has there been a subjective expectation exhibited; and b) is the expectation one which society recognizes as reasonable. Pen registers which record only telephone numbers dialed do not require a warrant because there is no expectation of privacy. [7]

In *Oliver v. United States*, the U.S. Supreme Court held that open fields are not protected by the Fourth Amendment because there can be no legitimate expectation of privacy in an open field, regardless of fences, no trespassing signs or its secluded location.[8] (See also *Hester v. United States*, 265 U.S. 57, 44 S. Ct 445, 68 L. Ed 898, (1924)). The fact that the officers were trespassing did not invalidate the search. Effects abandoned in a hotel room vacated by the defendant could be seized without a warrant because there could be no reasonable expectation of privacy. Similarly, the Fourth Amendment does not prohibit the search and seizure of opaque plastic garbage bags left outside the curtilage of a home because there is no expectation of privacy in garbage which has been thrown away.[9]

A state government employee has a reasonable expectation of privacy in private property in his office, drawer and files within the office and, therefore, the Fourth Amendment protects against searches by his employer or supervisor. However, a work-related search by government employers does not require a warrant and does not violate the Fourth Amendment if it was reasonable in its inception and was reasonably related to the justification. Reasonableness seems to depend upon the reason for the search. A search for work-related materials or to investigate violations of workplace rules would seem to be valid. Validity is not dependent upon probable cause, but on a general standard of reasonableness. [10]

Fourth Amendment protection extends to administrative searches and seizures, but strict probable cause need not be shown to obtain an administrative warrant.

In the *Marshall v. Barlow* case, the U.S. Supreme Court held that OSHA inspectors require a warrant to conduct their inspections, notwithstanding the fact that Congress had provided in the statute that OSHA inspectors did not need a warrant.[11] However, probable cause to believe that violations exist is not required to justify the warrant. Probable cause may be based on reasonable legislative or administrative standards. However, in *Donovan v. Dewey*, the Court held that no warrant was required for inspection for mine safety when Congress has determined warrantless searches are necessary for regulation.[12]

PROBABLE CAUSE

The Fourth Amendment says that no warrant is to be issued except "upon probable cause." What is probable cause? It is difficult to define. The courts will define it on a case-by-case basis.

One federal judge described it as follows: "Probable cause is a flexible, common sense standard. It merely requires that the facts available to the officer would warrant a man of reasonable caution in the belief that certain items may be contraband or stolen property or useful as evidence of a crime. A practical, nontechnical probability that incriminating evidence is involved is all that is required." [13]

Another federal judge described it this way: "Probable cause is a fluid concept — turning on the assessment of probabilities in particular factual contexts — not readily, or even usefully, reduced to a neat set of legal rules. It is clear that only the probability, and not a prima facie showing, of criminal activity is the standard of probable cause."[14]

The Fourth Amendment generally allows the seizure of any and all contraband, instrumentalities, and evidence of crimes, upon probable cause, even without a warrant in certain circumstances.

A detached, objective magistrate, not the investigator, is the one who determines if there is probable cause on which to base a warrant. The reason that a warrant is required, even if there is probable cause, is to protect the rights of individuals under the Fourth Amendment by getting the approval of a neutral and detached magistrate.

Hearsay may provide the probable cause to justify a warrant if it is credible.[15] However, not only must the informant be credible, but the basis for believing him/her must be shown and the basis of the information must also be reliable. Information given by unknown persons on the street is not sufficient.[16] A drug- or inflammable-liquid sniffing dog can provide probable cause.

In *Aguilar v. Texas* it was held by the U. S. Supreme Court that an affidavit to show probable cause to support a warrant (or evidence presented under oath to a magistrate) must not only state the conclusions of the affiant and the informer, but must also include facts from which the magistrate can find that the conclusions are warranted.[17] In addition, facts must be included to allow the magistrate to conclude that the informer's information is reliable and that the informer is credible. Strict application of this two-pronged test regarding "informer information" was replaced by the "totality of circumstances" test in the *Illinois v. Gates* case.[18] Under the latter test, the magistrate makes a practical common-sense decision whether, given all the circumstances before him/her, including the credibility of the informer and the basis of his/her knowledge, there is probable cause. Deficiency in one prong may be compensated for by a strong showing of the other. The Supreme Court said that reviewing courts should confine themselves to determining whether the magistrate had a substantial basis for his/her finding. An anonymous letter plus corroboration was held to be sufficient.

Police may not enter the suspect's house without a warrant to arrest unless exigent circumstances exist. An arrest warrant carries with it the limited authority to enter the suspect's dwelling when there is reason to believe that the suspect is within the house.[19]

SEARCHES

Not only may the fruits or implements of the crime be seized, but weapons and property, the possession of which is itself a crime, may be seized. Mere evidence can be seized provided that there is probable cause to believe that the evidence sought will aid in a particular apprehension or conviction. A search cannot be justified by what it produces nor can an arrest be justified by the fruit of an illegal search. A defective warrant is not made good by finding counterfeit items, heroin, opium or other evidence of a crime. A search is good or bad when it starts. A search which is good or reasonable when it is begun may still violate the Fourth Amendment if the scope or intensity authorized are exceeded. General exploratory searches are forbidden by the Fourth Amendment.

All searches without a valid warrant are unreasonable unless it is shown that they fall within one of the exceptions to the rule that a search must rest upon a valid warrant and the burden is on the government to show that the search comes under one of the exceptions.

The six basic exceptions to the warrant requirements are: consent, incident to a lawful arrest, with probable cause to search but with exigent circumstances, in hot pursuit, stop and frisk, inventory searches and other warrantless searches.

EXCEPTIONS FOR EMERGENCIES

Because the law recognizes that emergency situations require emergency action, one may not be required to obtain a warrant to search property or comply with due process requirements in emergency situations. The public policy underlying this exception is that the authority is given to government to prevent a threat to the public safety. It must be a threat of irreparable injury or imminent danger. For example:

- A locked exit door in a crowded night club imposes such an immediate peril to the occupants that it cannot await the obtaining of a warrant.
- A building being built in violation of the code must be given an immediate stop-work order before the foundation is covered or the walls are closed in.
- A building about to collapse must be evacuated at once.

However, even during an emergency situation, you must still have the authority to act. If the statutes do not give you that authority, legislation should be sought to grant you that power. No matter how well intended your actions might be, if you are acting illegally, you are open to a possible liability suit.

The due process clause of the U.S. Constitution requires that the person against whom you are proceeding has an opportunity as soon as possible to protect his/her interests. Due process requires notice and a hearing before you act, but during an emergency situation this is frequently impractical. However, immediately after the emergency is over, the person should be given the opportunity to have his/her rights protected. Due process might be satisfied by an informal meeting in the fire inspector's office or it might be a formal condemnation hearing in court.

A stop-work order on a construction project must be supported by a discussion with the owner, explaining the reasons for the stop-work order. The owner must also be afforded an opportunity to challenge the action as soon as possible. When you act during an emergency, the actions taken must be the minimum necessary to avert the danger.

Some jurisdictions give individuals a legal right to public utility service such as electricity and gas. If your emergency action interrupts these services, make sure you follow all requirements of the law.

Firefighters on the scene for suppression have the legal right to search bedrooms for victims or the attic for smoldering sparks so any evidence they find in plain view during these legitimate searches could be legally seized.

A warrantless entry or an entry pursuant to an administrative warrant must be related to cause-and-origin determination. The search must be limited to fire-damaged areas, and it should be limited to areas where evidence of cause-and-origin is reasonably expected to be found.

The proper procedure would be for the investigators to obtain an administrative warrant to enter and search the fire-damaged areas of the house to determine cause-and-origin. Then, once cause-and-origin had been determined, and they suspected arson, they should have relied on the evidence discovered up to that point to establish the probable cause to justify issuance of a criminal search warrant to authorize a search of the rest of the house.

CONSENT

Consent must be voluntary under the totality of circumstances. Mere submission to authority is not sufficient. The burden is on the government to prove that the consent was granted. When an officer told the suspect's grandmother that he had a warrant, but did not, and she allowed him to search, this was not a voluntary consent. Consent is invalid if it is obtained by fraud. Consent granted during an illegal detention is also invalid.

Who can consent? The accused, one authorized by him, or a third person who has common authority or other sufficient relationship to the premises or effects can consent. A landlord cannot consent for a tenant. Officers cannot use the right to inspect to justify a search. A hotel manager can consent to the search of a room which has been vacated, but not one which is still occupied.[20]

The person who grants permission must own or possess authority over the premises to be inspected.[21] If the inspectors reasonably believe that the person giving consent has the authority over the premises, it may be enough to justify the search even if it is later discovered the person had no such authority.[22] A landlord cannot give consent to search an apartment. [23] The consent must be freely and voluntarily given.[24] Consent which is given after the inspector's intimidation or harassment is not a valid consent.[25] If the person, after initially granting consent, later withdraws it, the inspectors must cease their search.

If the accused has less than ownership or possessory interest, the owner can consent and bind the accused, but the accused can contest the validity of the consent. A search incident to a lawful arrest may be made without a warrant, but if there is no probable cause for the arrest, the search is invalid. If the search is not contemporaneous with the arrest, or if it goes beyond what is under the immediate control of the defendant, it is not valid.

After making a lawful arrest of a person at his home, officers may, incident to the arrest, look in closets and other places immediately adjoining the place of arrest from which an attack against the officers could be launched. Beyond that, there must be facts and rational inferences which would warrant a reasonably prudent officer in believing that the area to be searched might harbor someone posing a danger to the arresting officers. This searching to look for potential assailants must be limited to a protective sweep. It cannot be a full search. It is a cursory inspection of places where someone could possibly be hiding.[26]

A limited search may be made, even if no arrest is made, when there is probable cause to make an arrest and there is reason to believe that the evidence will be destroyed.

PLAIN VIEW DOCTRINE

The so-called "plain view" doctrine allows evidence that is in "plain view" to be seized. There can be no expectation of privacy if the evidence is left in the open for anyone to see. The only basic requirement is that the person who finds the evidence must have had a lawful right to be where s/he was when the evidence was found.

If an officer is legally searching in connection with another crime or for another purpose, or who is otherwise where s/he has a right to be, items which come into his/her view may be seized and used in a prosecution of the crime to which they relate.

For example, a firefighter who is in a building to suppress a fire may seize evidence of arson which s/he sees in plain view.

Explosive devices and other evidence of arson found by firefighters during suppression activities can be seized because the firefighters did not need a warrant to enter the property to extinguish the fire. The emergency situation gave them the right to enter lawfully to put out the fire. Since they were in the building legally, the evidence they found there in plain view could be seized and later used in court. The same rule applies for investigators who have properly and lawfully entered the property to conduct a cause-and-origin investigation with respect to seized evidence which was in plain view.

The plain view doctrine is limited to evidence which is discovered inadvertently (except contraband, weapons, stolen items etc.). Objects in plain view inside the house, while providing the basis for probable cause, will not, without exigent circumstances, authorize entry to seize items without a warrant. Officers may seize evidence within the scope of a search incident to an arrest, even if not inadvertent.[27]

The plain view doctrine is based on destruction of the right of privacy once the item is viewed by the officers.

If the officer is not lawfully in an area protected by the Fourth Amendment where the evidence is open to being seen, he may not enter to seize it absent some other justification. His observation may, however, provide the probable cause basis to obtain a warrant, to arrest the person or to enter the premises if exigent circumstances exist.

What power does an investigator have to seize evidence not in plain view?

The answer lies in the answer to another question: What is the purpose of the search? Is it to determine cause-and-origin of the fire? If so, what is found in plain view may be legally seized. If the purpose of the entry is to obtain evidence of arson, a search warrant must be obtained before evidence is seized.

If the fire department personnel are on the premises in response to a fire call or for any other lawful purpose, information or evidence obtained incidental to that other purpose is legally admissible.   Any information or evidence gained through inspection or investigation which is conducted with the consent of the owner (or person in possession of the property) is also generally admissible. The only limitation is that the investigators may not misrepresent who they are in order to get permission to enter the premises.

FOURTH AMENDMENT PROTECTION NOT ABSOLUTE

There is no absolute right to be free from government searches or arrest. This amendment merely forbids UNREASONABLE searches and seizures. The courts always face the problem of balancing the rights of the individual against the government's intrusions.

The criminal justice system always balances the individual's right to privacy against the government's legitimate need to gather information to obtain evidence to be used against those who commit crimes against society.

WHAT IS A "SEIZURE"?

Seizure of a person is analogous to arrest. If, in view of all the circumstances, the person does not believe he is free to leave, he is seized.[28] The U.S. Supreme Court has said that, to constitute seizure of a person, there must be either the application of physical force, however slight, or submission to an officer's showing of authority to restrain the person's liberty.[29]

UNREASONABLE SEARCH AND SEIZURE

When an expectation of privacy which society considers reasonable is infringed, that is an unreasonable search.[30]

"The capacity to claim the protection of the [Fourth] Amendment depends not upon a property right in the invaded place, but upon whether the area was one in which there was a reasonable expectation of freedom from governmental intrusion."[31]

In seeking a search warrant, it is necessary to state with specificity the items to be seized and the place to be searched. It is not enough to say, for example, "to find evidence of arson."

PERSONNEL SHOULD KNOW SEARCH AND SEIZURE PROCEDURES

Because the law relating to searches and seizures is complex, it might be useful to have an attorney prepare forms to be used by fire department personnel in preparing requests for search warrants. Criminal procedures vary from one jurisdiction to another. Make sure all appropriate fire department personnel know the requirements in your jurisdiction.

The person serving the search warrant may use reasonable force to gain entry if necessary. Following service of the warrant, it is necessary to make a "return" of the warrant listing all the items which were seized under the warrant and other details of the search, including the time, location etc. Full attention should be made to this requirement because failure to make a proper "return" may later cause serious problems in prosecuting the case.

FIFTH AMENDMENT

The Fifth Amendment to the U.S. Constitution protects an individual from being forced to testify against him/herself which might tend to leave him/her open to criminal prosecution. If the person is granted immunity, s/he must testify.

The protections of the Fifth Amendment apply to the states through the fourteenth amendment. The privilege "protects an accused only from being compelled to testify against himself, or otherwise provide the state with evidence of a testimonial or communicative nature." [32]

"In order to be testimonial, an accused's communication must itself, explicitly or implicitly, relate a factual assertion or disclose information. Only then is a person compelled to be a witness against himself." [33]

It is permissible to obtain items from the accused. The only evidence which the accused cannot be forced to give is testimony against him/herself.

Placing a person in a lineup or asking him/her to repeat a phrase so a witness can listen to the voice are not prohibited by the Fifth Amendment. A person can be required to give a handwriting or voice sample or blood or fingerprints without violating the Fifth Amendment because these are identifying characteristics outside the privilege's protection.

SIXTH AMENDMENT

The Sixth Amendment to the U.S. Constitution requires that the defendant in a criminal case be given a speedy trial. It also provides that, in a criminal case, the accused is entitled to have a lawyer defend him/her against the charges and, if s/he can't afford to hire a lawyer, the government will hire one to handle the defense of the accused. Under our system, the accused — regardless of how heinous the crime — is entitled to the strongest, most vigorous defense possible, and his/her attorney has a legal and ethical obligation to provide it by presenting the law and the facts in the most forceful, persuasive manner possible to reflect most favorably on the client's position.

CHAPTER FOUR

CRIMINAL LAW

L aws, both civil and criminal, are enacted by federal, state and local governments. Within their respective jurisdictions, these laws must be obeyed.

CRIME

A crime is an offense against society which violates the law. It may be something the law forbids us to do, or requires us to do, to protect life and property, to preserve our freedoms, to allow us to live in a peaceful, harmonious society, to uphold our moral values, or to preserve our system of government. For example, we must not steal and we must file our federal income tax return every year.

It covers the firefighter who steals valuables from a burning building, the businessman who improperly disposes of hazardous wastes, the citizen who cheats on his/her expense account or income tax return, the government worker who sells defense secrets to a foreign nation as well as murderers, burglars, pickpockets, rapists, drug dealers/users and fire code violators.

Crime is a major problem for all of us in society. Well over 12 million crimes are reported every year while only about 20 percent result in arrests. Still fewer result in convictions. If those statistics were not bad enough, we know that most crimes are not even reported to the police.

Obviously, the victims of crime suffer the most. The price which victims of crime pay or the psychological effect which fear of crime has on us cannot be measured, but viewed strictly from a monetary point of view, crime costs us over $100 billion per year in losses, insurance costs, and costs of administering the criminal justice system.

People between the ages of 15 and 24 commit more violent crimes than any other age group. Males commit six times as many crimes as females, but the crime rate among females has been growing at the fastest rate.

CODE VIOLATION COULD BE A CRIME

Many fire codes define non-compliance with the code as a criminal act, subjecting the defendant to criminal penalties. Some make them civil infractions and others give the government the option of proceeding under either criminal or civil law.

BURDEN OF PROOF

The standard of proof in a civil case is much easier than in a criminal case. In a civil case the issue must be proved "by a preponderance of the evidence," while in a criminal case it must be proved "beyond a reasonable doubt."

"Beyond a reasonable doubt" does not mean that the proof must overcome all doubts, but only to overcome reasonable doubt. The jury must be firmly convinced of the defendant's guilt.

PROSECUTORS HAVE DISCRETION

For the fire department to proceed with a criminal prosecution, the concurrence of the prosecutor is required. Often s/he may perceive fire code violations as unimportant, particularly when measured against the many other crimes which s/he is faced with prosecuting. This requires education to get his/her attention focused on why fire code violations — matters of life and death — are important.

INVESTIGATORS MUST KNOW ARREST PROCEDURES

Fire Department investigators must understand proper investigative and arrest procedures. If they do not, a wrongdoer might go free as a result, and the fire department might face a civil suit.

For example, to make sure that statements made by an accused will be admissible in a subsequent trial, the defendant must be advised of his/her constitutional rights before questioning begins about the crime. In *Miranda v. Arizona*, the U.S. Supreme Court said, "The Fifth Amendment privilege is so fundamental to our system of constitutional rule and the expedient of giving an adequate warning as to the availability of the privilege is so simple, we will not pause to inquire in individual cases whether the defendant was aware of his rights without a warning being given. Assessments of the knowledge the defendant possessed, based on information as to his age, education, intelligence, or prior contact with authorities, can never be more than speculation; a warning is a clear-cut fact. More important, whatever the background of the person interrogated, a warning at the time of the interrogation is indispensable to overcome... pressures [of the arrest] and to insure that the individual knows he is free to exercise the privilege at that point in time.... Prior to any questioning, the person must be warned that he has a right to remain silent, that any statement he does make may be used as evidence against him, and that he has a right to the presence of an attorney, either retained or appointed.   The defendant may waive effectuation of these rights, provided the waiver is made voluntarily, knowingly and intelligently. If, however, he indicates in any manner and at any stage of the process that he wishes to consult with an attorney before speaking, there can be no questioning. Likewise, if the individual is alone and indicates in any manner that he does not wish to be interrogated, the [investigator] may not question him. The mere fact that he may have answered some questions or volunteered some statements on his own does not deprive him of the right to refrain from answering any further inquiries until he has consulted with an attorney...."[34]

CONFESSIONS

As stated above, confessions are not admissible unless they are voluntary. There is a correlation between the Fourth and Fifth Amendments. A search for what is forbidden under the Fifth Amendment is generally unreasonable under the Fourth Amendment.[35]

A statement made following an illegal arrest is not admissible unless the original taint is removed. Miranda warnings alone are not sufficient. Lapse of time, intervening events and flagrancy of the official misconduct may also be considered. The burden, as always, is on the prosecution to prove that the confession was voluntary and untainted.[36]

In respect to criminal law, the fire department should ask a number of questions about constitutional law and its own state's laws and, when the answers are learned, they should be incorporated into the department's training programs.

- What police powers do fire prevention and arson investigators have?
- Do they have arrest powers?
- Can they carry weapons?
- Are there guidelines on the use of deadly force?
- Can they serve subpoenas?
- What are the "Miranda" warnings and when should they be given to a suspect? (Many investigators carry a little card in their wallets to remind them of the "Miranda" warnings about which the accused must be advised before being questioned. These warnings are: "You have the right to remain silent. Anything you say can be used against you in court. You have the right to a lawyer and to have one present while you are being questioned. If you cannot afford a lawyer, one will be appointed for you before any questioning begins." Any questioning of a defendant during which he might confess, raises the issue of whether or not the defendant was advised of his/her rights, and whether the information derived during the questioning was voluntary and therefore admissible as evidence.)
- When is a confession voluntary?

- How is a line-up conducted?
- What is evidence? How can it be legally acquired? How must it be preserved?
- When and how should an arrest be made?
- If investigators can carry firearms, is there a policy as to when they may draw their weapons?

Not only should you ensure that fire department personnel are properly trained regarding these matters, but someone should maintain continuing liaison with the prosecutor's office to stay current with policies and legal requirements in these areas.

EVIDENCE — THE HEART OF EVERY INVESTIGATION

The heart of all criminal investigations is obtaining sufficient, legally admissible evidence to establish that a violation has occurred and that the defendant is the person who committed it. Regardless of the purpose of an investigation, the collection of evidence is critical to a successful conclusion.

Not only should investigators have an understanding of evidence, but all firefighters should have this understanding. They should know that evidence in plain view can be seized so they can prevent it from being accidentally destroyed before investigators arrive on the fire scene. They should also know how important it is to preserve the chain of custody of evidence. Firefighters should know they might be called upon to testify personally as to what they observed at the fire scene.

EVIDENCE — QUANTITY AND QUALITY

There are two general considerations in evidence. The first is the quantity of the evidence needed. This is especially important in a criminal investigation. To secure a conviction the prosecutor must prove every element of the crime. To do this, there must be adequate evidence to work with. If your investigators understand what evidence the prosecutor needs, the investigation can be conducted accordingly.

The second consideration in gathering evidence is the quality of the evidence. This means that the evidence collected must be the best and most reliable possible. For example, if the chain of custody of evidence cannot be proved, a defendant may successfully claim that it has been tampered with. If evidence is not properly collected and preserved, laboratory analysis could be impossible or inconclusive, and evidence illegally obtained is worse than no evidence at all.

ELEMENTS OF A CRIME

The law prohibiting or mandating something defines the crime and the elements which are necessary to commit that crime. In prosecuting an accused, the government must prove that all of the required elements for that particular crime have been met.

For example, assume your State statute defines robbery as "the unlawful taking of personal property in the possession of another against his/her will by force or by putting him/her in fear." The elements of the crime are: 1) "the unlawful taking of personal property," 2) "in the possession of another," and 3) "by force or by putting that person in fear." If all the elements are not present, it is not robbery.

The news media frequently reports that someone's home "was robbed last night while the family was out of town on vacation." From that statement alone, we know that it could not possibly be robbery because the victim was not present at the time of the crime so whatever was stolen was not taken from that person by force or by putting him/her in fear. This is sloppy reporting by the news media. If the government's prosecutor did, in fact, try to prosecute the burglar for robbery, the prosecution would fail because the elements of robbery could not be proved.

CRIMINAL INTENT

Virtually all crimes require an act and criminal intent (mens rea). This means that the person intended to commit a crime. It involves "knowing" and "wilful action." If the person performs the act through mistake or other innocent reason or because s/he is insane, there can be no criminal intent and the prosecution of the accused for that crime would fail.

STRICT LIABILITY

Exceptions to this rule are those crimes where the state legislature in passing the law imposed strict liability. In other words, the legislature made an act or omission a crime, even though the wrongdoer has no criminal intent. It puts liability on a non actor. In these cases, the act itself creates the crime whether the person knew s/he was committing it or not. A strong argument can be made that this is not just. Some environmental crimes fall into this category. Sometimes the doctrine of respondeat superior (which makes an employer liable for the wrongdoing of an employee) also applies to crimes. Some environmental laws make the president or the corporation itself liable for the criminal acts of employees, even though they did not know about them. This might be important for the fire department to consider in some hazardous material cases.

MOTIVE

Motive is different from intent. Motive is why someone does something. A good motive does not excuse a criminal act.

STATUTE OF LIMITATIONS

The Statute of Limitations specifies the time within which a legal action must be initiated or it is barred forever. These time restrictions apply in civil as well as criminal cases. There are different lengths of time for various matters. In criminal law, indictment of the accused stops the running of the time under the Statute of Limitations. In a civil case, filing a suit stops the running of the Statute of Limitations.

SAME ACT CAN BE BOTH CRIME AND TORT

The same act can be a tort (a legal wrong or breach of duty under civil law for which the law allows the victim to bring an action against the perpetrator and be compensated for his/her loss) and a crime (an offense against society prosecuted by the government, the object of which is to punish the wrongdoer and to deter others from committing the crime.)

Suppose a disgruntled male employee sets fire to a tavern. The owner of the tavern can bring a civil suit against the employee seeking compensation for the loss suffered from the fire and the State can prosecute him for arson. In the civil case, the law wants to try to make the victim whole again with money damages to make up for what has been lost. In the criminal case, the government brings the action against the man on behalf of all of us in society to punish him and to discourage others from doing the same thing.

CRIMES MAY BE FEDERAL OR STATE

Some crimes are federal offenses; some are state offenses. Some are violations of both state and federal law and can be prosecuted by either or both. The same act might also be a crime in more than one state.

FELONIES AND MISDEMEANORS

Crimes may be felonies (a more serious offense for which the punishment is greater, such as imprisonment for more than a year) or misdemeanors (less serious offenses for which the punishment is less than one year in jail.)

ACCESSORIES

An "accessory before the fact" is someone who participates in some way in the planning of the criminal act or in preparing for it, but was not personally present when it was committed.

An "accessory after the fact" is someone who somehow assists the perpetrators after the crime has been committed such as helping them escape or harboring them, or buying goods known to be stolen.

CONSPIRACY

A criminal conspiracy is an agreement between two or more individuals with intent to commit an unlawful act, or a lawful act in an unlawful way, and one or more of the conspirators does something to further the conspiracy. Under federal law and in most states, an essential element of the crime of conspiracy is that one of the conspirators must perform some overt act in furtherance of the conspiracy. For example, three people plan to burn a building. The crime of conspiracy has not yet been committed. One of them then goes out and buys gasoline to be used in the crime. All three of the plotters have now committed criminal conspiracy, because buying the gasoline is an overt act which fulfills the requirements of the crime.

The elements of the crime of conspiracy consist of 1) the existence of a conspiracy, either express or implied; (2) the defendant knew the essential objectives of the conspiracy; (3) the defendant voluntarily and knowingly participated in the conspiracy and (4) one or more acts are done in furtherance of the illegal purpose. The act in furtherance of the conspiracy by one of the conspirators does not have to be an illegal act.

SOLICITING

In some states it is a crime to solicit a person to commit a crime. For example, hiring someone to torch a building, or to commit murder are crimes. (Some states also call propositioning someone for prostitution "soliciting.")

ATTEMPT

In most states it is a crime to attempt to commit a crime, even though it is not carried out. The elements of the crime include, not only the intention to commit the crime, but also the taking of some step toward committing the crime. For example, a man begins to set a building on fire, but is scared off by something or someone before igniting it. He has committed the crime of attempt even though the building never burned. Attempt usually means any act which is a substantial step toward the commission of the offense and is done for the purpose of committing such an act. Unlike conspiracy, one person alone can commit the crime of attempt.

HOMICIDE

Homicide is the killing of one human being by another.

Noncriminal homicide is not a crime. It is killing which is justifiable. For example, a soldier killing an enemy soldier during wartime, the executioner who throws the switch at the electrocution of a convicted killer, a police officer who shoots and kills someone to save a victim's life, or a person who kills someone in self-defense.

CRIMINAL HOMICIDE

Murder is killing someone with malice. In other words, the perpetrator acts with the intent to kill or seriously hurt the victim.

First-degree murder is a deliberate, premeditated killing which is done with malice. The perpetrator thinks about it and plans it in advance.

Second-degree murder is a killing committed with malice, but without premeditation. The intent to murder the victim did not occur until the act itself was committed.

Felony murder is a killing which occurs during the commission of certain felonies. The prosecutor does not have to prove the intent to kill, because the killing might be accidental, but the fact that it happened during the commission of a serious crime, (arson, rape, robbery etc.) the perpetrator is held responsible for the results of his/her action. In most states felony murder is treated as first-degree murder.

Involuntary manslaughter is an unintentional killing which results from reckless conduct.

Voluntary manslaughter is an intentional killing, but it is committed under mitigating circumstances which, while they do not justify the act, they lessen the guilt.

Negligent homicide is killing someone through criminal negligence. Negligence is the failure to exercise the care which a reasonable and prudent person would exercise under the same circumstances. For example, driving a motor vehicle negligently and causing an accident in which a person is killed would be negligent homicide.

ASSAULT

Assault is intentionally putting someone in fear of his or her safety.

BATTERY

Battery is an intentional, unpermitted, offensive contact or touching of another. (When a man fondles a woman without her consent, that is battery, as well when someone strikes another.)

ARSON

State statutes define arson differently. At common law, arson was the malicious burning of the house or outhouse of another. It included structures appurtenant to the dwelling. Generally, it had to be a dwelling.

LARCENY

Larceny is unlawfully taking and carrying away the personal property of another with intent to steal it. Most states divide it into grand and petty larceny, depending on the value of the thing stolen.

EMBEZZLEMENT

Embezzlement is the taking of property which was in the embezzler's possession in trust.

BURGLARY

Burglary is defined in most states as the unauthorized entry into any structure with intent to commit a crime. At common law, it was restricted to entering dwellings at night. Some states have a separate crime of "store breaking" or they call burglary "breaking and entering."

ROBBERY

Robbery is the unlawful taking of personal property in the possession of another against his/her will by force or by putting him/her in fear.

FORGERY

Forgery is falsely making or changing a writing or document with intent to defraud.

EXTORTION

Extortion (sometimes called blackmail) is using threats to wrongfully obtain the property of another.

DRUNK DRIVING

All states have made drunk driving a crime. Some call it "driving while impaired," some "driving under the influence," or "driving while intoxicated." Society has been getting much more strict in prosecuting this offense and handing out stiffer sentences. Sentences for conviction of this offense can include suspending or revoking the driver's license, jail, fine, attendance at a special school, and community service. The sentence can be one or a combination of these punishments.

DRUGS

Distribution, sale or possession of illegal drugs is a crime which may violate both state and federal laws.

In some states a single possession of a small amount of a drug constitutes a serious felony. Some statutes allow major drug peddlers to be sent to prison for life without possibility of parole.

Drug use has become an increasingly serious problem for our society which affects all of us. It is estimated that illegal drug use costs our society more than $60 billion per year. In addition to drug crimes themselves, drug use has substantially increased crime in general since addicted users frequently commit crimes to obtain money to sustain their drug habit.

More and more people are being arrested, convicted and sent to prison for drug offenses, placing a tremendous burden on our justice system. More than half the inmates in federal prisons are serving drug-related sentences.

Drug use by fire department personnel is a very serious matter and should be dealt with vigorously.

EXCLUSIONARY RULE

When a court finds that government's search was unreasonable, the evidence uncovered in that search may not be used against the accused. This is called the exclusionary rule. The defendant can still be tried, but the illegally obtained evidence cannot be used.

OBTAINING A SEARCH OR ARREST WARRANT

In order to obtain a search warrant, the investigator must go before a judge, magistrate, justice of the peace or other judicial officer and explain what evidence s/he is seeking and show probable cause that it is related to a crime which has been committed.

Probable cause, of course, must also be shown in order to obtain an arrest warrant as well as a search warrant. Probable cause is defined as a reasonable belief than a person has committed a crime. This reasonable belief may be based on much less evidence that is necessary to prove a person's guilt at a trial. Probable cause may be established from information provided by citizens, victims, witnesses, informants or the investigators themselves. The officer seeking the warrant must go before the judicial officer and convince him/her that there is probable cause. Probable cause requires more than a hunch or mere suspicion.

When arresting without a warrant, investigators must use their own judgment as to what is reasonable under the circumstances.

EXCEPTION TO WARRANT REQUIREMENTS

The general rule is that to enter another's property, you must have a warrant. However, there are a number of exceptions when the government does not have to get either an arrest or search warrant. For example, investigators may legally search incident to a lawful arrest or with the owner's consent, or under "exigent" (emergency) circumstances, or when something is in a public place, or in plain view. Generally, if the investigator has a legal right to be where s/he is, s/he can seize evidence which can be seen.

TRAINING SHOULD COVER SEARCH AND SEIZURE

Fire Department training programs should alert personnel to the basic issues involved in search and seizure to ensure that those who are likely to be in a position to observe evidence during the course of their duties know what they should and should not do.

In searches and seizures the general rule is: if the searcher has the right to be where s/he is, anything found there is admissible evidence.

Evidence seized in violation of the Fourth Amendment not only cannot be used in a criminal trial, but could also result in a civil action against the individual and the department for violating someone's constitutional rights.

Unless there is a legal reason to do so — with a warrant, consent or under one of the exceptions — one may not enter upon the property of another. Any evidence observed or obtained while on the property illegally will not be admissible in court. Furthermore, any evidence which results from a lead followed up from the illegally obtained material is also barred from use. This is known as the "fruit of the poisonous tree" doctrine.

For example, in an arson case, suppose a key is found at the scene of the fire with the defendant's fingerprints on it. The key would be useful to establish evidence that the defendant was at the scene of the fire. Subsequent investigation reveals that the key fits a safe deposit box. A search of the box reveals cash and documents proving that the defendant had intentionally bankrupted his company and had set the fire to cover up the crime. If the key had been seized improperly in the first place, for instance, if it had been found during an illegal entry, it would be inadmissible. Furthermore, the fingerprint and the evidence obtained from the safe deposit box would also be inadmissible because they are the "fruit" of the earlier, illegal seizure, which would be the "poisonous tree."

AN INDICTMENT IS NOT A CONVICTION

Unfortunately, indictment for a crime is usually associated in the public's mind with guilt. This is completely inaccurate and highly unfair to the accused. The prosecutor controls the grand jury which returns indictments, and it almost always does exactly what the prosecutor wants it to do. Neither the defendant nor his/her attorney is present at grand jury deliberations, so there is no cross-examination of witnesses, hearsay evidence is permitted and the rules of evidence do not apply. The grand jurors hear only the prosecutor's slanted view of the case. They hear no defense from the defendant and no exculpatory information.

The news media publicizes an indictment in such a way that, not only does the uninformed public equate it with guilt, but often the publicity itself makes it difficult for the defendant to get a fair trial.

HOW CAN A LAWYER DEFEND THE GUILTY?

Frequently, in a criminal case, lay persons will ask, "How can that lawyer defend a client he knows is guilty?" The question reflects a lack of understanding of our system of justice.

It is not properly the attorney's concern whether the client is guilty or not. Under our adversary system of justice, everyone, no matter how heinous the crime with which s/he is charged, is entitled to the strongest, best possible legal defense.

When the adverse party is the government, with its massive power and vast resources, it is more important than ever that the defendant have a competent, aggressive attorney defending him/her to guarantee that his/her rights will be protected. The defendant's lawyer must make sure that the prosecutor follows all of the prescribed rules and that the defendant gets a fair trial.

It is very important to keep in mind that, when a person is charged with a crime, there is a strong presumption in the law that s/he is innocent until proved guilty beyond a reasonable doubt. It is not only a question of whether or not the defendant performed the act involved, but whether or not the government has enough legally admissible evidence to prove beyond a reasonable doubt all the elements of the crime and that the accused is guilty of having committed the crime with which s/he is charged. That is a major, important distinction. If the defendant is found guilty, it is then the attorney's job to attempt to get him/her fair punishment.

There are numerous safeguards in our U.S. Constitution and in our criminal laws to protect defendants against the government. Rightly so. Even with all of these safeguards, however, innocent people still get convicted and sent to prison. It is unfair to the defendant and to our system of justice for lay people to make judgments about the guilt or innocence of a person on the basis of sensational media coverage. If protection of rights and presumption of innocence can be denied to any defendant, they can be denied to any of us.

CHAPTER FIVE

ARSON

Arson in general is the willful burning of another's property. In most states by statute it is now arson to burn any building or structure whether it is owned by the person doing the burning or someone else. Usually, the statutes create a separate crime for intentionally burning any property for the purpose of defrauding an insurance company, no matter what kind of property it is or who owns it.

At common law, arson was defined as the malicious burning of the house or outhouse of another. It included structures appurtenant to the dwelling. Generally, the common law required that the house be a dwelling. In other words, if the house were abandoned, the person burning it would not be guilty of the common law crime of arson and burning other non dwelling buildings was not arson. Neither was it arson for a person to burn his/her own house. It had to be the house "of another."

State statutes vary in their definition of arson. If your state's arson statute refers to "dwelling house" rather than "house," it may be important to determine whether or not occupancy is an essential element which is required to constitute the crime of arson. Some states have specifically eliminated occupancy as an element of the crime and have broadened arson to include burning any real or personal property, including vehicles.

Some states have divided arson into first degree, second degree and third degree.

First degree arson is burning an inhabited dwelling house in the night time.

Second degree arson is the burning at night of a building other than a dwelling house, but close enough to a dwelling house to put it at risk.

Third degree arson is the burning of a building or structure not covered by first and second degree arson, or the burning of a property, the arsonist's own or of another, with intent to defraud or prejudice an insurer.

ELEMENTS OF THE CRIME OF ARSON

In general, the elements of the crime of arson are:

(1) There must have been a burning with damage to property. The slightest damage will qualify[37] or in Utah even the charring of fibers.[38]

(2) There must be criminal intent. (Willful and malicious burning, not necessarily the specific intent to set a particular building on fire.)

3) Conduct of the defendant in starting, conspiring to start, aiding and abetting in the starting of the fire or otherwise being involved in the fire. (Motive is not an element, but it might help to explain what happened.)

(4) The fire was not accidental.

(5) Incendiary origin of the fire.

MICHIGAN v. TYLER AND THE FOURTH AMENDMENT

The Supreme Court stated in the case of *Michigan v. Tyler*, that the basic purpose of the Fourth Amendment "is to safeguard the private security of individuals against arbitrary invasions by government officials."[39]

In the *Tyler* case a fire broke out during the night of January 21, 1970 at Tyler's Auction, a furniture store. When the Fire Chief arrived at 2 AM, firefighters were watering down smoldering embers. He was informed that two plastic containers of flammable liquids had been found in the store. The Chief and another officer, using portable lights, began an investigation. The Chief suspected that the fire was caused by arson. He summoned a police detective who arrived at about 3:30 AM and took photographs of the interior of the store and the plastic containers. At 4 AM, after the fire had been extinguished and all the firefighters and police had left the scene, the investigation was discontinued because smoke, steam and darkness restricted visibility. Four hours later, at first light, the Chief, along with an Assistant Chief, returned to the scene. It was the Assistant Chief's job to determine the origin of fires. After a cursory examination, they left. The Assistant Chief returned at 9 AM with the police detective. During that investigation, they found additional evidence of arson and seized it.

On February 16, nearly four weeks later, a Michigan State Police arson investigator searched the building without a warrant. Several additional warrantless entries were made after that date and various items of physical evidence were seized. The arson investigator, after observing the fire damage, expressed the opinion that it had been arson.

Tyler and another person were arrested and charged with conspiracy to commit arson. They were convicted. At their trial they objected to the admission of various items of physical evidence and the testimony of the investigating officers because none of the searches had been conducted with permission or pursuant to a warrant.

The prosecution argued that, once firefighters have entered a person's property to put out a fire, the owner has no reasonable privacy rights.

The U.S. Supreme Court did not agree. The Court said, "That innocent fire victims have no protectable expectations of privacy in whatever remains of their property... is contrary to common experience. People may go on living in their homes or working in their offices after a fire" and personal property often remains in the fire-damaged building. The Court said that it was not possible "to justify abandonment by arson when that arson has not yet been proven, and a conviction cannot be used" after the fact to validate the introduction of evidence used to secure the same conviction.

The Court said, as a general rule, fire investigators must obtain a search warrant before entering private property. It was pointed out that an administrative warrant was adequate to determine fire cause-and-origin, and any evidence found during that search under the administrative warrant could be "used to establish probable cause for the issuance of a criminal investigative search warrant or in prosecution." However, the courts said, if the cause of the fire is already known, and "authorities are seeking evidence to be used in a criminal prosecution, the usual standard [probable cause] will apply."

The High Court said that clearly a burning building presents an emergency situation to render entry without a warrant reasonable and, once the firefighters were in the building for this purpose, they could seize evidence in plain view. The Court said, "Fire officials are charged, not only with extinguishing fires, but with finding their causes. Prompt determination of the fire's origin may be necessary to prevent its reoccurrence, as through the detection of continuing dangers such as faulty wiring or a defective furnace. Immediate investigation may also be necessary to preserve evidence from intentional or accidental destruction. And, of course, the sooner the officials complete their duties, the less will be their subsequent interference with the privacy and the recovery efforts of the victims. For these reasons, officials need no warrant to remain in a building for a reasonable time to investigate the cause of a blaze after it has been extinguished. And, if the warrantless entry to put out the fire and determine its cause is constitutional, the warrantless seizure of evidence while inspecting the premises for these purposes is also constitutional." In other words, the two plastic containers which were found on the floor of one of the showrooms, were seized legally.

The Michigan Supreme Court had held that the officers re-entry four hours after they had left at 4 AM required a warrant. The U.S. Supreme Court did not agree on this point. The High Court said, "On the facts of this case, we do not believe that a warrant was necessary for the early morning re-entries.... As the fire was being extinguished [the Chief] and his assistants began their investigations, but visibility was severely hindered by darkness, steam and smoke. Thus, they departed at 4 AM and returned shortly after daylight to continue the investigation. Little purpose would have been served by their remaining in the building, except to remove any doubt about the legality of the warrantless search and seizure later that same morning. Under these circumstances, we find that the morning entries were no more than an actual continuation of the first, and the lack of a warrant thus did not invalidate the resulting seizure of the evidence."

But the Court said the entries several weeks later should have been conducted pursuant to a warrant. The Court said, "The entries occurring after January 22, however, were clearly detached from the initial exigency and warrantless entry. Since all of these entries were conducted without valid warrants and without consent, they were invalid under the Fourth Amendment, and any evidence obtained as a result of those entries must, therefore, be excluded at [the defendants'] retrial."

Dissenting justices in the *Tyler* case cited the practical difficulties in determining when a warrantless re-entry was a proper continuation of the prior legal search.

MICHIGAN v. CLIFFORD

In the *Michigan v. Clifford* case, the Supreme Court attempted to clarify this point.[40]

In the Clifford case, decided January 11, 1984, the Supreme Court, by a 5 to 4 decision further restricted fire investigators' access to fire-damaged property without a search warrant.

The Detroit Fire Department had responded at 5:42 AM on October 18, 1980 to a fire in a two-and-one-half-story, brick-and-frame private dwelling. At 7:04 AM firefighters and police left the scene when the fire was extinguished. Fire damage in the basement and lower interior structure was extensive, but the exterior and upstairs rooms suffered only minor smoke damage. During suppression and overhaul, firefighters discovered a Coleman fuel can in the basement. They left it outside by the basement side door.

The case was assigned to an arson investigator at 8 AM who, along with his partner, did not arrive on the scene until 1 PM.

The neighbor who had reported the fire advised the Cliffords, owners of the property — who were away on a camping trip — about the fire. They asked him to notify their insurance agent "to send out a crew to secure the house." When the investigators arrived on the scene, the insurance company's crew was already working and pumped nearly six inches of water out of the basement. The investigators found the Coleman fuel can and marked it as evidence. After the basement was pumped out, they entered the basement without a warrant. It was the Arson Division's policy that such entries were authorized "when the owner was not present, the premises were open to trespass and the search occurred within a reasonable time of the fire." The investigators confirmed that the fire had started under the basement stairway. They detected a strong odor of fuel throughout the basement and found two additional Coleman fuel cans under the stairs. In the debris, one of them found a crock pot with attached wires leading to an electrical timer that was plugged into an outlet a few feet away. The timer was set to turn on at approximately 3:45 AM. This evidence was seized and marked.

They then searched the remainder of the house, still without a warrant, and called in a photographer to take pictures throughout the house. They searched through dresser drawers and found them filled with old, worthless clothes. They found nails on walls, but no pictures hanging on them. They found the wiring and cassettes for a videotape machine, but no videotape machine.

The Cliffords were arrested and charged with arson. They moved to exclude all evidence and testimony on the grounds that the search was "conducted without a warrant, consent, or exigent circumstances." The trial court rejected their motion, but allowed them to appeal before going to trial.

The Michigan Court of Appeals reversed and the State of Michigan appealed. The Supreme Court of the United States agreed to hear the case "to clarify doubt that appears to exist as to the application of our decision in *Tyler*."

The U.S. High Court held that all the evidence seized was inadmissible except the Coleman fuel can which had been found first by the firefighters in plain view during suppression and overhaul.

The Supreme Court said, "At least where a homeowner has made a reasonable effort to secure his fire-damaged property after the blaze has been extinguished, and the fire and police units have left the scene, we hold that a subsequent post-fire search must be conducted pursuant to a warrant, consent or the identification of some new exigency." The emergency had ended when police and firefighters left the scene at 7:04 AM with no threat of the fire rekindling or immediate destruction of the evidence.

The Court said that, even if the initial entry into the house had been legal, a criminal warrant would still have been required to search beyond the basement area.

Once the arson investigators' entry was unconstitutional all the evidence they seized was inadmissible.

The Court went on to outline new restrictions on the scope of permissible searches into the cause-and-origin of a fire. The Court stated, "Even if the mid-day basement search had been a valid administrative search, it would not have justified the upstairs search. The scope of such a search is limited to that reasonably necessary to determine the cause-and-origin of a fire and to ensure against rekindling. As soon as the investigators determined that the fire had originated in the basement and had been caused by the crock pot and timer found beneath the basement stairs, the scope of their search was limited to the basement area. Although the investigators could have used whatever evidence they discovered in the basement to establish probable cause to search the remainder of the house, they could not lawfully undertake that search without a prior judicial determination that a successful showing of probable cause had been made. Because there were no exigent circumstances justifying the upstairs search, and it was undertaken without a prior showing of probable cause before an independent judicial officer, we hold that this search of a home was unreasonable under the Fourth Amendment, regardless of the validity of the basement search."

The court continued, "The warrantless intrusion into the upstairs regions of the Clifford house presents a telling illustration of the importance of prior judicial review of proposed administrative searches. If an administrative warrant had been obtained in this case, it presumably would have limited the scope of the proposed investigation and would have prevented the warrantless intrusion into the upper rooms of the Clifford home. An administrative warrant does not give fire officials license to roam freely through the fire victim's private residence."

The scope of an administrative warrant is limited to searching what is reasonably necessary to determine the cause-and-origin of a fire and to make sure it does not rekindle. Cause-and-origin searches must be restricted to those areas where evidence of the cause of the fire reasonably could be expected to be found. With the basement having been established as the place where the fire started, the arson investigators in the *Clifford* case had no legal justification for searching the upper rooms and most certainly not dresser drawers. Once the cause of the fire has been determined, the search must be limited to that area of the building.

Four dissenting justices would have required only an administrative warrant to search the remainder of the house. Justice Stevens concurred because, although he would require a criminal warrant rather than an administrative warrant to search, he would require no warrant if notice were given to the owner. Four members of the Court apparently would not require any warrant for entry within a reasonable time to determine cause of the fire, but might require notice in some circumstances not present in this case when the owners were out of town.

This case illustrates how difficult it is to determine what the law is when the Nation's top justices disagree among themselves as to what the law is. It can readily be seen by this case that a change of personnel on the U.S. Supreme Court could very well change the outcome of these type of fire department cases.

Whenever it appears that the fire is the result of arson, a criminal search warrant should be obtained immediately. That is the best way to make sure that evidence seized will be admissible at a subsequent trial.

Both the *Tyler* and *Clifford* cases point up the importance for fire departments to have standard operating procedures, forms and procedures, and training programs so that firefighters will know the limits of their authority to search and seize evidence at the fire scene, and so that investigators will conduct their investigations and searches in accordance with the U. S. Constitution's restrictions.

COURT CLARIFIES GEORGIA'S ARSON LAW

In a 1987 Georgia case a man was indicted on two counts of arson in the first degree. The superior court granted a directed verdict on one count and the jury found him guilty of criminal damage to property in the second degree.

The defendant appealed on the basis that the State of Georgia had failed to show damage to "the dwelling house of another" as the statute required, and the superior court, therefore, erred in failing to give him a directed verdict of acquittal.

He also argued that the trial court erred in not instructing the jury correctly concerning the elements of the crime for which he was charged. The State of Georgia countered by arguing that the mobile home where the defendant had allegedly started the fire had smoke damage and this is enough to satisfy the damage requirement of the charge against him.

The appellate court said that, although the State had failed to show damage to "the dwelling of another," the evidence was enough to convict the defendant of the lesser offense — criminal damage in the second degree. The court said a crime is a "lesser included offense" when it is established by the same proof or less proof than for the greater offense. The court said there was evidence of damage to a bed and bed clothes inside the mobile home and this damage was established by proof of the same conduct as the first degree arson, even though it required proof of a less culpable mental state. The guilty verdict was affirmed.[41]

STATE COURT UPHOLDS CONVICTION

A fire in a store was determined by investigators to be arson. A man, who was arrested on other charges, confessed to the arson and implicated a second man. The first man received a lighter sentence than the second man. The second man received ten years for second degree arson. He appealed.

The appellate court affirmed his conviction.

There were two statutes at issue: (1) burning insured property and (2) second degree arson. The court said the wording of the indictment was similar to the wording of the statute for second degree arson in that any person who willfully and maliciously sets fire to, or causes to be burned, any building shall be guilty of arson in the second degree.

This statute on second-degree arson carried a maximum penalty of ten years while the statute on insured property arson carried a maximum penalty of only five years.

The appellate court said it was clear from the language in the indictment that the second man was being prosecuted under the second degree arson statute and he could easily have determined this by looking at the indictment and the statute.[42]

While arson is usually a state crime, there are a number of federal statutes related to burning.

ANTI-ARSON ACT

The Anti-Arson Act (18 U.S. Code, Section 844 (i,j)) provides that anyone who "maliciously damages or destroys, or attempts to damage or destroy, by means of fire or an explosive, any building, vehicle, or other real or personal property used in interstate or foreign commerce or in any activity affecting interstate or foreign commerce" has violated the Act.

A security guard, who burned 12 units of a condominium under construction, was prosecuted under this act. Although the building was not occupied, the federal court held that the building met the "used" requirement of the Act. The court also found that the housing construction had an effect on interstate commerce.

NATIONAL FIREARMS ACT

The National Firearms Act (26 U.S. Code Section 5845) bans use of destructive devices. Destructive devices are defined as any explosive, incendiary, or poison gas, bomb, grenade, rocket with a propellant charge of more than four ounces, a missile having an explosive or incendiary charge of more than one-quarter ounce, a mine or similar device.

A homemade explosive was held to be a device covered by the Act[43], as was a device consisting of six trash bags, each containing a five-gallon can of gasoline, suspended and connected by overlapping paper towels.[44]

CHAPTER SIX

CIVIL RIGHTS AND WRONGS

MAJOR FEDERAL EMPLOYMENT DISCRIMINATION LAWS

Race or Color	Civil Rights Act of 1866
	Civil Rights Act of 1964, Title VII; Section 1983
	Civil Rights Act of 1871
Sex	Equal Pay Act of 1963, Title VII (1964)
	Pregnancy Act of 1978
	Civil Rights Act of 1991
Religious	Title VII (1964)
National Origin	Title VII (1964)
Age	Age Discrimination in Employment Act of 1967, as amended 1986
Handicap	Rehabilitation Act of 1973
	Americans with Disabilities Act of 1990
General	State and Local Fiscal Assistance Act, 1972 (revenue sharing for public safety)
	Intergovernmental Personnel Act of 1970
	Executive Order 11246 (1965)

Many states have anti-discrimination laws of their own and some are more stringent than the federal law. Some go beyond the federal law in prohibiting employment discrimination based on a person's marital status, physical appearance, sexual orientation, political affiliation, etc.

Title VII of the Civil Rights Act of 1964 in Section 703 provides that "It shall be an unlawful employment practice for an employer: 1) to fail or refuse to hire or to discharge any individual, or otherwise to discriminate against any individual with respect to his compensation, terms, conditions or privileges of employment, because of such individual's race, color, religion, sex or national origin or 2) to limit, segregate, or classify his employees or applicants for employment in any way which would deprive any individual of employment opportunities or otherwise adversely affect this status as an employee because of such individual's race, color, religion, sex or national origin."

Discrimination by labor unions or management-labor committees is also prohibited by the Act.

WORKPLACE RIGHTS AND RESPONSIBILITIES

What is the workplace? It may be a government agency — federal, state or local — or a not-for-profit entity, or one of a number of different forms of business organizations.

WORKPLACE HAS BECOME LEGAL BATTLEGROUND

In addition to civil rights and employment discrimination cases, in recent years the workplace has been the situs for other emerging constitutional rights issues, such as drug testing and sexual harassment.

DISCRIMINATION UNDER THE CIVIL RIGHTS ACTS

The Civil Rights Act of 1866, enacted after the Civil War, was designed to protect the rights of newly freed slaves. In recent years, Section 1981 of this act has been used to attack employment discrimination based on race and national origin. The U.S. Supreme Court has said that it protects all identifiable classes who suffer discrimination solely because of their ancestry or racial characteristics.

Title VII of the Civil Rights Act of 1964 is broader than the Civil Rights Act of 1866, but plaintiffs using Section 1981 of the Civil Rights Act of 1866 do not have to comply with Title VII's complex procedural requirements before filing suit. Also, higher damages are allowed under the old law. The Supreme Court has restricted use of the 1866 Act, holding that it applies only to the making and enforcement of private contracts. This restricts its use primarily to hiring cases. Some courts have stretched its use to include discriminatory firings.

CIVIL RIGHTS ACT OF 1964

Title VII of the Civil Rights Act of 1964 covers any employer which engages in an activity affecting interstate commerce and has five or more employees.

The term "employer" includes individuals, partnerships, corporations, labor unions, employment agencies and state and local governments. It does not cover independent contractors, Indian tribes or certain tax-exempt private clubs.

Title VII prohibits virtually all the ways an employer might discriminate against an employee because of race, color, sex, religion or national origin.

Someone alleging violation of Title VII must follow certain administrative procedures before filing suit. S/he must first file a complaint with the EEOC or a state human relations commission to allow these agencies to try to settle the dispute or to file suit in their own names.

Possible defenses which an employer might use include claims that the employment decision was made for business necessity, it was job related, that it was based on a bona fide occupational qualification, that it was based on a seniority system or a merit system or on quality or quantity of production, etc., or pursuant to a professionally developed ability test.

FIRING OVERWEIGHT FIREFIGHTER HELD NOT RACIALLY MOTIVATED

A Federal appeals court in Texas upheld a fire department's weight standard. The court found no evidence that disciplinary action taken against an overweight firefighter was motivated by the plaintiff's race.[44]

RELIGIOUS DISCRIMINATION

A Jewish Deputy U.S. Marshal in Washington, D.C. filed suit, complaining that some black deputies routinely ridiculed white deputies and a supervisor had joked about the Holocaust. He also said he was suspended for a week because an inmate escaped, but black deputies received no punishment for similar errors.

The U.S. District Court enjoined the U.S. Marshal's Service from harassing and disciplining the Jewish deputy and ordered back pay and removal of disciplinary notations in the Jewish deputy's file.[45]

AFFIRMATIVE ACTION

Vallejo, California, firefighters applied for promotion to a position of firefighter/engineer and took a competitive written examination and three practical examinations. The person ranking first was white and the fire department recommended that he be hired, but the City, in furtherance of its affirmative action plan, promoted a black who had finished third in the scoring.

The firefighters' union filed a grievance on behalf of the passed-over firefighter, alleging that the promotion violated the labor agreement between the City and the union. An arbitrator found that no contract violation had occurred.

The passed-over white firefighter then sued in U.S. District Court, alleging violation of Title VII of the Civil Rights Act, local statutes and the California Constitution.

The City moved for summary judgment and the district court granted it. The firefighter appealed.

In a de novo review, the U.S. Court of Appeals affirmed the lower court's granting of the summary judgment, holding that the promotion did not violate the City Charter, civil service commission rules or the affirmative action plan since race is an additional factor which can be considered in the selection of qualified candidates for promotion. The court said that the plaintiff had failed to prove a violation of Title VII of the Civil Rights Act since the City showed that it had considered race as a legitimate, nondiscriminatory reason for its decision. The Court also held that the plaintiff had failed to show any violation of his constitutional right to equal protection. The City's affirmative action plan was narrowly tailored to achieve racial balance and was similar to the Harvard Plan cited in the *University of California Regents v. Bakke* case[46], and the City properly considered race as an advantage in the applicant's favor.[47]

The current U.S. Supreme Court, as well as the Congress, is at this writing (March, 1995) weighing the conflict between affirmative action and reverse discrimination.

CONSENT DECREE HELD VIOLATION OF EQUAL PROTECTION CLAUSE

Under a consent decree, females were to be hired as firefighters in the same proportion to their numbers in the applicant pool, not by strict ranking based on their performance on the selection tests. Prospective male firefighters filed suit challenging the arrangement. An Ohio appellate court agreed that this arrangement violated the equal protection of law clause of the U.S. Constitution.[48]

HIGH COURT MAKES DISCRIMINATION CASES HARDER TO PROVE

In June, 1993, the U.S. Supreme Court, by a 5-4 decision, made it more difficult for employees to prove employment discrimination.

A black prison guard sued St. Mary's Honor Center, alleging that his firing was motivated by racial bias.

A U.S. District Court judge ruled that the purportedly non-discriminatory reasons given by prison officials for the firing were false because white employees had not been disciplined for committing more serious offenses. However, the judge said that the plaintiff had failed to meet his ultimate burden under Title VII of proving that his firing was motivated by some racial animosity.

The U.S. Court of Appeals for the 8th Circuit reinstated the plaintiff's lawsuit, ruling that he should win the case because he had disproved the prison officials' stated defense. The appellate court said he should not be required to come up with direct evidence of racially motivated treatment.

There had been disagreement among the federal circuits on this point so the U.S. Supreme Court agreed to hear the case.

The High Court reversed the 8th Circuit decision. Writing for the majority, Justice Antonin Scalia said, "Title VII does not award damages against employers who cannot prove a nondiscriminatory reason for adverse employment action, but only against employers who are proven to have taken adverse employment action by reason of, in the context of this case, race." He said, "Nothing in law would permit us to substitute for the required finding that the employer's action was the product of unlawful discrimination the much different (and much lesser) finding that the employer's explanation of its action was not believable."

Scalia was joined by Chief Justice William H. Rehnquist and Justices Sandra Day O'Connor, Anthony Kennedy and Clarence Thomas. Justices David H. Souter, Byron R. White, Harry Blackmun and John Paul Stevens dissented.[49]

SEX DISCRIMINATION

Title VII's prohibition against sex discrimination in employment covers men as well as women, but most of the cases have been brought by women. The ban applies only to gender-based discrimination and does not forbid discrimination based on sexual orientation. However, state statutes often are broader than the federal law and may prohibit discrimination on the basis of sexual orientation.

EQUAL PAY ACT

The Equal Pay Act of 1963 only forbids sex discrimination regarding pay. It basically requires that men and women performing equal work should receive equal pay. The Act was passed as an amendment to the Fair Labor Standards Act and its coverage is similar to that of other FLSA matters.

The Act prohibits pay discrimination against both men and women, but the cases brought under this act usually involve women. To prevail, she must show that she has received lower pay than a male employee who performed substantially equal work for the same employer. To be substantially equal, the two jobs must involve each of the following:

1) Equal effort (The amount of physical or mental exertion required.)
2) Equal skill (The experience, training, education and ability required.)
3) Equal responsibility (The importance of decisions or the degree of supervision required, the amount of discretion involved.)
4) Similar working conditions (Quality of physical surroundings, hazards, inside or outside work, exposure to heat or cold, fumes and toxic substances. Working conditions need only be similar, not identical.)

The defendant must justify the pay disparity in order to win the suit by showing justification under one or more of the Act's four defenses:

1) Seniority
2) Merit
3) Quality or quantity of production
4) Any other factor other than sex (Shift differentials, part of a training program.)

Employers cannot rely on discretionary, subjective employee evaluation systems which might allow sex-based criteria to influence their decisions. They must show that they used an organized, objective, precise set of criteria that apply equally to males and females and which have been communicated to all employees.

Successful plaintiffs generally receive back pay which was lost because of the discrimination, but they may also receive an equal amount in liquidated damages.

The EEOC enforces the Act. Plaintiffs do not have to submit their claims to the EEOC prior to filing a lawsuit.

CHAPTER SEVEN

TORTS

A tort is a legal wrong for which the law provides a remedy such as payment of money damages. It does not include breaches of contract or crimes which are, of course, also "legal wrongs."

The word, "tort," is a French word which comes from the Latin word, "torquere" meaning twist, bend or distort.

Torts involve a breach of duty coupled with a commission of an improper act or the omission of an act which should have been done.

It is important to remember the differences between civil law and criminal law. Torts fall under civil law. However, remember that the same act can be both a tort and a crime and give rise to two separate legal processes — a criminal prosecution and a civil suit.

As noted in earlier chapters, the government, through its prosecutor, initiates a criminal action to punish the wrongdoer for his/her offense against society and to deter others from committing the crime. A person injured by the same act, can initiate a civil action in his/her own name for the purpose of being compensated for the damages sustained through the defendant's wrongful action.

CATEGORIES OF TORTS

Torts can be divided into four categories:

- INTENTIONAL TORTS (Require a state of mind to intentionally injure the plaintiff. Assault and battery — which are also crimes — are examples of intentional torts. The plaintiff can often recover both compensatory and punitive damages.)

- NEGLIGENT TORTS (Involve no moral wrongdoing, but the defendant has failed to use reasonable care. A firefighter who, through inattention or lack of skill, causes injury to another or another's property might be liable for a negligent tort.)

- STRICT LIABILITY TORTS (Involve liability without fault. The defendant has done nothing wrong, was not negligent, had no intention of causing harm, was not reckless, but the law imposes strict liability because the defendant had control over the thing, condition or circumstances which gave rise to the injury. Examples are manufacture of a product, keeping wild animals, explosives or other hazardous materials or something else which is inherently dangerous. The Fifth Circuit U.S. Court of Appeals and the Louisiana Supreme Court have held that storage and disposal of hazardous wastes are ultrahazardous activities for which strict liability will apply.)

- CONSTITUTIONAL TORTS (Cases where a person infringed a right of another which is protected by the U.S. Constitution. Most torts are derived from state law, but constitutional torts are federal matters involving U.S. statutes and the U.S. Constitution. Examples are discrimination related to a person's race, religion, disability, gender or national origin. Successful plaintiffs can recover attorneys fees as well as damages. Immunities found in state law provide no protection from federal suits based on constitutional torts.)

The Civil Rights Act of 1871 creates tort liability for any public official or employee who injures a person by depriving him/her of constitutionally guaranteed rights. The Supreme Court has extended this to local governments which support such behavior by "custom, practice or policy."

The Fifth Amendment, which was made applicable to states by the Fourteenth Amendment, provides that a person shall not be deprived of life, liberty or property without due process of law.

Some possible areas where constitutional torts might arise include: unfavorable recommendations about former employees, use of zoning for political or discriminatory purposes, failure to award contracts to the lowest bidder in some cases, wrongful discharge of employees, failure to adequately train employees, use of permits, inspections and licensing for political, harassing or discriminatory purposes.

30-MINUTE MEETING WITH CHIEF WAS DUE PROCESS FOR FIRED EMPLOYEE

A fired firefighter filed a civil rights suit in U.S. District Court alleging that his pre-termination hearing was not adequate due process. The U.S. Court of Appeals said he was given notice of the proposed dismissal, informed of the reasons for his firing and given an opportunity to respond and, although the meeting between him and the fire chief lasted only thirty minutes, it satisfied due process requirements. The post-termination hearing included a neutral and detached fact finder, the right to call favorable witnesses and to cross-examine adverse witnesses, to be represented by counsel and to obtain a record of the proceedings. The fact that the plaintiff failed to avail himself of the second hearing was held to be irrelevant. The court said, "One who fails to take advantage of procedural safeguards available to him cannot later claim that he was denied due process."[50]

The termination hearing does not have to be formal or elaborate but the procedure must only be sufficient to serve as an initial check against wrong decisions.[51]

Officials can be held personally liable, even if the actions were official. In 1991 the U.S. Supreme Court, by an 8 to 0 decision said that state officials can be held personally liable under Section 1983 even if their actions were taken as part of their officials duties. A suit was brought against the Pennsylvania State Auditor for firing employees because of their political affiliation and their support of her opponent in the election. She was found personally liable and had to pay the judgment out of her own pocket.[52]

KNOW WHAT DUTIES ARE IMPOSED BY LAW

It is important to know what duties are imposed on firefighters and fire departments by state law. Some states hold that local governments do not have a duty to respond and fight fires, so a negligence action for failure to do so in those cases would be rejected by the courts. Other states, on the other hand, do impose on firefighters and local governments the duty to fight fires in a non-negligent manner.

You should learn what immunities are provided to firefighters and their governments by state statutes and court decisions. This is of vital importance because, depending on the parameters of immunity in your state, a suit against you and the department might be barred.

NEGLIGENCE

Legal liability is becoming an increasing concern to fire service personnel. Justifiably so. As was stated earlier, there has been a proliferation of law suits in all aspects of our society. Coupled with this, is a steadily expanding role for the fire service, more complicated hazards, as well as new laws and court interpretations which increase the opportunity for making a mistake and getting into legal trouble.

Negligence is a major area of the law requiring the attention of fire service personnel. The elements necessary to establish negligence differ from state to state. Negligence in general will be covered here.

NEGLIGENCE DEFINED

Negligence is the failure to exercise that degree of care which the reasonable and prudent person would exercise under the same circumstances.

WHAT PLAINTIFF MUST PROVE IN NEGLIGENCE CASE

In order for a plaintiff to recover, the attorney must prove the following:

1) That the defendant owed the plaintiff a duty;
2) That there was a standard of care for that activity;

3) That the defendant breached that standard of care;
4) That the plaintiff was damaged; and
5) That the defendant's action was the proximate cause of that damage.

DEFENSES TO A NEGLIGENCE ACTION

- Assumption of risk
- Statute of limitations
- Contributory or comparative negligence
- Immunity

DUTY OF VOLUNTEERS

If a person has no duty to do something and yet undertakes to do it voluntarily, s/he has the obligation to perform that service in a non negligent manner. If a person sees an injured person on the road, an ordinary citizen has no legal duty to render assistance (except in a few states such as Vermont and Minnesota). S/he may have a moral duty, but not a legal duty. However, if s/he undertakes to assist that person, then there is imposed on that volunteer rescuer a duty to act carefully. If I see someone drowning in the river, I have no duty to try to rescue him/her, but if I do, I must use due care. A lifeguard at a beach, on the other hand, would have a duty to attempt to rescue the drowning person. (Your state's Good Samaritan law, if you have one, may affect liability.) Your state law might also impose a special duty on firefighters to respond.

REASONABLE, PRUDENT PERSON STANDARD

In negligence cases, the legal standard of care question is asked: "Would the reasonable and prudent person performing the same duties under the same circumstances have acted in the same manner?"

Consensus standards such as those of the NFPA 1500 may be used as the yardstick against which your department's performance will be measured. The NFPA 1500 may establish what the reasonable and prudent department similarly situated would have done.

The violation of a standard of care established by statute is negligence per se (as a matter of law). Under this concept the plaintiff need only prove the applicability of the statute and that the defendant's violation of it caused the injury. The statute establishes the duty and standard of care.

For example, state law might require that nursing homes be inspected periodically. If the fire department fails to inspect a nursing home when it is supposed to be inspected and fire injury occurs as a result, the department could be liable for negligence per se. You must know the statutes in your state which affect your responsibilities.

STANDARDS OF CARE OF PROPERTY OWNERS

The property owner or person in charge of the property owes varying standards of care to those who enter onto their property and are injured. It depends on the legal status of the one who enters.

TRESPASSER

If the person is a trespasser, the property owner is only liable if his/her behavior is reckless, willful or wanton. The owner is not free to inflict wanton or intentional injury on anyone, not even a trespasser, but no special duty of care is owed to a trespasser.

LICENSEE

A licensee is one who enters onto the property with the owner's permission, usually for the visitor's own purposes. The licensee must assume all the risks s/he finds on the property. The property owner owes the licensee no special duties other than to not inflict willful or wanton injury on him/her. The owner must exercise reasonable care to assure the safety of the licensee. If there is an unsafe condition on the property and the owner fails to warn the licensee about it and had an opportunity to give the warning, and the licensee, not being aware of it, is injured, the owner may be held liable.

INVITEES

Invitees are workers, customers in stores, patrons at a movie theater, etc. The owner is liable if the owner knew, or should have known, of the dangerous condition and should have realized that it might cause harm to invitees. The owner should assume that they would not discover the danger on their own, or that they would fail to protect themselves against it. If the owner fails to exercise reasonable care to protect invitees against the danger, s/he is liable.

STATUS OF FIREFIGHTERS WHILE ON PROPERTY

Most states generally consider firefighters to be licensees, in spite of the fact that it could be assumed that they have an "invitation" from the property owner, explicit or implied, to enter onto the property to extinguish the fire.

The property owner has no duty to use reasonable care to make the premises safe for the firefighters. S/he does, however, have a duty to warn them of known dangers, hidden risks, when s/he is aware of them and has an opportunity to warn the firefighters about them.

In those jurisdictions which consider the firefighter an invitee, the property owner has the duty to exercise reasonable care. In other words, an owner who negligently creates a fire hazard or causes a fire may be held liable to a firefighter-invitee who is injured, but that is generally not the law in most states.

It should be noted that these exceptions are made on the basis that the firefighters enter the premises to discharge their duties. These duties include facing fire hazards. This fact might afford a defendant the defense of assumption of risk in an action for negligence. Some courts state that the owner must only provide safe access onto the premises for firefighters, nothing more.

A firefighter, while fighting a fire at a house under construction, fell from a balcony on which the railing had not yet been installed. The fire had been caused by an unattended space heater. The injured firefighter sued the owner for damages. The firefighter won at the trial court level, but the appellate court reversed the decision and ordered that judgment be entered for the defendant. The firefighter appealed. The Supreme Court of New Jersey held that the evidence did not establish wanton misconduct on the part of the owner-defendant so the appellate court's decision was affirmed.

RESCUE DOCTRINE AND THE FIREMAN'S RULE

The Rescue Doctrine holds that a wrongdoer who commits a tort — intentionally or through negligence — is liable for all injuries to a rescuer which directly result from the wrongdoer's action. However, the Fireman's Rule is an exception to the Rescue Doctrine. Under the Fireman's Rule, a firefighter or police officer cannot sue a negligent person for injuries which grow out of the incident which created the need for the firefighter or police officer's presence at the scene. In other words, the emergency responder cannot recover when his/her cause of action is based on the same conduct which created the need for him/her to be present at the emergency scene.

Various reasons are used to justify the rule:
- Self-help measures might be taken in an emergency by negligent individuals out of fear they might be sued;
- Property owners or occupiers might be discouraged from calling the emergency responders;
- Liability would place too heavy a burden on property owners to keep their premises safe from the unpredictable arrival of firefighters or police officers in an emergency;
- Since a large percentage of fires are caused by the negligence of the owner or occupier of the property, it is unreasonable to hold the property owner or occupier liable for damages sustained by firefighters who are employed and trained to fight such fires;
- The risk of firefighters' and police officers' injuries should be spread to society as a whole through worker's compensation systems, pay, fringe benefits, etc., rather than to individuals;
- A proliferation of litigation would result;
  Firefighters and police officers know they are in an occupation which is dangerous and they assume the risks;
- They are hired and trained to handle these risks.

PROFESSIONAL RESCUER KNOWS THE RISKS, COURT SAYS

In the *Young v. Sherwin Williams* case, the court said, "While the reality may be that a professional rescuer's pay and benefits are often inadequate to compensate for a given injury, the fact remains that a professional rescuer knows before accepting the employment both what the risks of the job are and what the compensation and benefits will be."[53]

EXCEPTIONS TO THE FIREMAN'S RULE

There are exceptions to the Fireman's Rule. If a firefighter is called to a site where there is a pre-existing hidden danger about which s/he should have been warned by the property owner who knew about it, or if the property owner's willful or wanton conduct causes the injuries, the Fireman's Rule will not be applied. Explosives are a good example of a hidden risk. The law does not require firefighters to assume ALL possible risks If the injury results from arson, or willful misconduct or gross negligence, recovery will be allowed in most states. If negligent acts occur after the firefighter arrives on the scene which are outside those risks which firefighters usually anticipate facing during the course of their duty, it will not apply.

WANTON, WILLFUL MISCONDUCT NEGATES FIREMAN'S RULE

A firefighter, while fighting a chemical company fire, was injured. He sued the chemical company, its parent and the company which had supplied the combustible chemical. The trial court applied the Fireman's Rule and granted the defendants' motion for summary judgment. The firefighter appealed. The Supreme Court of New Jersey reversed and remanded the case back to the lower court.

The court said to accord immunity under the Fireman's Rule to one who deliberately and maliciously creates the hazards that injures the firefighters stretches the policy underlying the Rule beyond the logical and justifiable limits of its principle. The court said wanton and willful misconduct by the chemical manufacturer sufficient to deprive the company of immunity under the Fireman's Rule was adequately asserted by allegations that the manufacturer was fully aware that the chemical increased the flammability of combustible materials such as the fiber containers which it began using for shipments, that it had communicated privately and publicly its conclusion that the danger of spontaneous combustion was so great that the use of fiber containers should cease and that, even knowing this fact, it nonetheless still shipped 100 fiber containers of the chemical to the warehouse where the fire occurred which injured the firefighters.[54] New Jersey has since repealed the Fireman's Rule legislatively.

ARSON NEGATES FIREMAN'S RULE

Three firefighters died fighting a construction site trailer fire set by an arsonist. The trailer contained ammonium nitrate and fuel oil. Their estates filed suit against several defendants, including the State Highway Department. The Highway Department filed a motion for summary judgment on the basis of the Fireman's Rule and the plaintiffs appealed. The appellate court said, under the Fireman's Rule, a firefighter while performing firefighting duties cannot recover against the person whose ordinary negligence created the emergency. A firefighter assumes all risks incidental to his/her firefighting activities except for hidden risks which are known to the landowner. Explosives are a hidden risk. The law does not require firefighters to assume all possible lurking hazards and risks. There is no valid reason why they should be compelled to assume the extraordinary risk of hidden peril of which they might easily be warned. Affidavits submitted in support of the motion for summary judgment were in conflict concerning whether the firefighters knew about the explosives at the construction trailer. Because of this ambiguous information, the court said there was a genuine issue of fact which should be decided at trial so the summary judgment was not appropriate. The survivors of the firefighters were entitled to a trial on their claim.[55]

FIREFIGHTER LOSES SUIT AGAINST GE FOR COFFEE MAKER FIRE

A firefighter responded to a residential fire caused by a coffee maker. In jumping off the roof of the house to avoid an explosion, he was injured. He knew from experience that explosions often occur at fires. He began collecting worker's compensation benefits following the accident. He filed suit against General Electric, manufacturer of the coffee maker. The court granted GE's motion for summary judgment based on the Fireman's Rule and the firefighter's assumption of risk. The appellate court affirmed this decision. Even if the Fireman's Rule did not apply, the issue of product liability was barred. Since no defect in the coffee maker was the proximate cause of the firefighter's injuries, it could not be based on a negligence or a strict liability claim.[56]

FIREMAN'S RULE HELD APPLICABLE TO EMT'S

A paramedic who worked for a county hospital and provided emergency medical services injured his back when extracting a man from a truck which had crashed into a ditch. Tests showed that the driver of the truck had a blood alcohol level above the legal limit for operating a motor vehicle. The paramedic and his wife brought suit against the driver and his employer, a coal company which owned the truck. The defendants moved for a summary judgment based on the Fireman's Rule, but the court denied their motion stating that it did not apply to paramedics.

On appeal, the court said that the Fireman's Rule created an exception to the Rescue Rule and that Indiana had adhered to the Fireman's Rule for 100 years. The court said the rule basically provides that professionals, whose occupations by nature expose them to a particular risk, may not hold another liable who is negligent in creating the situation to which they respond in their professional capacity. The court pointed out that other courts had extended the rule to public safety officers and police. The court said the Fireman's Rule rests on three distinct, but related, theoretical pedestals: the law of premises liability, the defense of incurred risk and the concerns of public policy. By choosing work as a public safety professional, the court said, the paramedic entered into a relationship with taxpayers and members of the public. A paramedic knows the work involves certain risks and has, thereby, implicitly agreed to relieve the negligent persons of responsibility when the risks lead to foreseeable harm. The court said, in essence, the paramedic consented to the harm risked.

The court pointed out that the paramedic is already compensated for the hazards and inherent risks of his work through pay, medical, disability, pension schemes and worker's compensation and that allowing him to recover in tort would amount to a double recovery.

The court held that the driver's intoxication was irrelevant because the statute barring alcohol-impaired drivers from operating a motor vehicle was designed to protect the general public on the highway, not paramedics performing rescue of an intoxicated driver. The case was reversed in favor of the truck driver and his employer.

IOWA SUPREME COURT ADOPTS FIREMAN'S RULE

Two police officers responded to a tavern where a fight was underway. A drunken patron assaulted them and the officers sued the tavern for damages. The tavern owner raised the Fireman's Rule in defense, arguing that their recovery was barred because the injuries occurred while they were acting in their official capacity. The district court found in favor of the police officers, stating that Iowa had not yet adopted the Fireman's Rule. The tavern owner appealed. The Iowa Supreme Court in reversing the lower court said that almost all jurisdictions recognize that firefighters and police officers cannot recover on a complaint that is based on the same conduct which initially created their presence in their official capacity. If the negligence which created the risk which resulted in the injuries was the reason for their presence, they cannot recover.

FIREMAN'S RULE NOT APPLICABLE IN OREGON

The State of Oregon by statute abolished the doctrine of assumption of risk as a defense in an action for negligence and, since assumption of risk is the foundation for the Fireman's Rule, the Fireman's Rule will no longer bar recovery of emergency responders in Oregon.

In Oregon the key question in these negligence cases now is what legal duties did the defendant owe to the emergency responders.[57]

Some other states have also abolished assumption of risk as a defense in negligence actions. Some states have also abrogated the Fireman's Rule. Some have repealed it legislatively or have overruled it by court decision. Some never had it, but most still recognize it. Again, know what the law is in YOUR state.

FIREMAN'S RULE HELD APPLICABLE IN MARYLAND IN SUIT BY VOLUNTEER

A Maryland volunteer firefighter, while fighting a fire in an apartment building, fell twelve stories down an elevator shaft. He sued the building owner and the elevator manufacturer for his injuries. The circuit court sustained the defendants' demurrer on the basis of the Fireman's Rule. The firefighter appealed and the Court of Appeals, Maryland's highest court, affirmed. The Fireman's Rule was held to be applicable to both defendants.

The Court of Appeals restated the law regarding the Fireman's Rule: If there is negligence in failing to warn firefighters of a pre-existing danger where the defendant knew of the danger and had an opportunity to warn them, the Fireman's Rule would not apply. Also, acts occurring subsequent to the firefighters' arrival on the scene which are outside their anticipated occupational hazards will not be protected by the Fireman's Rule. The Fireman's Rule does not apply in arson cases, but it does apply in "fires of suspicious origin."[58]

WISCONSIN COURT DENIES RULE'S PROTECTION IN STRICT LIABILITY CASE

In *Hauboldt v. Union Carbide Corp.*, a Wisconsin court held that a manufacturer of an acetylene tank which exploded could not take advantage of the Fireman's Rule because strict liability applies. The tank did not start the fire, but exploded during the fire because of defective safety devices. The court said the Fireman's Rule does not apply to manufacturers of defective products which are the proximate cause of injury to firefighters who are unprepared for the danger which a defective product presents. In explaining why the property owner would not be liable, the court said, "Public policy considerations led to the conclusion that imposing liability would place too great a burden on homeowners and other occupiers of real estate and would permit the law of negligence to enter a field that has no sensible or just stopping point. Based on these public policy considerations, we held that one who negligently starts or fails to curtail the spread of a fire that causes injury to a responding firefighter is immune from liability."[59]

Incidentally, the court, in an interesting bit of *obiter dicta*, commented that it agreed with the Rhode Island Supreme Court that the Fireman's Rule should be re-named in view of the number of female firefighters.

FIRE CODE VIOLATORS NOT LIABLE BECAUSE OF THE FIREMAN'S RULE

Firefighters were seriously burned while suppressing an automobile fire. Even though the defendants were criminally charged with using a defective gasoline pump which had caused gasoline to overflow and for allowing the driver of the car to smoke, these fire code violations did not create liability for the defendants because of the Fireman's Rule.[60]

PRIVATE RIGHTS MUST USUALLY YIELD TO PUBLIC NECESSITY

At common law, where public convenience and necessity come into conflict with private rights, the latter must yield. Public safety permits the doing of acts to prevent imminent loss or destruction of property or life and those acts will not be deemed to be tortious because of the element of necessity. However, a subsequent entry exceeding one's authority might make the person a trespasser ab initio (from the beginning.)

The courts have trouble resolving issues concerning a firefighter entering on another's property without the owner's consent. They are not trespassers because they have a duty and privilege to enter upon the property. However, the privilege they have is not derived from the owner's consent. Whether or not they are invitees or licensees is resolved by the duty or care owed to them by the owner and the resultant liability of the owner to them.

COSTS OF FIRE SERVICES ARE GENERALLY BORNE BY PUBLIC AT LARGE

The common law rule in force in most jurisdictions provides that, absent authorizing legislation, the cost of public services for protection from fire or other safety hazards is to be borne by the public as a whole, and not assessed against the tortfeasor whose negligence created the need for the service. However, the courts appear open to allow cost recovery from negligent parties where statutes or local regulations provide for such recovery.[61]

VOLUNTEER COMPANY HELD LIABLE FOR DAMAGES FROM BRUSH FIRE

An appellate court in Maryland held that a volunteer fire department could be liable for negligence in failing to control and extinguish an outdoor brush and trash fire adjacent to a warehouse.[62]

DEPARTMENT LIABLE FOR FAILURE TO PROVIDE NOISE PROTECTION

The City of Baltimore was held to be negligent for not providing additional protective equipment for fire boat crew members who suffered significant hearing loss due to diesel engine noise. The department should have furnished sound-reducing "ear muffs" to crew members.[63]

CITY FOUND LIABLE FOR AMBULANCE DELAY

A Honolulu jury awarded $1.9 million against the city for a two-hour delay in ambulance arrival.[64] In a similar case, (and there have been many) a Chicago court awarded $2.5 million against the City of Chicago and the School Board to a little boy who was kicked in the head during recess because of the delay in getting him treatment. After the school made three calls to 911, the ambulance finally arrived at the school ninety minutes after the first call. The School Board had a policy of not transporting sick or injured children to the hospital so they waited for the ambulance. The hospital was located across the street from the hospital! This points up the importance of keeping your policies flexible.

FIREFIGHTER AWARDED $2 MILLION FOR CITY'S NEGLIGENCE

A firefighter who was injured in a fire in a city-owned building, recovered $2 million for the city's negligence.[65]

IMMUNITY

Several years ago in Waverly, Tennessee a train carrying several propane gas tank cars derailed. The fire department responders assessed the situation and determined that, if they did not promptly off-load the gas, the rising temperature would cause it to explode. Several propane gas companies were contacted and asked to assist in the off-loading. They all refused.

Why would these companies not cooperate?

Because they were concerned about liability.

The firefighters were correct in their assessment. The tank cars exploded, injuring a number of people. If the propane gas companies had assisted in the off-loading, they would most assuredly have been sued.

If you want propane companies or chemical companies to help you in a hazmat incident or other companies with specialized expertise to assist you, urge your state legislature to enact a "Good Samaritan" law which immunizes companies which come to the assistance of government in an emergency and are subsequently sued.

DON'T RELY ON IMMUNITY

Most state laws and court decisions afford fire departments varying degrees of immunity, but don't rely on immunity. It might not be there.

SOVEREIGN IMMUNITY

Sovereign immunity was described at common law as: "The King can do no wrong." When our government was set up by our founding fathers, they adopted the legal system in use in England with which they were familiar as British colonists. Although we did not have a king, we did adopt the concept of sovereign immunity so that government could not be sued no matter how negligent and no matter how egregious its behavior was. State legislatures and courts began to recognize the inequity of this and started a gradual process of eroding, and in some instances totally abolishing, sovereign immunity. This trend continues today.

Most states did not jettison sovereign immunity completely, but created distinguishing principles in addressing the issue. Some elements of these treatments of immunity exist in some fashion in various states. It is very important for you to find out how the legislature and courts treat immunity in your state.

GOVERNMENTAL ACTIVITY VS. PROPRIETARY ACTIVITY

One test employed in some jurisdictions is: "Whether the activity which gave rise to the suit was a governmental activity or a proprietary activity."

Firefighting, police matters, public health, etc., are governmental, but collecting trash and garbage, running bus companies, water and electric utilities, etc., are proprietary. The courts are not unanimous as to what constitutes governmental activity and proprietary activities.

In Utah, the test is whether or not the function that government is carrying out is a function that MUST be carried out by government. If the private sector COULD do it, government is not immune. This approach will cause serious problems as some local governments continue the trend to contract out jails, fire service, etc. Private companies do not enjoy governmental immunity. The state legislature, however, could give it to them by statute.

DISCRETIONARY VS. MINISTERIAL

Another test used is whether or not the activity which caused the injury was discretionary or ministerial. A ministerial function means a routine action which does not require the choosing between various alternatives, one that is required to be performed without regard to one's own judgment. A discretionary act requires exercise of reason and judgment in determining how or whether to act and the choosing among various alternatives.

The higher someone goes on the management ladder, the more likely his/her activities will be deemed to be discretionary. The on-scene fire commander who must decide among various options in deploying equipment and manpower to fight a fire will usually be deemed to be engaged in discretionary acts.

The decision whether or not to grant a variance from the fire code is discretionary, but actually executing the paper work for it after the decision has been made is ministerial.

A dog pound master did not destroy a vicious dog and, as a consequence, a person was injured. When the dog pound master was sued, the court held that he was immune because he had broad authority and discretion with respect to destroying dogs.

In an unreported Florida case involving the rehabilitation of an historic hotel, the owner requested a variance from the requirement of enclosing an ornamental wooden staircase. The owner agreed to take other fire prevention measures on the strength of which the city granted him a variance from the code's requirement to enclose the staircase. A fire occurred and those injured sued the government. The court held the city was immune because granting the variance was a discretionary act.

A police officer carelessly discharging his gun or negligently driving his car in a routine situation has been held to be ministerial.

The decision to order an evacuation or selecting the evacuation route, or adopting emergency medical policies are discretionary acts, but failure to carry out established evacuation plans or to take action contrary to medical policies would probably be deemed ministerial and, therefore, not immune.

PUBLIC DUTY DOCTRINE

Another limitation on the liability of government falls under what is called the public duty doctrine. Government does not guarantee that a citizen's house will not burn down or that a person will not be the victim of crime. The government has only a general public duty to provide services, such as fire protection, health service, police protection etc., but there is no guarantee that, in spite of the provision of these services that a person will not be damaged by the government's failure to provide preventative services. However, if there can be shown to be a special relationship between the government and the victim, the public duty doctrine will not apply.

A New York State case is illustrative. A woman called 911, complaining that a prowler was outside her house. The dispatcher told her to stay right where she was, that help was on the way. He did not repeat the address she had given him and dispatched the police cruiser to a wrong address, a vacant lot. Assuming it was a false report, the police officers went back in service and took no further action. The prowler entered the woman's house, raped and murdered her. Her heirs sued the city which defended on the basis of the public duty doctrine. The court rejected this defense, stating that the telephone conversation between the dispatcher and the victim formed a special relationship between the victim and the city government, rather than a general one, and the city was held to be liable.

In *Archie v. City of Racine*, the U.S. Court of Appeals for the Seventh Circuit found a special duty (rather than a general public duty) when a dispatcher refused to provide rescue services requested by a woman who subsequently died. The court said this was not mere negligence. The employee's conduct was reckless and made with a disregard for the natural and foreseeable consequences of his inaction. The court said it was a constitutional deprivation of life without due process. This case illustrates that there may be immunity for negligent actions of employees under state law, but when the actions go beyond simple negligence there could be liability under federal law, such as a constitutional tort as in this case.[66] Furthermore, immunity under state law does not bar a suit on a federal legal basis.

IMMUNITY STATUTES MAY HAVE SPECIAL PROVISIONS

In addition to general immunity, there may be special provisions in your state statutes such as giving blanket protection for all emergency activities.

In some emergency situations, individuals who have been injured or who have suffered property damage as a result of the government's negligence, may file suit against the agency causing the harm. Even though the government responding to the disaster may have immunity during the response phase, it might be liable for its actions prior to the incident which caused the injury or created the hazard, such as failure to train personnel adequately or failure to prepare adequately for the emergency or for failure to keep equipment in proper working order, or for not having an adequate supply of water in the pumper.

For example, if a state statute requires the local government to have a current emergency preparedness disaster plan and the local government fails to carry out this duty, this could lead to liability. It would also be negligence per se.

To repeat again: know what your state provides by way of immunity. Sometimes a statute gives blanket immunity coverage, including preparation of plans, but in others it only applies during the emergency event itself.

IMMUNITY DISALLOWED

A woman called the Gary, Indiana, Fire Department Ambulance Service stating that her husband could not catch his breath. The dispatcher told her an ambulance was on its way. She later called 911 again and was told that the ambulance should be there soon. She later called a third and fourth time. When she called the fifth time, the dispatcher said the ambulance would be there momentarily. In fact, he said, it was probably outside already.

The ambulance service was apparently short-staffed when the woman called. The dispatcher tried to contact one ambulance crew, but the crew was not where it was supposed to be stationed. When he finally located them, the crew was asleep. The crew logged the call but it took them quite a while to leave the station. It arrived 42 minutes after the wife had first called. The EMT's failed to adequately assess the husband's condition, failed to call promptly for a paramedic and failed to start CPR quickly enough. The man died of heart failure and cardiac arrest about fifteen minutes after arriving at the hospital.

The wife sued the city for wrongful death, arguing that the city was liable because it negligently responded to her calls.

The city claimed it was immune from suit and asked for a judgment in its favor. The trial court denied the city's motion and ruled in favor of the wife. The city appealed.

The appellate court affirmed the lower court's decision in favor of the wife. The court said the city owed her husband a special duty, not a general duty and it breached that duty by failing to respond quickly enough to the wife's emergency call. The repeated calls and the assurances of the dispatcher, the court said, were enough to establish the special duty rather than a general public duty. The court held that the ambulance crew also owed the husband a special duty which they breached by arriving late and providing inadequate care, as a result of which the patient died. The court held the city liable for the death.[67]

OPERATORS OF FIRE APPARATUS HELD NOT PUBLIC OFFICIALS

A Maryland court ruled that operators of fire apparatus and ambulances are not considered public officials covered by official immunity while operating emergency vehicles. The Court laid down four guidelines regarding immunity:

(1) The position was created by law and involves continuing and not occasional duties.

(2) The holder performs an important public duty.

(3) The position calls for the exercise of some portion of the sovereign power of the State.

(4) The position has a definite term for which a commission is issued and a bond and an oath are required.[68]

This case was superseded by statute on another point as stated in *Prince George's County v. Fitzhugh.* [69]

MICHIGAN COURT FAILS TO FIND GROSS NEGLIGENCE, OTHERS MIGHT

The Michigan Supreme Court ruled that the City of East Detroit could not be held liable for its firefighters' negligence unless they were grossly negligent.

The City of East Detroit had two trained emergency medical technicians on its staff, but sent two untrained firefighters to the home of a heart attack victim. While the victim was vomiting into the bag valve mask, the firefighters kept pumping, despite the appeals of a bystander who urged them to clear the device. The victim suffered brain damage because of their action.

The Court upheld a Michigan statute which reads: "When performing services consistent with training, an ambulance attendant is not liable for acts or omissions in the treatment of a patient unless the acts or omissions were the result of gross negligence or wilful misconduct." The Court held that the action of the firefighters was not gross negligence.[70] (Another court might certainly have held otherwise.)

BEFORE CLEANING DEBRIS FROM PROPERTY, GIVE NOTICE TO OWNER

The Village of Chebanse tried for two years to get the property owner to clean debris off his property. He was cited for a code violation, but still he did nothing. The Village hired someone to clean up the property.

The owner sued the Village Board of Trustees and recovered $7,150, the value of the goods removed from the property and $3,500 in attorney's fees.

The Village should have notified him before it removed the property to give him an opportunity to present his case. The Due Process Clause of the U.S. Constitution requires this. The Village could also have obtained a court order authorizing the cleanup. Under a state statute the property could have been declared a "common law nuisance" which would have given the Village authority to remove the hazard.[71]

USE OF WARNING DEVICES IN NON-EMERGENCY LEADS TO LIABILITY

Suppose a firefighter, in the absence of an actual emergency, uses his fire department vehicle's flashing lights and siren to allow him to ignore traffic rules, and while doing so, an accident occurs damaging another vehicle and injuring the driver. Will his actions be immune from liability? Probably not. Most immunity statutes do not provide protection for gross negligence or willful misconduct. In addition, if his/her government normally provides an attorney and pays judgments when personnel are sued, they will not do so for willful misconduct or gross negligence. The general rule is that an employer will not be held liable for the torts of an employee unless they are committed within the scope of employment. Some courts have held, however, that, if the employer allows the employee to use an employer-owned vehicle for personal business, the employer will be held liable for accidents which occur in the vehicle, even if the employee is not working at the time.

CHAPTER EIGHT

VOLUNTEERS AND ORGANIZATION OF FIRE DEPARTMENTS

BASIS OF AUTHORITY

The basis of the fire service's authority is the police power of the state. Police power does not relate to law enforcement matters, although that function falls within the police power. It is broader than that. It means the power of government to take actions to promote and protect the health, safety and welfare of its citizens.

Fire departments owe their very existence to laws and the law impacts on every fire service activity.

Why does it matter if the department is properly and legally established?

Because state laws relating to immunity, exemption from traffic regulations while responding to an emergency etc. only protect departments which conform to statutory standards, a fire department which is not legally organized operates outside the scope and protection of the laws, regulations and immunities and exposes itself and its members to legal risks.

It is very important to know the extent and limits of your authority because if you exceed that authority, you may be personally liable. An act beyond your authority or beyond your department's authority is "ultra vires" and will put you in the no-man's-land of personal liability. Understanding the law is essential for fire personnel — volunteer and career.

Usually volunteer firefighters and volunteer departments are affected on the same basis as paid firefighters and governmental departments.

Both governmental and volunteer fire departments must have a legal basis for their existence. One is established by enactment of a law while the other is usually created by incorporation according to applicable laws.

The U.S. Constitution created the states, but not the counties or cities or other subdivisions of the states. These jurisdictions were created by the State governments. It is important to remember that local governments have no power unless it was delegated to them by the states.

One of the first grants of power to local governments was the power to provide fire protection. This grant of power carried with it state regulation.

The activities of both volunteer firefighters and career firefighters are regulated by law. Fire districts, because they have broader powers, including the power to tax, are usually even more closely regulated. They also must be concerned about director liability and would be wise to obtain errors-and-omissions insurance to protect their directors from the consequences of personal liability.

ARE FIREFIGHTERS PEACE OFFICERS?

Fire personnel investigate arson and other violations so the question as to whether or not they are peace officers is important. Do they have the authority to carry concealed weapons, make arrests, serve subpoenas, arrest and search warrants, and are they required to complete the training required for police officers (use of firearms, search and seizure, arrest procedures etc.) To answer this question, you must consult your state statutes.

VOLUNTEERS AND THE LAW

Volunteer fire departments are older than the Nation itself, but the organization of these departments was not regulated until the 20th century.

Nine out of ten firefighters in this country are volunteers. Some volunteer companies have been established under state laws authorizing the formation of fire districts, some have been organized as individual fire companies within the departments of incorporated communities.

Generally, volunteer companies have the same rights, duties, powers and privileges as paid firefighters, assuming that the department has been properly organized and maintained, and that the volunteers' names are properly recorded on the volunteer company's register as required by law. Individuals must also be in full compliance with all the requirements for retention of member status and training. If so, they have all the rights and duties available to their department.

The following are some of the areas where laws regulate the behavior of volunteer firefighters:

- Response to fires in private cars
- Worker's compensation
- OSHA and other safety regulations
- Training requirements
- Federal tax liability
- Compliance with civil rights/anti-discrimination laws
- Hiring and recruitment practices
- Exercise of due care in firefighting
- Financial reporting

The standard of care for the volunteer firefighter is the same as that for the paid firefighter.

Volunteers are also vulnerable to liability suits. The fact that someone is working as a volunteer does not exempt that person from liability for willful misconduct. S/he is not protected from liability merely because the call was responded to in a private vehicle, equipment was purchased with private funds and the person was not paid for firefighting.

Usually, volunteers are eligible for worker's compensation and their dependents are usually entitled to all the available benefits in the event of death or injury while acting within the scope of their "employment." Check your state's law to determine what the status of volunteers is. The federal law which gives a lump sum payment to the widows of police and firefighters killed in the line of duty also includes volunteers.[72]   (This bill was sponsored by the author while he was a Member of Congress.)

Many states have statutes which give worker's compensation benefits to anyone who volunteers to help suppress a fire when requested to do so by an officer of a public agency.

Usually state statutes and local ordinances spell out the status of the fire department as to powers, duties, composition training, etc. Unless state laws gives the volunteer immunity, the volunteer will be liable for negligence. If provisions in the laws are felt to be inadequate, contact should be made with the state legislature to seek amendments to the statutes.

Some states have enacted special provisions in law to give the volunteer firefighter the right to install a warning light, siren etc. on his or her personal car and have been granted privileges to waive traffic laws under emergency conditions. In the absence of such specific authority in the law, however, the volunteer may not use emergency signals and must obey all traffic laws.

No firefighter should ever respond to a fire call when he or she is under the influence of alcohol, because not only will this potentially impair his or her own life and those of his/her fellow volunteers, but will also potentially expose the volunteer and his department to liability. The volunteer company should have a strict policy on this and it should be stringently enforced.

In this regard, volunteer associations frequently serve alcoholic beverages at social functions held at the volunteer fire company's facilities or allow outside organizations renting the facilities to serve them. If so, it would be wise for the volunteer company to obtain adequate liability insurance to cover any accidents on the premises or from an automobile accident which might result from the actions of a person who became intoxicated at the volunteer company's facilities.

Volunteers must not only be given all the training required by laws, but must be given all the safeguards (protective clothing, self-contained breathing apparatus, etc.) which are provided to paid firefighters.

Some states prohibit an employer from firing a volunteer for missing work while responding to a fire call. Other states exempt volunteer companies from paying property or sales taxes.

IS A VOLUNTEER AN "EMPLOYEE"

Federal appeals courts have held that volunteer firefighters can be "employees" and privately incorporated fire companies can be liable under 42 U.S.C. Section 1983 for civil rights violations. The court found that the fire company received some public funds and that compensation need not be wages.[73] In *Krieger v. B.C.C. Rescue Squad* another federal court in Maryland held that a member of a volunteer rescue squad was not a public employee because he operated an emergency light truck ancillary to, but not part of, the "public function of firefighting."

ADMINISTRATIVE AGENCY MUST DECIDE VOLUNTEER'S RIGHTS

During a volunteer firefighter's initiation rites, he was required to wear an old uniform, travel to bars with fellow firefighters and consume large amounts of alcohol. At one location, the recruit was blindfolded and the captain informed him that he was two stories above the pavement and must jump into a fireman's net. In fact, he was only about twelve inches above the pavement. The recruit slipped and broke his leg.

Without his knowledge, the captain told the city clerk to file a worker's compensation claim on behalf of the recruit. His medical expenses were all paid and he cashed the worker's compensation checks.

The recruit sued the city, the fire department, the chief and other individuals personally for negligence. The city defended on the grounds that worker's compensation was the exclusive remedy and all other claims should be dismissed. He responded that, since he was a volunteer, he was not entitled to worker's compensation coverage.

The court decided in favor of the plaintiff, stating that the question of whether or not worker's compensation was the appropriate remedy should have been decided by the Department of Industry, Labor and Human Relations. Until that decision was made, the negligence claims would not be dismissed, but they would be held in suspension. The court said that agency also has jurisdiction over the decision as to whether or not the plaintiff's acceptance of benefits precludes his negligence suit.[74]

FIRE DISTRICTS

In many areas, fire protection is provided by special territorial sub-jurisdictions called fire districts. There are approximately 3,000 of these districts in the United Sates. A fire district is a political subdivision of a state, a municipality, a county or other entity. A fire district must be formed in conformity with all applicable state laws. These statutory provisions will cover the formation, elections of boards of directors, financial and taxing authority, boundaries, procedures for annexation, etc.

Fire districts have only those powers granted to them by statute, so it is essential to know the authorizing statute so that the powers and limitations of the fire district are well understood.

These powers are often extensive and include qualifications for firefighters as well as the power to tax, the right of eminent domain, the right to sue and be sued, to own property, to establish a corollary water district so that an adequate supply will be available for firefighting purposes, to operate ambulances as well as firefighting equipment, to enter into mutual aid agreements, establish a merit personnel system for employees, provide insurance etc. The fire district can have paid as well as volunteer firefighters.

The powers of the fire district are exercised through its board of directors which is usually elected.

The fire chief is an employee and usually the Board's top administrator. The Board sets the policies; the fire chief carries them out.

The members of the board of directors by virtue of their position take on special liability, so it behooves the fire district to insulate them with appropriate insurance and policies providing them with attorneys in the event they are sued for something they do in their official duties and paying any judgments which might be levied against them. Directors are liable for the wrongful expenditure of public funds. They cannot make private use of public property or use public facilities or funds to aid in their elections.

Normally the directors individually would not be personally liable for the torts of members of the fire department, but they may become liable if they direct the wrongful acts or ratify them. The Board itself as a body may very well be liable under the doctrine of *respondeat superior*.

Obviously, directors are subject to criminal and civil penalties for taking bribes or kickbacks on purchase of equipment or doing business with the fire department at inflated charges, even if done through straw parties. State conflict of interest laws will usually apply to fire district directors.

State law may require that the business of the fire district be conducted publicly. If the State has a open meetings law, it will usually be applicable to the fire district's board of directors. These laws usually contain exceptions to the open meeting requirements authorizing meetings in executive session under certain conditions, such as when personnel matters are being discussed.

As with all fire departments, the fire district should have competent legal counsel to guide it through legal matters.

CHAPTER NINE

CONTRACTS

A contract is a promise or a set of promises for the breach of which the law gives a remedy or the performance of which the law in some way recognizes as a duty.

We owe certain duties to others in society and the breach of these duties gives rise to an action in tort. In contracts, however, there is no duty prior to the contract. The agreement itself imposes duties on the parties to the contract. If it were not for the contract, there would be no duty.

For a valid contract, you must have mutuality of assent — a meeting of the minds — between the parties to the agreement.

ORAL OR WRITTEN CONTRACTS

Contracts can be oral or in writing. It is better to have them in writing because it makes the terms of the agreement easier to prove. However, an oral contract is perfectly legal and enforceable.

There are some contracts, however, which are required by the Statute of Frauds to be in writing. The Statute of Frauds sets forth those types of contracts which must be in writing, or must be evidenced by a written memorandum signed by the person being sued. For example, all contracts for an interest in real estate for a year or more must be in writing. Under the Uniform Commercial Code, contracts for the sale of goods valued at $500 or more must be in writing.

EXPRESS OR IMPLIED

Contracts can be express, which means all the terms are specifically spelled out whether the contract is oral or written, or it can be implied.

For example, the City of Philadelphia had a fire in a high rise building one night. As the firefighters were attempting to extinguish the fire, the windows began popping out of the building and the shards of glass came plummeting to earth with tremendous force, cutting the hoses. The firefighters went to a lumber company and acquired plywood with which to cover the water hoses. Suppose the Department later refused to pay for the plywood on the grounds that the purchase did not comply with the city's purchasing requirements. The law would imply a contract to pay from the circumstances surrounding the situation and the city would be liable for the payment.

Most governments have very specific regulations and procedures for issuing contracts for goods and services. Failure to comply with them could lead to voided contracts, court challenges, protests by unsuccessful bidders, disruption of work and perhaps even personal liability for the person making the purchase for exceeding his/her authority. But the government cannot deny payment when one of its agents — in the above case, a Philadelphia firefighter — makes a purchase for the city.

ELEMENTS OF A CONTRACT

There are certain essential elements which must be present to have a valid contract:

- Offer and acceptance. (One party makes an offer and the other accepts it. If that person changes the offer and does not accept it exactly as the first party offered it, there is no valid contract unless and until the first party accepts the counteroffer made by the second party.)
- Consideration. (In a contract, consideration means something given in return for something else to bind the contract which creates benefit and detriment to the parties.)

- The contract must be for a legal purpose. (An agreement to engage in prostitution for a fee, to murder someone for a fee, or a gambling debt are not valid contracts because they are for illegal purposes.)
- Parties to the contract must be legally competent. (Parties who enter into agreements with insane persons and minors cannot enforce them.)

OFFER AND ACCEPTANCE

The offer must be stated with the intent to enter into a contract, not merely to negotiate.

The offer must be communicated to the offeree.

The offer must be definite and certain.

The offer is invalid if offeror was joking. If the contractual intent is lacking, there is no offer.

The offer can be terminated by:

- expiration of the stated time (or, if no time is stated, after a reasonable time)
- revocation (by the offeror)
- rejection (by the offeree)
- counteroffer (by the offeree)
- death or insanity of either party
- illegality (a new law is passed)
- it is against public policy
- destruction of the subject matter of the offer.

EXECUTED AND EXECUTORY CONTRACTS

An executed contract is one where the parties have performed everything required of them under the contract. (The term is also used to designate the signing of a written contract.)

An executory contract is one in which only partial performance has been accomplished.

QUANTUM MERUIT

Quantum Meruit means literally, "as much as he deserves." It is an obligation imposed by law which gives a party the value s/he deserves, notwithstanding the lack of an agreement between the parties. In other words, one party innocently performs a service or does something of benefit to another party. Even though there was no contract, the law may force the enriched party to pay the reasonable value of the benefit received.

QUASI-CONTRACT

A quasi-contract is not really a contract. It is an obligation imposed by law to do justice or to prevent the unjust enrichment of one party at the expense of another. The obligation is inferred from the circumstances.

FRAUD

Fraud is the misrepresentation of a material fact with the intent that a person rely on it, that person does rely on the misrepresentation and is damaged thereby. Fraud can give rise to a civil suit and a criminal prosecution.

OPTION

An option is an agreement to hold an offer open for a specified time in return for a consideration.

DURESS

A contract entered into through duress is not valid. If someone holds a gun to your head and says, "Sign this agreement," it is not enforceable against you.

GIFTS

An agreement to give someone a gift is not enforceable because there is no consideration. In a gift, the intended recipient gives up nothing to bind the agreement, so there is no consideration and, therefore, no valid contract. However, if the intended recipient, relying on the person's promise to make a gift, changes his/her position to his/her detriment, the gift may be enforceable. For example, suppose a wealthy widow in your community offers to give you the funds to build a new fire station. Later, she changes her mind and tells you she will not give you the money. Generally, there is nothing you can do legally to force her to make the gift. However, if relying on her promise to make the gift, you entered into a binding agreement with a contractor to build the new station, or you had an architect draw up the plans, then the gift would probably be enforceable.

VOID AND VOIDABLE CONTRACTS

If a contract is void, there is no contract. For example, if it was for an illegal purpose, or against public policy, there is no contract.

In a voidable contract, the defects could make the contract unenforceable. For example, your sixteen-year-old daughter signs an agreement with a company to buy a certain number of compact discs per year. Even though she lacks the capacity to contract, as long as she wants to continue the arrangement, it is valid. In other words, she can enforce the contract against the company. However, if she changes her mind and does not want to buy any more compact discs, the agreement is not enforceable by the company against her because she is a minor. The contract is voidable by the minor, not by the non-minor.

MUTUAL AID

A mutual aid agreement is a contract. Make sure you have viable mutual aid agreements. Update them regularly. Make sure they meet your needs. Put them in writing so there is no subsequent confusion as to what was agreed to. Draft them with care, anticipating all of the problems which might arise so that parties to the agreement clearly understand what liabilities they are assuming and exactly what is expected of them.

What is the status of your mutual aid agreements? Are they up to date? When were they drafted? Are they still relevant and responsive? Have you provided for second and third tier responders? Do your mutual aid agreements provide the type of specialized help you might need? Are they fully authorized by law? Do the responders come as self-contained units or will you be required to provide housing, food, etc. for them. Whose worker's compensation applies if they are hurt? Are you responsible for paying them overtime if they are entitled to it? Who pays for what? Who is liable for what? How are volunteers protected? Do the mutual aid responders meld into your chain of command or do they operate as a separate entity under their own chain of command?

Suppose you have a hazardous spill at an interstate highway cloverleaf in your jurisdiction. One ramp is in the county, another is in the city and the state controls the interstate. Who's in charge? Who orders an evacuation? Who supervises the cleanup? These are the kinds of questions which should be resolved beforehand in your mutual aid agreements.

Too often fire departments operate for years without formal mutual aid agreements. This works fine until you have a problem. Avoid these future difficulties. Have written mutual aid agreements so that their terms will be clear to everyone involved.

CHAPTER TEN

EMPLOYEE RELATIONS

I n the field of employee relations, it is essential that management treat all employees with the same even-handed fairness.

When taking any disciplinary action it is very important for the manager to document his/her actions. This reduces the possibility of making rash, incorrect decisions based on a misinterpretation of facts which could have serious legal consequences. Documentation is also needed to justify later steps in a progressive disciplinary process and as proof that proper procedures as required by the personnel code, departmental policies and the union contract were followed. The U.S. Constitution requires that employees be accorded due process in all proceedings.

KNOW PROVISIONS OF UNION CONTRACTS

If the fire department has a union contract, it will probably specify how disciplinary matters must be handled and will provide for a grievance procedure. The union contract will usually also call for an arbitration process when management and labor reach an impasse in their dealings with each other.

The fire department is legally obliged to adhere to the terms of the union contract as long as it is in effect, so everyone in management should be familiar with its provisions.

TREAT ALL EMPLOYEES THE SAME WAY

Suppose the department has a trouble-maker employee. S/he is often on sick leave on Monday mornings, grouses about everything the department tries to do, has trouble getting along with fellow employees, has regularly committed minor breaches of the rules etc. and in general has been a difficult person to deal with. One day this employee commits a serious infraction of the rules.

"That's the last straw," management says, "S/he's fired!"

In the same department, there is a model employee who is always on time, never on sick leave, is always looking for extra work to do, is brave, loyal and in general is a completely dedicated employee.

But everyone is human. One day, this outstanding employee commits the same infraction of the rules as the above employee. Management says, "Well, s/he has such a wonderful record, we'll be lenient. We'll suspend him/her for one day."

If the first employee sues, the courts will undoubtedly overturn his firing because management failed to treat both of them the same way for the same infraction.

However, if you document in the personnel records each minor infraction of employee number one and each outstanding performance of employee number two, you will probably be able to justify the disparate action. Without this documentation, however, employee number one will probably be successful in overturning the dismissal.

RESIDENCY REQUIREMENTS

A number of jurisdictions have imposed residency requirements on personnel, especially executives and emergency workers. The rationale is based on a number of factors: government employees should pay taxes to support their own city and they should be involved in the affairs of the jurisdiction which employs them, they should be available to vote for needed bond issues, and, in the case of emergency workers, they should be able to respond quickly when their services are needed.

The U.S. Supreme Court and several state supreme courts have upheld the validity of residency requirements. In the case of *McCarthy v. Philadelphia Civil Service Commission*, a firefighter with sixteen years of service who lived in New Jersey challenged the residency requirement. The U.S. Supreme Court upheld the validity of Philadelphia's residency requirement and said it was not an infringement on the employees' constitutional right to travel in interstate commerce. It said further that the firefighter did not have a protected right in his job and that his refusal to move his residence was a valid basis for firing him.[75]

However, when residency requirements have the effect of discriminatiung against a protected class, they will be overturned.

FEDERAL COURT HOLDS FAMILY ASSOCIATION NOT INFRINGED

In *Hameetman v. City of Chicago*, a federal court ruled that a residency requirement does not violate a person's constitutional right of family association because it does not directly regulate family affairs.[76] If the restriction interfered with vital family needs such as access to required medical treatment or special education which is not available in the employing city, the residency requirement would interfere with an employee's protected rights.

WAIVER OF RESIDENCY REQUIREMENT DENIED

In 1981 the City of Clinton, Iowa, passed an ordinance requiring all police, firefighters and some other city employees to live within 10 miles of their place of work. A woman police officer, who began work in March, 1987, moved to Clinton to comply with the residency requirement. However, she later asked for an exception. She was planning to be married and her fiance lived on a farm about 18 miles from where she worked.

The Chief of Police recommended approval of the exception, but the Clinton City Council's employee relations committee voted to strictly enforce the 10-mile residency requirement.

She sued in district court, arguing that the residency requirement was unconstitutional and contrary to Iowa law. The federal court disagreed with her and dismissed the case. She appealed.

In affirming the lower court's decision, the appellate court said that, according to Iowa law, cities may set reasonable maximum distances outside their borders where police officers, firefighters and other critically important city employees must live. The court felt that ten miles was reasonable.

The burden was on the plaintiff to prove that the 10-mile distance was unreasonable which the court felt she failed to do.

She also alleged that enforcement of the residency requirement denied her equal protection of the law under the U.S. Constitution's Fourteenth Amendment. In determining whether she was denied equal protection, the court applied the rational basis standard which, the court said, means a statute is constitutional unless it is patently arbitrary. The court said a residency requirement bears a rational relationship to one or more legitimate purposes and is, therefore, constitutional under the traditional equal protection test. The court cited the legitimate purposes as: enhancing community safety by enabling emergency workers to respond more quickly, improving employee performance since the employee would have a stake in the community, enhancing the tax base and local economy by increasing the chances that the person's wages will be spent in the city, improving community attitudes and cooperation, increasing loyalty to the community and reducing absenteeism.[77]

FEDERAL COURT IN MICHIGAN REFUSES TO UPHOLD RESIDENCY RULE

A federal court in Michigan came to a different conclusion, holding that a pre-application residency requirement violated the Equal Employment Opportunity Act by discriminating against minority applicants.[78]

In *Gantz v. City of Detroit*, the Michigan Supreme Court upheld residency requirements.[79]

RESIDENCY RULE DID NOT REQUIRE CHIEF'S RESIDENCY BEFORE HIRE

Volunteer firefighters who were discharged by the fire chief brought suit against the City of Kennett, Missouri, claiming that the chief had no authority to terminate them, that only the mayor was permitted by statute to do so. They also claimed the chief did not have the authority to fire them since he was hired in violation of a statute which stated that "all officers elected or appointed to offices in the city government shall be voters under the laws and constitution of this state and must be residents of this city."

The fire chief was appointed on December 6, 1988 but did not move into the city until a few weeks later. They claimed that, since the city council appointed a nonresident in contravention of the statute, the appointment was illegal.

Judgment was entered without trial for the city. The court held that this interpretation of the statute would limit the pool of qualified applicants to persons already living within the city which would place an unreasonable restriction on the city's ability to recruit qualified people. The court construed the statute to mean that an appointed official would have to become a resident of the city at the time his duties began. Therefore, the court said, the fire chief was properly in office. The court also disagreed that only the mayor could terminate firefighters. The court said this would strip the city department heads of the basic responsibility for hiring and disciplining their subordinates. The court said the statutes do not require that the mayor make personal decisions regarding the hiring and termination of every city employee.[80]

OHIO COURT REQUIRES RESIDENCY TO BE COLLECTIVELY BARGAINED

An appellate court in Ohio held that the City of St. Bernard, which had enacted a residency requirement, was obliged to collectively bargain the matter with its unions. A state statute said that laws regarding civil rights, affirmative action, unemployment and worker's compensation, public retirement and residency requirements prevail over conflicting clauses in collective bargaining agreements. The court said that the legislature, if it had desired the enumerated subjects to be excluded, would have done so in the statute's enabling act, but the statute merely said that provisions in a law supersede any CONFLICTING provisions in union agreements.

WHAT IS A "RESIDENCE"?

What constitutes a "residence"? In *Fagiano v. Police Board of the City of Chicago*, the Illinois Supreme Court in upholding the city's residency requirement, said residence is synonymous with domicile, "the place where a person lives and has his true, permanent home to which, whenever he is absent, he has an intention of returning."[81] When a Detroit firefighter had an apartment in the city, but his family lived outside the city, he was told he did not meet the city's residency requirement. He refused to move his family into the city and was fired. The local court said he could not be considered to be a resident of the city when his family resided outside the city. The Michigan Court of Appeals, however, held that it was an error for the lower court to rely exclusively on the residence of the employee's family in determining his residence. In the case of *Civil Service Commission of Pittsburgh v. Parks*, the firefighter's family lived outside the city, he received his mail there, his vehicle registration and his voter registration were all outside the city and his apartment in the city was not inhabited. The court said, as a matter of law, his residence was not in the city.[82]

STATE LAW TAKES PRECEDENCE

If a state law prohibits local governments from imposing residence requirements on their workers, such requirements are unenforceable. Where there is a conflict, state law takes precedence.[83]

USE OF POLYGRAPH

A U. S. District court in Ohio upheld the Cincinnati Fire Department's background investigations which included credit and employment checks, questions about unusual sexual conduct and criminal convictions. It also allowed polygraph testing. The court disallowed questions about arrests which were prohibited under a federal consent affecting the Fire Department requiring its job selection procedures to be job-related. Several unsuccessful black applicants had challenged their rejection and the selection procedures.

The federal court said that, even though there was uncontradicted testimony at the trial that the polygraph examination intimidated many applicants to reveal further background information, its use was job related because it allowed the city to make a more informed choice in its hiring.[84]

A Colorado Court of Appeals also upheld the use of polygraph examinations and the firing of a police officer who refused to take one. A Colorado Springs police officer had received two speeding tickets while off-duty. He was questioned by internal affairs officers and was ordered to take a polygraph exam. He refused. On his behalf, the union objected on the grounds that the subject of the inquiry did not directly relate to the performance of the officer's official duties.

The trial court held that the officer had been horsing around with another officer and drove at high speeds after drinking alcohol and that this conduct affected his fitness for public service. The Colorado Court of Appeals affirmed the officer's dismissal for failure to follow a lawful order.[85]

PROMOTIONAL TESTS

While awaiting a court's decision on its request for a preliminary injunction, the Black Firefighters Association of Dallas moved to prohibit promotions to second driver position and the use of allegedly biased promotion tests. The court granted the motion, but denied the injunction and the Association appealed.

The appellate court affirmed the lower court's ruling, stating that in order to be entitled to a preliminary injunction, the plaintiff must show irreparable harm, substantial likelihood of success, favorable balance of hardships and no adverse effect on the public interest. The trial court had ruled that the Association had failed to meet those requirements. In an effort to show that it could meet those requirements, the Association, among other things, presented statistics showing that in the previous two years, blacks performed worse than whites on the tests. Citing a U.S. Supreme Court ruling that a bare statistical comparison does not establish a basic case of discrimination, the appellate court affirmed the lower court's denial of the preliminary injunction. [86]

"AT-WILL" FIRE CHIEF FIRED, REINSTATED BY COURT

The fire chief of a fire protection district approved twelve-hour shifts starting at 6 P.M. and ending at 6 A.M. for full and parttime firefighters. When a schedule was set for part-timers which provided no training for them, the chief spoke out against the lack of preparedness this caused. He was fired. He sued the district, alleging that his discharge was retaliatory for his speaking out against the lack of preparedness for the part-time firefighters.

The trial court dismissed the suit for failure to state a claim and the chief appealed.

The appellate court held that in Illinois, the tort of retaliatory discharge is a limited and narrow cause of action recognized as an exception to the general rule that at-will employees can be fired at any time for any reason or for no reason. To establish a cause of action in retaliatory discharge, the plaintiff must show that he was fired in retaliation for his activities and that the discharge was in contravention of a clearly mandated public policy. The court said that the Illinois Fire Protection Training Act clearly establishes that the public policy of the state is to maintain a high level of training for its fire protection personnel. The discharge of the fire chief for speaking out against the lack of training contravened the clearly mandated public policy of upgrading and maintaining a high level of training for fire protection personnel. Consequently, the court said, the lower court should not have dismissed the chief's cause of action for retaliatory discharge.[87]

The fire chief could also have brought an action for violation of his First Amendment rights. Although he was arguing about an internal matter, it had great public import — fire safety — and therefore would probably be protected under the First Amendment.

MISSISSIPPI RECOGNIZES EXCEPTIONS TO "AT-WILL" DISCHARGES

The Mississippi Supreme Court has modified the State's "at-will" employment doctrine. Mississippi has a statute which says that municipal employees serve at the pleasure of the government and court decisions have upheld these terminations "at the employer's will for good reason, bad reason or no reason at all, excepting only reasons independently declared legally impermissible." Mississippi does not have a whistle-blower protection act, but the State Supreme Court unanimously recognized two exceptions to at-will discharges: 1) when the employee refuses to participate in an illegal act and 2) when an employee is terminated for reporting an illegal act of the employer.[88]

GROOMING

Can a fire department forbid its employees from having beards and long hair?

Probably, although cases have gone both ways on this issue. Clearly, if a beard interferes with the tight seal on a breathing apparatus mask, the ban can be enforced. Some cases have also been decided in the department's favor on the grounds that in a quasimilitary organization, such as police and fire departments, grooming standards are permissible because they relate to the esprit de corps and are part of the uniform.

Since new protective equipment has made it possible for firefighters to have longer hair than in the past and still be safe and since SCBA face pieces and flame-resistant hood and helmet liners insure that all surfaces of the head are protected, maybe the safety considerations regarding length of hair are not as relevant as they once were.

Some fire departments have adopted a safety standard which states that no hair shall be exposed during firefighting activities. This avoids the need to regulate hair length and merely bases the restriction on considerations of firefighter safety.

The courts and the EEOC have not provided a clear answer to the question of grooming standards.

Some in the fire service think that fire departments should move away from the paramilitary regimen and, as such, they feel grooming standards are no longer as important as they once were.

The courts have generally held that prohibiting long hair for male employees is not sex discrimination if grooming standards for both genders are related to community standards and are applied in an even-handed way. A safety regulation is gender neutral and does not discriminate against either sex. Some courts have pointed out that Title VII of the Civil Rights Act was not intended to interfere with employers who desire to maintain reasonable standards of appearance for their personnel as required by their business. EEOC, however, has treated different grooming rules for men and women as constituting sex discrimination. But, the EEOC has upheld reasonable safety concerns as they relate to grooming.

Some employees have maintained that their appearance is part of personal liberty and that right should not be infringed. The courts have generally declined to overturn reasonable grooming regulations that can be justified by the need for discipline, uniformity and esprit de corps, even where no safety considerations are present. This was the ruling when a male firefighter challenged the department's hair standards in the case of *Quinn v. Muscare.*[89]

FIREFIGHTERS LOSE LONG HAIR

In 1989 four firefighters with long hair were found to be in noncompliance with the East Providence, Rhode Island, Fire Department's grooming standards which were established in 1974. The firefighters' union claimed that the department was violating its contract with the union, and that the general order was not the correct way to establish grooming standards. On the latter point, the Court said the union was too late in complaining. The order should have been contested in 1974 when it was first promulgated.

Judgment was entered for the city. The court pointed out that the firefighters were not disciplined or threatened with discipline so the Department did not violate that part of the union contract which said that "general, orders and memoranda will not be issued to promulgate what should be rules and regulations."[90]

LOUISIANA COURT FINDS GENDER BIAS IN GROOMING RULE

On the other hand, a Louisiana court interpreting the Louisiana constitution held that the City of Shreveport could not restrict firefighters' hair to shoulder length or shorter. Women on the department had been allowed to pin their hair up to conform with the regulation. The court issued an injunction against Shreveport finding that the regulation was not "gender neutral" because men and women were treated differently and this had the effect of classifying individuals "on the basis of sex." The court said, "Similarity of appearance can be and was in fact achieved by requiring fire personnel on duty or in uniform to have their hair, if longer than a regulation length, pinned to meet a length requirement established by their employer. The esprit de corps and equal treatment as to grooming standards are easily achieved by uniform enforcement of hair length while on duty or in uniform with recognition of an individual's right to have whatever length hair s/he desires as long as while on duty or in uniform it is kept to a level set forth in the employer's regulations...While similarity of appearance has been recognized as an appropriate and rational goal in a 'paramilitary' civilian service, commonality may not and should not be required at the expense of reason and purpose."[91]

EMT'S LOSE BEARDS

When three Arkansas EMT's were fired for having facial hair, they filed suit in U.S. District Court under the Civil Rights Act. The lower court granted the county's motion for a summary judgment. Upon appeal, the U.S. Court of Appeals affirmed the U.S. District Court's decision, citing the U.S. Supreme Court's holding in the *Kelley v. Johnson* case.[92] The Court said the grooming standard promoted esprit de corps, presented a uniform and professional public image and minimized concerns that one's hair would impede emergency duties. The Court said that County officials "have the power to treat hair, or the lack of it, as part of the EMT's uniform," even if the policy is mistaken or even silly, and it does not violate the Fourteenth Amendment.[93]

FIREFIGHTER HAS NO CONSTITUTIONAL RIGHT TO SHOW CHEST HAIR

In June, 1992, a federal judge ruled that the Montgomery, Alabama, Fire Department did not violate a firefighter's freedom of expression by ordering him to trim his chest hair which protruded from his collar. The firefighter had been ordered to either cut his chest hair or wear a T-shirt under his uniform. He refused to do either and filed a grievance. He was disciplined by being assigned to 25 consecutive night shifts. He filed a lawsuit in federal court, alleging that the Fire Department had violated his First Amendment right to freedom of expression. The U.S. District Court judge did not agree with him. The Court said that his refusal to wear a T-shirt or trim his chest hair was not deserving of First Amendment protection. A public employer may require its employees to be neat in appearance, the Court said.[94]

NO MUSTACHE RULE IN MASSACHUSETTS UPHELD

A Massachusetts State Police no-mustache rule was upheld by a federal Court of Appeals, citing the Supreme Court's *Kelley v. Johnson* decision. When a number of police agencies were merged with the State Police, the head of the consolidated force extended the existing rule against mustaches to the newly merged other agencies.

The rule was challenged by employees who sought an injunction to prevent its enforcement. The lower court rejected their plea and they appealed to the U.S. Court of Appeals. The appellate court affirmed by a 3 to 0 vote, stating that the Supreme Court's *Kelley* decision was based on esprit de corps which the court said was a "sufficiently rational justification" for the Massachusetts State Police rule. The Court said the head of the State Police had the responsibility for "melding a cohesive unit instilled with a 'common purpose' and 'shared mission' from disparate parts" and consistency of appearance promoted those goals and "would enhance the chances of a successful consolidation."[95]

PFB NO JUSTIFICATION FOR BEARDS, GEORGIA COURT SAYS

The Atlanta Fire Department had a rule prohibiting beards. The rule was challenged in federal court by 12 black firefighters who suffered from PFB (pseudofolliculitis barbae). Police officers have won similar suits in the past, but the Court held that firefighters need a tight fit on their breathing apparatus masks which are regularly worn at fires. The Court said, while blacks are more vulnerable to PFB, the policy itself is racially neutral and is imposed for employee safety considerations. The U.S. Court of Appeals observed that the U.S. Congress had created a "business necessity" exception to Title VII of the Civil Rights Act and the city fell under this exception with its no-beards rule. Atlanta had introduced an affidavit from an occupational health and safety expert claiming that any amount of facial hair could interfere with the seal on the breathing apparatus masks and expose the wearer to risk from toxic fumes.[96]

MICHIGAN COURT FINDS IN FAVOR OF FIREFIGHTER WITH PFB

Coming to an opposite conclusion, a Michigan court held that a firefighter with pseudofolliculitis barbae was handicapped under Michigan law and could not be discharged for failure to shave his beard unless the fire department could prove that he was unable to perform the functions of his job with the help of adaptive devices or aids. If no such device or aid is available, the firefighter might be entitled to a light duty assignment or disability retirement.[97]

D.C. COURT OF APPEALS UPHOLDS RIGHT OF MAN TO HAVE BEARD

The District of Columbia Court of Appeals (*Kennedy v. District of Columbia*) held that a firefighter who had been fired for wearing a beard and handlebar mustache was discriminated against on the basis of appearance and was entitled to back pay and reinstatement. The fire department, which had a regulation that forbade mustaches and beards (except for those firefighters who suffered from pseudofolliculitis barbae), argued that facial hair interfered with the seal on the firefighter's face mask and was, therefore, a safety hazard. Relying on the record established by the EEO director, the appellate court found that the beard and mustache did not interfere with the seal of the face mask.[98] Other cases have come to an opposite conclusion.

GROOMING STANDARDS MUST BE PROMULGATED PROPERLY

Grooming standards may be overturned when they have not been lawfully promulgated. One board of fire commissioners failed to properly adopt a rule, regulation or by-law regarding grooming standards and consequently the court said they had no power to enforce them. Only when the grooming standards have been adopted in conformity with due process (or union contract requirements) will they be enforceable.[99]

MANNING ISSUES MAY OR MAY NOT BE BARGAINABLE

A labor contract may provide for negotiation and/or arbitration of various issues with labor unions. This Oregon case, for example, held that manning levels on apparatus was subject to mandatory bargaining.[100]

Decisions regarding minimum staffing levels for a city fire department were held to be a management prerogative and not subject to bargaining.[101]

Neither was it held to be a mandatory topic of bargaining in New York[102] or in Maine.[103]

A Pennsylvania court, however, ruled that the number of firefighters per rig and the number of firefighters per shift are mandatory subjects of bargaining.[104]

A Michigan appellate court also required management to bargain over the number of firefighters per shift,[105] but a city was not required to bargain with the union on the number of police officers in a patrol car or on duty each shift.[106]

PENNSYLVANIA COURT SAYS CITY NEED NOT BARGAIN OVER LIGHT DUTY

When York, Pennsylvania, adopted a policy on light duty assignments for firefighters who are temporarily sick or injured, the union filed an unfair labor practice charge. The State's Labor Relations Board dismissed the charge and an appeals court affirmed that decision. The court said that a decision to implement a light duty policy is within the City's management discretion and is not a mandatory subject of collective bargaining.[107]

STRIKES

Many states prohibit strikes by public employees. Firefighters who participate in an illegal strike can not only be fired but they can be arrested. In several cases, courts have ruled that individual strikers and the union are both liable for damages caused by the strike. The strike would be both a crime and a tort.

COMMAND OFFICERS EXCLUDED FROM BARGAINING UNIT

A Michigan appellate court excluded command officers from the basic firefighters bargaining unit.[108]

EMPLOYEE REFERENCES

A potential source of liability is the matter of employee references, both verifying and checking information provided by applicants and responding to inquiries from prospective employers about the work performance of former employees.

In considering applicants for hire, the fire department has an obligation to exercise due diligence in checking them out to ensure that nothing in the applicant's background would indicate that s/he would pose a threat to the department or its employees. Applicants often exaggerate and even falsify educational and other achievements so their applications should be thoroughly checked. Failure to do so could leave the department open to liability. The fire department should learn as much as possible about an applicant before making a job offer. In trying to gather such information, however, the department might find that former employers will limit the information they will provide to dates of employment, pay rates and positions held. This makes it difficult to obtain meaningful and helpful information about the applicant's suitability. The department may want to ask the applicant to sign a written authorization allowing the department to contact former employers and waiving any legal action against such employers for responding truthfully to reference inquiries. This may or may not loosen the reluctance of the past employers.

The department should check to see if the applicant has a criminal record in any of the places s/he has worked or resided.

With respect to employment inquiries which the fire department receives about former department employees from prospective employers, a policy and procedure should be established for handling them. The persons authorized to respond to these inquiries should be limited and these authorized individuals should be mature, experienced officers who have been instructed in the possible legal consequences of responding to these inquiries. They should limit their responses to job-related information, the more objective the better, and restrict their comments to the questions asked so they do not exceed the limited privilege that comes from truthfully responding to inquiries from prospective employers.

A federal district court in Washington, D.C. held that an ex-employee could sue her former employer for improperly divulging that she had filed discrimination charges against the employer. The plaintiff claimed that such a disclosure was retaliatory in nature and caused her new employer to terminate her when learning this information.[109] This case points out the importance of channelling all employment inquiries to designated individuals whose background and training have prepared them to respond in a prudent, legally defensible manner.

CHAPTER ELEVEN

FAIR LABOR STANDARDS ACT

BACKGROUND

I n 1938 the Federal Fair Labor Standards Act was enacted, providing that employers must pay their workers at one and one-half times their regular rate of pay for hours worked in excess of 40 hours per work week.

State and local government workers were exempt from the act, but in 1974 the law was amended to include them. Two years later the U.S. Supreme Court in the *National League of Cities v. Usery* case held that it was unconstitutional to apply the FLSA to State and local governments.[110]

In 1985 in the *Garcia v. San Antonio Metropolitan Transit Authority* case, the U.S. Supreme Court, by a 5 to 4 decision reversed the *National League of Cities* decision.[111]

The impact of the Garcia decision was to bring approximately 13.8 million State and local employees under FLSA coverage.

Employers who willfully violate the FLSA could face criminal prosecution. The Statute of Limitations is two years for unwillful violations and three years for willful.

Employees who file complaints under the FLSA cannot be discriminated against.

CONGRESS GRANTS POLICE AND FIRE PARTIAL EXEMPTION FROM FLSA

Because of the unusual working conditions of some state and local government employees such as firefighters, and the special burdens this would place on State and local governments, Congress added Section 7 (k) to the act as a special overtime provision for these employees. Instead of a 40-hour work week, a longer period was specified.

APPLICABLE STANDARD WORK WEEK FOR FIREFIGHTERS IS 53 HOURS

Instead of the 40-hour standard for most workers, the "applicable weekly standard" for firefighters is 53 hours. (The 212-hour exemption of Section 7 (k) for 28-day work periods divided by 4 weeks.) For police, the "applicable weekly standard" or "applicable maximum hours standard" is 43 hours. For all others it is 40 hours. In other words, for all state and local government employees other than police and fire, when the employee has worked 40 hours in a work week, all time in excess of that must be paid on the basis of one and one-half times the normal rate of pay. When a firefighter works 53 hours or a police office 43 hours in the work week, all time above that must be compensated on the time-and-a-half basis.

The public agency can claim the exemption allowed by Sections 7 (k) and 13 (b) (20), for "any employee...in fire protection activities." This means any worker who:

1) Is employed by an organized fire department or fire protection district;

2) Has been trained to the extent required by State statute or local ordinance;

3) Has the legal authority and responsibility to engage in the prevention, control or extinguishment of a fire of any type, and

4) Performs activities which are required for and directly concerned with, the prevention, control or extinguishment of fires, including such incidental non-fire fighting functions as housekeeping, equipment maintenance, lecturing, attending community fire drills and inspecting homes and schools for fire hazards.

The term includes such employees even if they are trainees and whether they are permanent or probationary, and irrespective of what their job title might be (for example, firefighter, engineer, hose or ladder operator, fire specialist, fire marshal, lieutenant, captain, battalion chief, deputy chief, chief) and regardless of their assignment to support activities. The term also includes rescue and ambulance service personnel if such personnel form an integral part of the public agency's fire protection activities.

It does not cover such agency personnel who do not fight fires on a regular basis. It may include such employees during emergency situations when they are called upon to spend substantially all (80 percent or more) of their time during the applicable work period on one or more of the covered activities.

CIVILIAN EMPLOYEES NOT EXEMPT

So-called civilian employees engaged in support activities such as dispatchers, alarm operators, apparatus and equipment repair and maintenance workers, clerks, cooks, etc., are not included in the term "employee in fire protection activities" and the agency may not, therefore, claim the FLSA exemption for these workers.

The attendance at a bona fide fire academy or other training facility when required by the employing agency constitutes engagement in activities under 7(k) only when the employee meets all the applicable tests. If the tests are met, the training is considered incidental to and part of the employee's fire protection activities.

DEPARTMENTS WITH FEWER THAN FIVE EMPLOYEES EXEMPT

Section 13(b)(20) of the FLSA provides a complete overtime pay exemption for any employee of a public agency engaged in fire protection or law enforcement activities, if the public agency employs fewer than five employees in such activities.

EMPLOYEES MAY NOT VOLUNTEER IN SAME JOB FOR SAME EMPLOYER

Section 39(e)(4)(A)(ii) of the Act does not permit an individual to perform hours of volunteer service for a public agency performing the same type of services which the individual is employed to perform for the same public agency. In other words, if a firefighter works for a county government in one part of the county as a career firefighter, s/he may not "volunteer" to work for the county government as a firefighter in another section of the county.

Employees may volunteer hours of service to their public employer or agency provided "such services are not the same type of services which they are paid to perform for the public agency." Employees may volunteer their services in one capacity or another without contemplation of pay for services rendered. The phrase "same type of services" means similar or identical services. An employee of the City Parks and Recreation Department, however, may serve as a volunteer firefighter without his/her volunteer hours being applied to his normal work hours for FLSA purposes.

Montgomery County, Maryland, a combined career and volunteer fire service, consolidated all of its fire companies to bring them under central county control. Many paid firefighters worked as volunteers in their home area. The U.S. Department of Labor ordered the county to pay overtime for all of the extra hours which these paid firefighters had worked as volunteers. The county vowed to challenge the ruling, but it appears to be an obvious FLSA violation. As a result, the county ordered all paid firefighters to cease working as volunteers in their off hours. Both the county and the volunteers were upset by the ruling, but unless the FLSA is amended, they will probably lose.

VOLUNTEERS WORKING FOR A PUBLIC AGENCY ARE NOT "EMPLOYEES" UNDER THE FLSA.

Volunteer firefighters will not be considered employees, even though they may receive reimbursement for any expenses they incur, including payments for meals, transportation, tuition, books, worker's compensation, insurance benefits and other fringe benefits, supplies or other materials essential to their training or services. Furthermore, volunteer firefighters may be paid a nominal fee or stipend such as under a paid-on-call system and still not be considered employees for purposes of the FLSA.

EMT'S MAY NOT BE EXEMPT FROM FLSA OVERTIME RULES

The exemption applies to rescue and ambulance service personnel if such personnel are an integral part of the public agency's fire protection activities. They are considered employees engaged in fire protection for purposes of the 7(k) exemption if their services are substantially related to firefighting (or law enforcement) in that:

1) They received training in the rescue of fire, crime and accident victims or firefighters or law enforcement personnel injured in the performance of their respective duties, and

2) They are regularly dispatched to fires, crime scenes, riots, natural disasters and accidents.

A federal Court in Baltimore (and in several other jurisdictions) ruled that fire departments' emergency medical personnel were not exempt from FLSA's 40-hour overtime provision. The court said that Congress could have exempted emergency medical response personnel, but it did not.[112] In order for EMT's to be exempt from the 40-hour overtime provisions, the criteria set forth above must be met.

APPEALS COURT EXPLAINS 80% / 20% RULE IN REMANDING CASE

Firefighters and EMS personnel made up Martin County, Florida's emergency response team. The EMS personnel worked in eleven paramedic stations and in six of the County's fourteen fire stations. Most EMS workers responded to emergencies in their assigned ambulances, but at one station they rode to emergencies on fire trucks. Their shifts were the same as for firefighters — 24 hours on duty and 48 hours off.

The EMS workers provided emergency treatment to victims of fires, accidents, crimes, hazmat incidents and other medical crises. All of their duties were related to emergency medical services.

Martin County classified them as firefighters and treated them under FLSA's Section 7 (k) exemption, paying them time and a half for overtime only after they had worked 53 hours, rather than the general FLSA requirement of 40 hours. Suit was brought against the county, challenging this designation. When the lower court granted summary judgment for the county, the EMS personnel appealed.

The appellate court held that the county must compensate an employee at an overtime rate for all work performed in excess of 40 hours during a work week unless the worker was eligible for an exemption. The court pointed out that the section 7 (k) exemption provides the employer with partial exemption for employees engaged in fire protection or police work. These workers must be compensated for overtime after they have worked 53 and 43 hours, respectively.

The Court said that, while the Fair Labor Standards Act itself does not specifically address emergency medical workers who work for public agencies, the Wage and Hour Division of the Department of Labor's regulations state that a public agency may treat EMS personnel as firefighters or police only if their services are substantially related to fire protection or police work. The court said the employer may lose the exemption if the covered employees work more than twenty percent of their time on non-exempt activities, and this rule should be applied to the EMS workers.

The county's computer printout of the EMS dispatches was confusing and the county could not show that 80% of the EMS workers' time was spent in activities which should be exempt. It appeared that the EMS personnel functioned autonomously on about half of their calls.

The court said that where ambulance and rescue personnel perform both fire and law enforcement duties, the applicable standard is the one which applies to the activity in which the employee spends the majority of his/her work time during the work period. The regulations require a comparison of the number of hours spent in each type of activity instead of the percentage of calls falling into each classification. Similarly, it is difficult to classify whether a car on fire is an EMS activity, a fire activity or a police activity.

The trial court had granted summary judgment in favor of the county, but, the appeals court said, such granting was inappropriate. The trial court could not have found as a matter of law that the EMS personnel fell into either the firefighter or police officer exemption because evidence was not heard regarding the number of hours spent in each type of exempt activity. For the county to place the employees in the section 7 (k) exemption category, it must show that the exempt employees worked at least 80% of their time in activities related to the exemption. The case was remanded back to the trial court to determine whether or not the EMS employees were exempt.[113]

King County, Washigton also lost a case brought by its EMS personnel. The court said in order for the 7 (k) exemption to be properly claimed for EMS workers, the county would have to show that their work was an integral part of firefighting or law enforcement activities. A computer printout showed that only one-half of one percent of the EMT's dispatches were for fire emergencies and less than five percent were related to law enforcement emergencies.[114]

Ambulance and rescue service employees of public agencies such as those who work for government hospitals or other institutions primarily engaged in care of the sick, the aged or mentally ill or defective are not covered under the exemption. Neither do such personnel employed by private organizations come under the exemptions even if their activities are substantially related to the fire protection and law enforcement activities performed by a public agency or if the private employer is operating under a contract with the public agency to provide these services.

EXECUTIVE, ADMINISTRATIVE AND PROFESSIONAL EMPLOYEES ARE EXEMPT

The act provides a complete exemption under section 13(a)(1) from the overtime and minimum wage requirements for any employee who is in a bona fide executive, administrative or professional capacity. The fire department or fire district can claim the exemption for such personnel only if they meet certain specified tests relating to duties, responsibilities and salary.

TO BE EXEMPT, EXECUTIVES' SALARY MUST BE FIXED AND IMMUTABLE

The Ninth Circuit U.S. Court of Appeals held in the *Abshire* case that Kern County, California, Battalion Fire Chiefs are not salaried within the meaning of the Fair Labor Standards Act and thus are not bona fide executives exempt from the overtime provisions of the FLSA. The court said, "The dispositive factor is that under the County's policy, the employee's pay is at all times subject to deductions for tardiness or other occurrences. Either pay is fixed and immutable, and not subject to such deductions, or it is contingent. Battalion Chiefs' pay is contingent."[115]

BUT EIGHTH CIRCUIT DISTINGUISHED RULING IN OMAHA CASE

Assistant fire chiefs in Omaha, who are paid a predetermined amount of money no matter how many hours they work, were held to be executive or administrative employees for purposes of the FLSA exemption in spite of the fact that their pay is subject to being docked for absences of less than one day unless accumulated leave can be reduced. This decision is in conflict with the *Abshire* decision. The Eighth Circuit distinguished the pay system of Omaha from that of Kern County. In Kern County battalion chiefs were paid overtime for every one-tenth of an hour beyond regularly scheduled periods and regular rates for training activities outside of normal duty hours. That scheme was inconsistent with salaried status, the court said. In Omaha assistant chiefs are paid an annual salary, no matter how many hours they work. While personal leave and compensatory leave are part of an employee's compensation package, they do not constitute "salary." The Court cited decisions in the Fifth and Eleventh federal circuits which did not follow the *Abshire* case. The Court said the *Abshire* approach "does not persuade us."[116]

Fire captains and lieutenants were found to be "executives" and, therefore, exempt from FLSA overtime provisions in an Ohio case[117] and deputy fire chiefs and captains were held to be exempt from overtime in Huntington, West Virginia, but not lieutenants.[118] Similarly, a federal court in Texas also ruled that district and deputy fire chiefs are "salaried executives" exempt from overtime.[119]

VIRGINIA COURT SAYS LIEUTENANTS NOT EXEMPT

A U.S. District Court in Virginia upheld the overtime claims of 78 present and former lieutenants in the Fairfax County Fire Rescue Department. The Court said they were not exempt from the overtime provisions of the FLSA. The Court said the question is not whether or not they are called "executive employees," but whether they are paid on an hourly or salary basis. The judge said a salaried employee is one who is paid his/her full salary for any week in which s/he performs any work no matter how many days or hours are worked. Under the FLSA's regulations, a compensation deduction for a salaried employee is allowed only 1) when the employee absents him/herself from work for a day or more for personal reasons other than sickness or accident or 2) for absences of a day or more occasioned by sickness or disability...if the deduction is made in accordance with a bona fide plan, policy or practice of providing compensation for loss of salary by both sickness and disability.

COMP TIME IN LIEU OF CASH

The fire department may offer compensatory time off in lieu of cash, but only under the following conditions:

1) It must be pursuant to a collective bargaining agreement, a memorandum of understanding or other type of written or oral agreement between the agency and employee's union or, if there is no union representation, through agreement with the employee prior to performance of the work;

2) If the employee has not accrued compensatory time in excess of the limit applicable to the employee.

Concluding that an employment agreement was coercive, a federal appeals court approved a firefighter's claim for overtime pay.[120]

This compensatory time must be at the rate of not less than one and one-half hours of compensatory time for each hour of overtime worked.

PUBLIC SAFETY EMPLOYEES MAY ACCRUE 480 HOURS OF COMP TIME

Public safety employees may accrue 480 hours of compensatory time. Non-public safety employees may accrue only 240 hours. In other words, the 480 hour limit on accrued compensatory time represents not more than 320 hours of actual overtime worked.

The 480-hour accrual limit will not apply to office personnel or other civilian employees who may perform public safety activities only in emergency situations, even if they spend substantially all of their time in a particular week in such activities. For example, a maintenance worker employed by a public agency who is called upon to perform firefighting activities during an emergency would remain subject to the 240-hour limit, even if s/he spent an entire week or several weeks in a year performing public safety activities.

The employee must be regularly engaged in "emergency response" activities to be covered under the 480-hour limit. A city worker who may be called upon to perform rescue work in the event of a flood, snowstorm or hurricane would not be covered under the higher limit since such emergency response activities are not a part of the employee's regular job.

If the agency decides to pay the employee for the accrued compensatory time, it must pay at the rate of pay the employee is earning at the time the payment is being made.

Upon termination of employment, any unused compensatory time must be paid at a rate of compensation not less than the average regular rate received by the employee during the last three years of his/her employment or the final regular rate received by that employee, whichever is higher.

The agency must grant the employee's request to take compensatory time within a reasonable time after making the request unless it would disrupt the operations of the department.

MUTUAL AID AGREEMENTS AND THE FLSA

Suppose someone works as a paid firefighter for a jurisdiction which has a mutual aid agreement with a nearby community where that person serves as a volunteer. That firefighter, while in a volunteer capacity, responds under that mutual aid agreement to extinguish a fire in the community where s/he works as a paid firefighter. Must those hours be counted as part of his/her hours of employment for FLSA purposes? The answer is "no". The mere fact that services volunteered in one jurisdiction may in some instances involve performance in the firefighter's jurisdiction of employment does not require that the volunteer's hours are to be counted as hours of work with that employer.

TOUR OF DUTY

For those combined police and fire agencies where the employee performs both functions, the applicable standard is the one which applies to the activity in which the employee spends most of his/her work time during the work period.

"Tour of duty" applies only to police and fire personnel who have the 7(k) exemption. It means the period of time during which an employee is considered to be on duty for purposes of determining compensable hours. It may be a scheduled or unscheduled period. Such periods include shifts assigned to employees, often days in advance of the performance of the work. Scheduled periods also include time spent in work outside the shift which the public agency employer assigns. Such unscheduled time, for example, might include time spent in court.

Compensable hours of work generally include all of the time during which an employee is on duty on the employer's premises or at a prescribed workplace, as well as other time, including all pre-shift and post-shift activities which are an integral part of the employee's principal activity or which are closely related to the performance of the principal activity. For example, attending roll calls, writing reports, washing and re-racking fire hoses.

OFF DUTY MUST MEAN OFF DUTY

Periods during which an employee is completely relieved from duty and which are long enough to enable him/her to use the time effectively for his/her own purposes are not hours worked. An employee who is required to remain on call on the employer's premises or so close thereto that s/he cannot use the time effectively for his/her own purposes is working while on call. An employee who is not required to remain on the employer's premises but is merely required to leave word at his/her home or with the employer where s/he may be reached, or is required to wear a pager, is not working while on call.

Brief rest periods may not be deducted from work time. They must be counted as time worked.

Restrictive standby time was held compensable in Kansas by a federal court which awarded double damages to firefighters.[121]

A requirement that firefighters wear a pager or provide a telephone number did not convert off-duty periods to time worked.[122]

Normal home-to-work travel time is not compensable, even when the employee is expected to report to work at a location away from the employer's premises. This is so even if the employee is driving a government vehicle and is required to keep the radio on to be able to respond to an emergency. Of course, the time spent responding to such emergency calls is compensable.

The fact that employees cannot return home after work does not necessarily mean that they continue on duty after their shift. For example, firefighters working on a forest fire may be transported to a camp after their shift in order to rest and eat. As a practical matter, the firefighters may be precluded from going to their homes because of the distance.

Section 7(k) of the FLSA provides partial overtime pay exemption for fire protection and law enforcement personnel who are employed by public agencies on a work-period basis. This section of the Act formerly permitted public agencies to pay overtime compensation to such employees in work periods of 28 consecutive days only after 216 hours of work. However, the 216-hour standard has been replaced by 212 hours of work for fire protection employees and 171 hours for law enforcement employees. In the case of such employees who have a work period of at least 7 but less than 28 consecutive days, overtime compensation is required when the ratio of the number of hours worked to the number of days in the work period exceeds the ratio of 212 (for firefighters) or 171 (for police) to 28 days.

SLEEP TIME

Sleep time can be excluded from compensable hours of work in the case of firefighters who are on a tour of duty of 24 hours or more, but only if there is an express or implied agreement between the employer and the employees to exclude such time. In the absence of such an agreement, the sleep time is compensable. In no event shall the time excluded as sleep time exceed 8 hours in a 24-hour period. If the sleep time is interrupted by a call to duty, the interruption must be counted as time worked. If the sleep period is interrupted to such an extent that the employee cannot get a reasonable night's sleep (which for enforcement purposes means at least 5 hours) the entire time must be counted as hours worked. It would seem that firefighters should be entitled to 5 hours of UNINTERRUPTED sleep, but FLSA regulations do not require this. The regulations say the employee is entitled to a "reasonable" night's sleep.

MEAL TIME

The fire department may exclude meal time from hours worked, provided that the employee is completely relieved from duty during the meal period. If the employee is required to remain on call at the fire station or similar facility, s/he is not considered completely relieved from duty and any such meal periods must be compensable.

WORKERS MUST AGREE TO DEDUCTION FOR SLEEP AND MEAL TIME

A U.S. District Court in North Carolina disallowed deductions which Catawba County had made for meal periods and 8 hours of sleep time from 24-hour shifts of EMS personnel because they had never agreed to the deductions. The court held that in the absence of such an agreement the employees are entitled to overtime pay for those hours which had been deducted from work time. If the agreement is not in writing, the court said, it will not be implied if an employee asserts reasonable contemporaneous oral objections or protests the employer's actions. The fact that the employees continued to cash their checks was not proof of their agreement to the arrangement. Furthermore, the court held that because the EMS workers had to respond to emergency calls during meal periods, the County could not deduct 1.5 hours per day for meals. The Court awarded the EMS workers double pay for the hours deducted.[123]

When there has been no express or implied agreement regarding the deductibility of meal and sleep times by the employees or their representative organization, the sleep and meal times must be counted as work time.

An employee who is required to be on duty for less than 24 hours is working even though s/he is permitted to sleep or engage in other personal activities when not busy. If the person is required to be on duty, the time is work time.

As used in section 7(k) the term "work period" refers to any established and regularly recurring period of work which, under the terms of the Act and legislative history, cannot be less than 7 consecutive days nor more than 28 consecutive days. Except for this limitation, the work period can be of any length, and it need not coincide with the duty cycle or pay period with a particular day of the week or hour of the day. Once the beginning and ending time of an employee's work period is established, however, it remains fixed regardless of how many hours are worked within the period. The beginning and ending of the work period may be changed, provided that the change is intended to be permanent and is not designed to evade the overtime compensation requirements of the act. The employer may have one work period applicable to all employees or different work periods.

VOLUNTARY RELIEF TIME IS NOT TIME WORKING

It is common practice among fire personnel to relieve employees on the previous shift prior to the scheduled starting time. Such early relief time may occur pursuant to an employee agreement, either express or implied. This practice will not have the effect of increasing the number of compensable hours of work for employees employed where it is voluntary on the part of the employees and does not result, over a period of time, in their failure to receive prior compensation for all hours actually worked. On the other hand, if the practice is required by the employer, the time involved must be added to the employee's tour of duty time and must be treated as compensable hours of work.

TRAINING TIME

Time spend in attending training which is required by an employer is normally considered compensable hours of work. However, such training is not counted as compensable hours of work when attendance outside of regular working hours at specialized or follow-up training is required by law. If such training is required for certification by a higher level of government, it does not constitute compensable hours of work. For example, where a state or county law imposes a training obligation on city firefighters, this would not be compensable time, even if all or part of the cost of the training is paid for by the employer.

Firefighters who are in attendance at a fire academy or other training facility are not considered to be on duty during those times when they are not in class or at a training session if they are free to use such time for personal activities. Such free time is not compensable.

NO OVERTIME FOR PERIODS OF AT LEAST 7 DAYS BUT LESS THAN 28

For fire service employees who have a work period of at least 7 days but less than 28 consecutive days, no overtime compensation is required under 7(k) until the number of hours worked exceeds the number of hours which bears the same relationship to 212 as the number of days in the work period bears to 28.

The ratio of 212 hours to 28 days for employees engaged in fire protection activities is 7.57 hours per day (rounded).

Public agencies can balance the hours of work over an entire work period. For example, if a firefighter's work period is 28 consecutive days and s/he works 80 hours in each of the first two weeks, but only 52 hours in the third week, no overtime compensation would be required since the total hours worked do not exceed 212 for the work period. If the same firefighter had a work period of only 14 days, overtime compensation would be due for 54 hours (160 minus 106 hours) in that 14-day work period.

ARSON INVESTIGATORS HELD NOT EXEMPT

The City of Minneapolis classified its arson investigators as employees engaged in fire protection activities, making them eligible for overtime pay only for hours worked in excess of 53 during a seven-day work period. The investigators, however, argued that they were engaged in law enforcement activities and they should be entitled to overtime pay after only 43 hours of work per week. The arson investigators brought an action for a declaratory judgment, seeking back pay for unpaid overtime hours. Both the plaintiffs and the defendant filed motions for judgment without trial. The arson investigators' motion was granted and the city appealed. The appellate court said how these employees are to be classified depends on their responsibilities and tasks, not on their job title or their place of work. They worked for the Minneapolis Fire Department and they were assigned to fire stations. They responded to fire calls and spent time at fire scenes, but, the court said, they did not perform activities required for, and directly concerned with, the prevention, control or suppression of fires. They investigate arson, look for evidence of arson at fire scenes, investigate leads, compile evidence and assist in the arrest and prosecution of suspected arsonists. The court found that these activities were closer to law enforcement activities than to fire protection activities and upheld the decision of the lower court in favor of the arson investigators.[124]

CITY'S RECALCULATION OF HOURS REJECTED BY COURT

Firefighters in Murfreesboro, Tennessee, worked twenty-four hours and then had 48 hours off. Overtime was paid after 159 hours in a 21-day period. Certain FLSA rulings convinced city officials that much of the firefighters' time should be considered overtime. Officials told the firefighters that their hours were being cut by excluding meal times and their hourly wage was being reduced. The officials felt that these new provisions would give the firefighters the same annual income and bring the city into compliance with FLSA regulations. Firefighters brought suit against the city objecting to the arrangement and the court granted a summary judgment for them against the city.

The court said that, while unilateral recalculation was not specifically prohibited by the FLSA, it was inconsistent with the legislative intent of Congress. In a joint explanatory statement, both Houses of Congress had said that any attempt to nullify the overtime provision by compensatory recalculation of wages would negate the premium wages required by the FLSA. The firefighters were given back pay and attorney fees.[125]

CONGRESS TO STUDY FLSA

The new Republican majority which took over Congress in January, 1995, indicated that the FLSA would be revised. GOP members on the House Economic and Educational Opportunities Committee said that early in the session they would introduce legislation to make the rules more flexible and attuned to the needs of the modern work force. While they were not specifically referring to the fire service, the GOP, in an internal memorandum, called the FLSA a "barrier to flexible scheduling and compensation in the workplace." The memo said, "The demographics of the work force and the societal pressures influencing it have changed dramatically since 1938, yet the FLSA has not kept pace with these changes."

One of the changes they have in mind is the tremendous influx of women in the work force, many of whom would benefit from a more flexible work week.

CHAPTER TWELVE

AMERICANS WITH DISABILITIES ACT AND THE FIRE SERVICE

The Americans with Disabilities Act has many implications for fire departments, particularly in employment matters.

Someone in the fire department should develop expertise in the ADA (as well as other statutes which create special mandates.) This person should read everything possible about these subjects, attend seminars and be a resource person with ready answers for the department. The EEOC has issued a "Technical Assistance Manual" to assist employers in complying with the ADA which this designated expert should have readily available.

The Americans with Disabilities Act, which became effective in 1992, has two main parts: employment and public accommodations.

The ADA is designed to protect people with mental as well as physical impairment. The act requires employers to make reasonable accommodation for their employees' or applicants' handicaps.

It also requires those serving the public to make necessary adjustments to accommodate those with disabilities.

The Equal Employment Opportunity Commission oversees the Act's employment provisions and the Department of Justice enforces its public accommodations aspects.

In the first year after its enactment, nearly 12,000 complaints by individuals were filed with the EEOC alleging discrimination under the ADA. Eighty percent of these complaints were from current employees rather than from applicants.

The most common types of charges alleged in the complaints were as follows:

Discharged from work because of disability 48%

Employer failed to make reasonable accommodation 22%

Disability prevented worker from getting job 13%

Harassed because of disability 10%

Unfairly disciplined because of disability 7.2%

Denied benefits because of disability 3.6%

Most common disabilities for which complaints were filed:

Back impairment 18%

Mental illness 9.8%

Heart impairment 4.3%

Neurological 3.7%

Diabetes 3.6%

EMPLOYMENT DISCRIMINATION AGAINST QID'S VIOLATES ADA

The ADA states that employers shall not discriminate against a Qualified Individual with a Disability (QID) because of the disability of such individual in regard to job application procedures, the hiring, advancement, or discharge of employees, employee compensation, job training and other terms, conditions and privileges of employment.

Title I of the ADA prohibits employers from discriminating against qualified applicants or employees with a disability who can perform the essential functions of the job with or without "reasonable accommodation."

The Act does not require that employers give special preference to disabled employees or applicants, but only prohibits discrimination against them because of the disability.

Title II prohibits discrimination by a public entity, including state and local governments.

Title III requires that public accommodations and commercial facilities be accessible to individuals with a disability.

Title IV relates to communications common carriers, requiring them to provide accessibility for persons with speech or hearing disabilities.

Title V includes miscellaneous provisions of the Act.

EMPLOYMENT DISCRIMINATION

The law prohibits employers from discriminating against a Qualified Individual with a Disability (QID) regarding:

- Recruitment and advertising for applicants
- Hiring
- Firing
- Promotions
- Transfers
- Rates of pay or other compensation
- Job assignments
- Leaves of absence
- Training and conferences
- Social and recreational activities sponsored by employer
- Fringe benefits, even if they are administered by someone else.
- Shift assignments
- And any other terms, conditions or privileges of employment

DISABILITY DEFINED

The ADA defines disability as:

1) A physical or mental impairment which substantially limits one or more major life activities. (For example, walking, seeing, hearing, speaking, learning, etc.)

2) Having a record of such impairment. (Someone who had an impairment in the past such as a recovering alcoholic or drug user, someone who was confined to a mental institution, or someone who was out of work in the past because of a heart attack.)

3) Being regarded as having such an impairment. (The person actually does not have a disability, but people think s/he does. For example, someone who is rumored to have AIDS but does not.)

WHAT CONSTITUTES IMPAIRMENT?

The ADA lists things which do not constitute physical or mental impairments:

- Temporary impairment such as a broken limb
- Normal pregnancy
- Concussions
- Physical characteristics such as eye or hair color, baldness or left handedness
- Pyromania
- Kleptomania
- Compulsive gambling.
- Homosexuality
- Transvestism
- Exhibitionism
- Predisposition to disease or illness
- Poor judgment
- Quick temper
- Environmental or cultural disadvantages
- Having a criminal record
- Not having a good education
- Current illegal use of drugs (but reformed drug addicts ARE protected.)

TYPES OF DISCRIMINATION

The Act lists seven types of actions which are included under the term "discriminate":

1) Limiting, segregating or classifying a disabled applicant or employee in such a way that s/he is adversely affected.

2) Participating in a contractual or other relationship that has the effect of discriminating against an applicant or employee. (For example, a fire department using an employment agency or testing company which discriminates against the disabled in its screening methods, or a fire department holding a training program at a private facility which is not handicapped accessible.)

3) Using standards which discriminate on the basis of disability or perpetuating the discrimination of others. (For example, having a

policy in the fire department requiring all employees to have a driver's license would discriminate against a departmental secretary who was unable to drive because of his/her disability because driving is not an essential function of the job. However, such a policy would be acceptable with respect to firefighters because driving is an essential function of that job.)

4) Denying jobs or benefits to someone who has a relationship with, or associates with, a disabled person. (For example, not hiring a woman who takes care of her wheel-chair bound parent on the grounds that she might have to take an inordinate amount of leave to care for the disabled parent, or failing to hire someone whose roommate has AIDS.) The ADA does not require an employer to make a reasonable accommodation for an applicant or employee who has such a relationship with a disabled person as it does with the QID him/herself. In other words, the department would not have to make special shift or leave arrangements to accommodate such a person.

5) Not making a reasonable accommodation for the disabled applicant or employee with a disability. If the employer would suffer undue hardship by making the reasonable accommodation, it does not have to be made. (For example, if a firefighter requests leave to enroll in an alcohol rehabilitation program, the department may not deny the request unless granting it would cause an undue hardship in the department's operations. Such a request might be reasonable in a large department, but unreasonable in a small department.) If the disability prevents the disabled person from performing the essential functions of the job, but there is something the employer can do which would enable the disabled person to perform those essential functions of the job, the employer must make the accommodation unless it would create an undue burden or would jeopardize the safety of other employees or the public. Reasonable accommodations include providing physical access to the work site, restructuring the job or the work schedule, acquiring special equipment or devices, providing leave, modifying training materials or programs, etc.

6) Using tests or selection criteria which have the effect of screening out disabled persons. The tests must be shown to be job-related

and constitute a business necessity. (For example, departmental secretary applicants, who might have a disability, could not be required to qualify on the department's physical exertion tests which are used for firefighter applicants.)

7) Failing to administer tests to disabled persons which accurately reflect their skills. (Example, a firefighter with dyslexia should be given an oral rather than a written test.)

CURRENT DRUG USERS NOT PROTECTED; ALCOHOL USERS ARE

The ADA's provisions state that a "qualified individual with a disability" does not include applicants who currently use illegal drugs. The Act does cover applicants or employees who are currently abusing alcohol. Such a person is entitled to the ADA's protection. In other words, the act distinguishes between use of alcohol and illegal use of other drugs. However, the employer may hold alcohol users to the same qualification standards as for other workers.

The ADA does prohibit discrimination against someone who has successfully completed or is currently enrolled in a supervised rehabilitation program and is no longer using drugs illegally. The Act also protects those who are erroneously regarded as illegal drug users.

The department may prohibit use of drugs or alcohol at the workplace and may insist that employees be drug-free or alcohol-free at work and require employees to otherwise comply with the Drug-Free Workplace Act. Even though a person may be disabled because of addiction to nicotine, employers may prohibit smoking in the workplace or impose restrictions on it. Restrictions on smoking are becoming more and more prevalent.

The ADA has no prohibitions against employers maintaining a drug-testing program for applicants and employees and the employer may make employment decisions on the basis of the results of such tests.

SMOKING BANS

May a fire department prohibit firefighters from smoking on and off the job? The court would assess whether or not this restriction was reasonable. Considering the nature of the occupation, the courts would probably find that it is reasonable.

ESSENTIAL FUNCTIONS OF THE JOB

The disabled person's ability must be measured against the essential functions of the job. The department should set forth the essential functions of the job in a job description before recruitment of applicants begins. This will be persuasive evidence of the job's essential functions if the department is subsequently challenged.

The ADA distinguishes between essential functions and marginal functions of the job. The department may not discriminate against a person who can perform the ESSENTIAL functions, but cannot perform the MARGINAL functions. (For example, a firefighter capable of performing all the essential functions of the firefighter position may have an allergy to motor oil which would prevent him from changing the fluids of fire trucks. Since this is a marginal rather than an essential function of the job, the department could not refuse to hire him on this basis.)

PHYSICAL ACCESS

The ADA requires installation of ramps and other facilities to make your public space accessible to the handicapped.

All public buildings must now have made "readily achievable" changes in facilities to afford access to people with disabilities as long as the adjustments would not cause an "undue burden" or "fundamentally alter the nature of the firm's product or service." Accessibility must be afforded for emergency services, offices, meeting rooms, recreational facilities, etc., of any public place —unless this would create a threat to the health and safety of others.

The "reasonable accommodation" feature of the Act ensures that buildings will not have to be shut down if they do not comply. While the ADA is flexible for existing buildings, all new buildings and major renovations must adhere to ADA standards and requirements.

Existing fire stations are not necessarily exempt from the public access provisions of the Act. With respect to these older fire houses, it should be kept in mind that it is not necessary to furnish an accommodation which creates an undue hardship. An undue hardship is one which requires significant difficulty or expense when considered in the light of the nature and cost of the accommodation, the overall financial resources of the department, the total size of the department and its number of employees, the type of structure and so forth.

Some changes are easy. If an office which serves the public is on the second floor with no elevator access, the service could be moved to an office on the first floor. Ramps, restroom access, replacing door knobs are all relatively simple changes which can be made to accommodate the handicapped.

It should be emphasized that the Act relates not only to physical barriers, but access to communications and information such as teletype 911, interpreters at meetings, visual or audible alarms etc.

PROVISIONS OF ADA

To summarize, among the provisions of the Act are the following:

- New vehicles must be accessible to QID's, including those in wheel chairs. Vehicles must have ramps, lifts, wheel chair space, etc. unless manufacturer can show after good faith effort that equipment is not available.
- The law requires accessibility for emergency services, offices, meeting rooms, recreational facilities, etc. of any public place unless the accessibility would create a threat to health and safety of others without reasonable accommodation.
- It requires interpreters, readers, braille, etc. to provide the QID with equal opportunity for employment promotion, etc.
- It requires all new facilities to be readily accessible to the handicapped.
- Visual and audible alarms etc. must be modified to provide warning to those with disabilities.
- Telephone companies must provide telecommunications relay services for the hearing and speech impaired. This will impact on 911, TDY's and modems.
- Remedies available to those who have been discriminated against in violation of the ADA: reinstatement with or without back pay, attorney fees and court costs, compensatory, but no punitive damages.

MEDICAL EXAMS AND QUESTIONS PROHIBITED

The ADA includes the following other provisions:

- There can be no pre employment medical exam. (Alcohol and drug tests are not considered "medical")

- You can require a medical examination only if it is specifically job-related and consistent with business necessity, but only after the applicant with a disability has been conditionally offered a job.
- You can't ask if the applicant has disability or about the person's physical or mental health.
- You CAN ask if s/he can perform an essential function of the job. The person may be asked how the applicant would perform the function, but the disabled person may not be asked to demonstrate this ability unless all other applicants are also asked to do so.

IMPERMISSIBLE QUESTIONS

According to the EEOC, these are some of the questions which should not be asked:

Do you have: ... AIDS? ... asthma? ... a disability which would interfere with your ability to do the job? ...any health problems?

Have you ever been injured on the job? Have you ever filed a worker's compensation claim?

How much alcohol do you drink each week?

Have you ever had surgery?

Have you ever been hospitalized?

Have you ever been treated for mental health problems?

Have you ever consulted a psychiatrist?

How many days sick leave did you take last year?

What kind of exercises do you do?

PERMISSIBLE QUESTIONS

It is permissible to ask the following kind of questions:

Can you perform the essential functions of this job with or without reasonable accommodation?

Describe how you would perform these functions.

How did you break your leg?

Do you have a cold?

Have you ever taken Tylenol for a fever?

Can you meet the attendance requirements of this job?

Can you work X number of hours per week?

Can you arrive at work at 7 AM?

Can you drive a truck?

Can you perform CPR on an adult patient?

How many days annual leave did you take last year?

Do you use illegal drugs? Have you used illegal drugs in the past two years?

How much do you weigh? How tall are you?

Do you usually eat three meals per day?

Do you have all the required licenses required for this job?

MEDICAL EXAMS

As was stated above, giving a medical examination to applicants is prohibited, but a medical examination MAY be given after a conditional job offer is made and before work begins. There are some conditions attached to this:

1) ALL applicants must be subjected to the medical examination. The disabled person cannot be singled out for the medical examination.

2) Information obtained regarding a medical condition or medical history must be collected and maintained on separate forms and kept in separate, confidential, medical files. Supervisors may be informed regarding any needed restrictions of duties or necessary accommodations for the employee's safety, but not about the actual test results or a diagnosed condition. When appropriate, first aid personnel may be advised if the medical condition might later require treatment. EEOC inspectors can also be given access to this information.

3) The results of the medical examination must be used in compliance with the ADA.

WHEN AN EXAM IS "MEDICAL"

What is a medical exam? The following factors are used to determine whether or not a test is medical:

1) Whether the test is given by a health care professional or trainee.

2) Whether the results of the test are interpreted by a health care professional or trainee.

3) Whether the test is given for the purpose of revealing an impairment or the state of an individual's physical or psychological health.

4)   Whether the test is invasive (drawing blood, urine, breath etc.)
5)   Whether the exam measures physiological/psychological responses or whether it tests performance of a task.
6)   Whether the test is normally done in a medical setting.
7)   Whether medical equipment or devices are used for the test.

WHEN JOB OFFER MAY BE WITHDRAWN

Following the medical exam, the ADA allows withdrawal of a job offer:

1)   If the employer can show that the withdrawal is related to business necessity.
2)   If the applicant would pose a direct threat to health or safety.
3)   If no reasonable accommodation without an undue burden can be made to enable the applicant to perform the essential functions of the job.

The offer may NOT be withdrawn because the employer is concerned about possible future worker's compensation claims.

The examining physician should be given the job description and/or the essential functions of the job prior to the exam so s/he can advise the employer regarding the applicant's limitations. Once this information has been acquired, the employer must then consider making a reasonable accommodation.

PHYSICAL FITNESS AND AGILITY TESTS ARE ALLOWED

Physical agility or physical fitness tests designed to demonstrate the person's ability to perform actual or simulated job-related tasks, or to measure the person's performance of physical criteria such as the combat challenge course are not medical examinations. However, during the latter test it is not permissible to measure an applicant's blood pressure or heart rate after the test. Psychological tests which measure skills or tastes are also permissible as well as tests for the illegal use of drugs.

POLYGRAPH EXAMS ARE ALLOWED

The EEOC has said that polygraph examinations are not medical exams, but the polygraph operator may not ask prohibited questions which would elicit medical information. Even though polygraph tests record physiological responses of the person being tested, the EEOC's Guidelines say polygraph testing is not medical. It is not permissible to ask the person if s/he has a physical impairment which might be adversely affected by the stress of the polygraph exam. However, an employer may ask the applicant to assume responsibility and release the employer from liability for any injuries resulting from any physical or mental disorders. The employer may also invite the person to consult his/her physician about whether or not s/he may safely take the polygraph examination. The applicant may not be asked if s/he is taking any medication which might affect the results of the exam. If the employer learns after an offer has been made that the person was under medication which might have negated the results of the exam, the exam can be given again. If you plan to use the polygraph in your applicant program, make sure state law does not prohibit it.

TESTS TO MEASURE HONESTY, TASTES AND HABITS ARE NOT MEDICAL

The EEOC noted that employers sometimes examine applicants to see if they are suitable for shift work or are likely to respond appropriately in an emergency or whether they are likely to steal. Whether such tests are medical or not will be considered on a case-by-case basis. If a test is to determine whether or not an applicant is likely to lie, it is not a medical examination and may be given at the pre-offer stage. The EEOC has said that a test designed and used to measure only such factors as an applicant's honesty, tastes and habits is not medical. If a psychological exam discloses whether an applicant has a mental disorder or impairment as categorized by the American Psychiatric Association in its "Diagnostic and Statistical Manual of Mental Disorders" or in the "International Classification of Impairments, Disabilities and Handicaps" of the World Health Organization, it should be considered a medical examination. Included in the list would be excessive anxiety, depression and certain compulsive disorders.

HIRING POOLS ARE ALLOWED

The ADA allows the use of hiring pools. Offers may be made to more persons than there are job vacancies. The fire department could have only 25 positions open, but could make 50 offers so that it would have people ready to begin work when a vacancy occurs since many of the offers will be revoked following the post-offer medical tests or because some offerees may decline to accept the job. In setting the size of their hiring pools fire departments might find it helpful to keep statistics on how many applicants reject offers when they are made.

If your department creates such a hiring pool, each applicant must be informed of his/her ranking BEFORE any post-offer ranking and must be informed of any change in his/her ranking AFTER any post-offer ranking. If psychological profiles are used in re-ranking applicants, all applicants must be told their initial and final ranking and told that their placement has changed because of the post-offer examination. The burden is on the employer to show that the standard used to lower the applicant's ranking is job-related and consistent with business necessity.

The EEOC guidelines state that, if there are more offers made than positions exist, the employer must hire individuals from the pool based on pre-established, objective standards such as date of the application. If the pre-established, objective standards have an adverse impact on a class of disabled applicants, the employer must justify the standards as job-related and consistent with business necessity.

MENTAL ILLNESS IS A DISABILITY UNDER ADA

Mental illness is a disability under the Act. The issue of mental illness presents very complex compliance problems for fire departments because people with such conditions as schizophrenia, depression or anxiety disorders are afforded the same protection under the ADA as those with physical impairments. (The National Mental Health Association reports that an estimated 20 percent of all people will have some type of mental disorder in their lifetime.)

CAN PERSONS WITH INFECTIOUS DISEASES BE "FLAGGED?"

Are there restrictions placed on notification of emergency personnel when patients are diagnosed as having an infectious disease?

You can't have a list of people with infectious diseases in your computer or on your fire trucks, but you can have informal systems. It is proper for the dispatcher to ask about medical problems: heart condition, high blood pressure, breathing problems, AIDS, etc.

FIREFIGHTER WHO IS CARRIER OF HEPATITIS B WINS SUIT

A U.S. District Court judge in Washington, D.C. ruled that a firefighter who is a carrier of hepatitis B must be allowed to perform mouth-to-mouth resuscitation.

When it was learned that the firefighter was infected with hepatitis B, fire department officials placed him on leave without pay for 20 months. After officials consulted with medical specialists on infectious diseases, he was allowed to return to work. The city government agreed to give him back pay and damages of $100,000, but refused to rescind its order forbidding him from performing mouth-to-mouth resuscitation.

The federal court held that under the Rehabilitation Act of 1973, the D.C. Fire Department had discriminated against the firefighter by forbidding him to perform part of his job. The judge pointed out that there are no reported cases in which the virus was conveyed through mouth-to-mouth contact. The judge said the firefighter was legally disabled because the D.C. government treated his disease as an impairment. He was awarded post-settlement legal fees and costs.[126]

According to the federal Centers for Disease Control and Prevention in Atlanta, the virus which causes hepatitis B usually is transmitted through sexual contact or blood, but it is also present at very low levels in saliva.

OBESITY AS A HANDICAP

Looking ahead at the trends in the law, obesity as a disability and appearance in general are likely to be addressed more frequently by the courts and legislatures. (Facial scars are already included among the conditions where the ADA prohibits discrimination.)

In November, 1993 a U.S. Court of Appeals ruled that a 5-foot-2-inch, 320-pound woman was discriminated against when the Rhode Island Department of Mental Health refused to re-hire her.

The woman worked as an attendant at a state mental health center until she left to care for her ill son. When she re-applied for her job 18 months later, she passed a physical examination and a nurse concluded there was nothing limiting her ability to do the job. The medical director of the center, however, told her that her weight would be a liability and she would have to meet weight standards for her height before she could return to work. In other words, she would have to reduce her weight from 320 to 117 pounds. In defending their action, the medical director and other state officials argued that her weight put her at risk for heart disease and other ailments which could lead to worker's compensation claims.

She filed suit in U.S. District Court. Following a jury trial, she was awarded reinstatement, back pay, pension contributions, interest and damages for pain and suffering, totalling $100,000.

In an *amicus curiae* brief filed with the court, the EEOC adopted the view that morbid obesity is a federally protected "disability," defined as overweight by 100 pounds, or twice one's desirable weight.

When the State of Rhode Island appealed, the federal Court of Appeals affirmed the lower court's decision and stated, "In a society that all too often confuses 'slim' with 'beautiful' or 'good' morbid obesity can present formidable barriers to employment."

The plaintiff's lawyer commented, "I think this ruling establishes the concept that obesity deserves tolerance and understanding instead of prejudice and stereotyping." [127]

A similar case was adjudicated in California in September, 1993, recognizing obesity as a Disability. The California Supreme Court ruled that fat people can sue for job discrimination under a State statute, but only if the obesity was caused by a physiological disorder, such as faulty metabolism or systemic disorder rather than bad eating habits which are within the person's control. The plaintiff must prove a handicap within the meaning of the California discrimination law.

The plaintiff was a 305-pound woman who claimed she was dismissed from a health food store because her employer thought her weight would hamper her work. She offered no evidence that her weight was due to a physiological condition, or that her employer had thought that it was. She maintained that she was healthy and fit, notwithstanding her weight.

The court dismissed her claim because "she demonstrated neither an actual or a perceived handicap" within the meaning of the state's discrimination law.

The court did not explain how employers and courts are to distinguish between obesity caused by a medical condition and that caused by overeating that is not medically related.

The court declined to address the question that a person may be fat because of a compulsive behavior caused by a psychological or physiological dependency on eating. No one is fat because s/he wants to be. In other words, if overeating is due to a mental dependency, would that be protected by the law? This was a state statute under which the suit was brought, but it is likely that the protection under ADA might be extended to obesity as in the Rhode Island case cited above. [128]

BIAS AGAINST FIREFIGHTER WITH DIABETES FOUND

A U.S. District Court judge ordered a firefighter reinstated in his job without loss of seniority or longevity and awarded him $335,600 in back wages, holding that he was illegally fired from his job with the Sioux Gateway Fire Department because of his diabetic condition. The Fire Department, a division of the Iowa Department of Defense Military Division, employed the man with the airport-based fire team from November, 1987, to May, 1988. His letter of discharge stated he was being fired because of his "inefficiency, failure to perform assigned duties in a reasonable period of time and inadequacy in performance of assigned duties."

The firefighter claimed the real reason he was fired was because of his diabetic condition, a condition which his superiors did not know about until a month before his discharge when he inadvertently failed to take his glucose. This resulted in his behaving erratically on the job and missing four days' work. The court noted that the man had not been given any prior warnings or reprimands about poor job performance prior to the discovery that he was diabetic. The court was convinced that he would not have been fired if his superiors had not discovered his diabetic condition.

The judge said, "This court believes a firefighter who has diabetes can legally be discharged because of matters related to his condition. If such person has problems making him unable to at times perform in emergencies, an employer need not keep such an employee. An employer is entitled to ask that a diabetic do the things a diabetic should do to enable him to perform the duties required by his job. But an employer cannot legally discharge a person because he has diabetes." In other words, a fire department can fire someone who cannot do the job, but not because s/he has a disability.[129]

DEPARTMENT MUST JUSTIFY USING A DISEASE AS DISQUALIFIER

The Spokane Fire Department listed Crohn's disease as one of the diseases for which an applicant would be rejected. Crohn's disease, or regional enteritis, is an inflammatory disease of unknown origin which affects the esophagus, stomach and intestines. It is often controlled by medicine, but in some cases part of the digestive tract is removed. When an applicant challenged the rule in court, the U.S. Court of Appeals said that rejecting the firefighter applicant because he had Crohn's disease was improper because the department, in order to show a bona fide occupational qualification (BFOQ), would have had to prove that all or substantially all, of the applicants with the disease could not properly perform the duties of a firefighter.

An exception to ADA is that anyone who poses a direct threat to the health and safety of others (which cannot be eliminated by "reasonable accommodation") does not have to be hired.[130]

FIREFIGHTER WHO BLEEDS WHEN EXPOSED TO SMOKE MAY BE FIRED

An appellate court upheld the right of the New Orleans Fire Department to fire a firefighter who experienced episodes of unexplained bleeding from his nose and mouth when he was exposed to smoke. He was fired because he could not perform the duties of a firefighter. The firefighter had been exposed to burning butadene in 1987 and 1989. He appealed on the grounds that the Louisiana handicap laws required employers to reasonably accommodate an employee's handicap.

FAILURE TO REMOVE VERMIN, TRASH MAY BE CONSTITUTIONAL TORT

An example of the lengths to which the federal courts will allow plaintiffs' claims to be litigated under civil rights theories is a New York case where residents at a mental care facility alleged a civil rights violation because the facility was infested with vermin and littered with trash.

Inmates of a mental health facility sued under 42 U.S. Code Section 1985 (3), alleging a conspiracy to deprive them of the equal protection of laws.

The federal court recognized the right of disabled persons to sue on a civil rights conspiracy basis as well as under the Americans with Disabilities Act. The court said the ADA was an alternative, but nonexclusive remedy.[131]

FIRE DEPARTMENT FORCED TO HIRE STUTTERER

The burden of proof will always be on the fire department employer to show that its medical or physical standards are reasonably related to the requirements of the job. Unless the department can prove that an applicant would be unable to perform the essential functions of the job, or would jeopardize the safety of him/herself or others, the department must make a reasonable accommodation for the disabled person.

An interview board for the Columbus, Ohio, Fire Department rejected a firefighter applicant because of a speech impediment and he filed a complaint with the Ohio Civil Rights Commission. Because of the man's stuttering, the interview took three times as long as it normally would. The reason cited for the applicant's rejection was public safety on the grounds that at fire scenes visibility is often poor and verbal communication is essential.

For ten years the applicant had worked as a parttime firefighter for another community and had fought 20 to 30 fires without a problem. He had even served as second-in-command, supervising subordinates. Evidence showed that during emergencies the man's speech improved and during a training exercise, he had been the first to warn others of a flash-over.

A speech pathologist testified that the cause of stuttering is not known and that it is an individualized problem He said that it is not uncommon for the person to stutter in social situations, but not when the person's mind is intently focused.

The Ohio Civil Rights Commission ruled that the applicant was handicapped and had been unlawfully discriminated against. The Commission said the man's inability to speak without stuttering during an interview had no relationship to his ability to serve as a firefighter.

The City of Columbus appealed, but the appellate court affirmed the Civil Rights Commission's decision.[132]

MUST SHOW APPLICANT COULDN'T FUNCTION WITH IMPAIRED EYESIGHT

Similarly, when an applicant for Colorado's North Washington Fire Protection District was rejected because his uncorrected eyesight was impaired, the Colorado Supreme Court held that the fire district did not adequately substantiate that the applicant could not function safely and efficiently while wearing a Scott Air Pak.[133]

BLIND FIREFIGHTER HERO WRONGFULLY FIRED

Even before enactment of ADA, courts were recognizing and penalizing discrimination under the Rehabilitation Act. In Baltimore the court ordered the fire department to rehire a fire fighter who had gone blind and awarded him $108,000 in damages.

The firefighter had been named "Fireman of the Year" and Baltimore's "Best Firefighter" in 1983. A rare genetic disorder caused him to go blind in 1986 and he was ordered to retire. In 1987 he was allowed to try out for a job as dispatcher, but was not allowed to touch any of the equipment, and was denied the job. The court held that the department did not make sufficient efforts to make an accommodation for him so that he could hold a job in spite of his disability. The court said he could have become a dispatcher or performed other jobs in the department with the aid of "adaptive technology."[134]

COURT SAYS DEPARTMENT NEED NOT ACCOMMODATE ASTHMATIC

In another Maryland case, the federal district court held that a county need not accommodate an asthmatic firefighter.

The case was brought by a former firefighter recruit who suffers from asthma. He alleged that Howard County, Maryland, had unlawfully discriminated against him because of his handicap.

He had applied for and was accepted as a firefighter recruit at which time he informed the fire department of his asthmatic condition. The county's physician cleared him for training. However, during the training exercises, he experienced great difficulty breathing. In fact, on a number of occasions he had to discontinue the required one-and-a-half-mile run.

The recruit then was seen by three additional physicians, all of whom expressed the opinion that he could not perform the duties of a firefighter even with the aid of medication and that his asthma posed an extreme safety risk for him and other firefighters.

The county advised him that his employment was terminated because of his chronic asthmatic condition. He filed suit, alleging that the county had failed to accommodate his disability as required by the Rehabilitation Act. The county responded that he did not have a disability as defined by the Act and, even if he did, the county had not failed to accommodate him.

In making the initial determination whether the recruit qualified as a disabled person, the U.S. District Court found that asthma impaired his ability to breathe, a major life function. Next, the court had to determine whether he was an otherwise qualified disabled individual, depending on whether he could perform the essential functions of the job. The court held that he was not able to perform the essential functions of a firefighter, and pointed out that firefighters are required to work under extremely difficult conditions and that his asthma prevented him from maintaining the stamina needed to perform the functions of the job.

The court also noted that it would be unreasonable to allow him to use his inhaler at the scene of a fire, considering the extreme temperatures and the difficulty in using it while wearing protective equipment such as a face mask. The court said any other accommodation would be an undue hardship on the department.

The court said it was impressed by his sincere desire to serve his community, but it could not require the county to make a sacrifice to employ him as a firefighter, considering his inability to perform all the requirements of the job and the risks involved. Therefore, the court dismissed his claim, granting summary judgment for the county.[135]

CIVIL SERVICE COMMISSION BACKS COLORBLIND RECRUIT

A hearing officer held in 1992 that the Saratoga Springs, New York, Civil Service Commission acted illegally when it denied a job on the City's fire department to a man who was colorblind.

PUBLIC EMPLOYEE MAY FILE SUIT BEFORE FILING CLAIM WITH EEOC

A U.S. District Court in Wisconsin ruled that, when the employer is a public entity, it is not necessary for an employee to exhaust his administrative remedies by filing a complaint with the Equal Employment Opportunity Commission before filing suit in federal court for alleged discrimination under the Americans with Disabilities Act.

The plaintiff was an employee of the University of Wisconsin. The court said that the ADA allows public employees the remedies which were afforded to them under the earlier Rehabilitation Act of 1973 which did not require public employees to file a complaint with the EEOC before filing suit. The court pointed out that the ADA requires the U.S. Department of Justice to set up a procedure for resolving complaints, but it does not require complainants to exhaust those administrative remedies.[136]

DECREASED SENSATION IN FINGERS NOT AN IMPAIRMENT UNDER ADA

An Oklahoma federal appellate court said that rejection of a firefighter applicant who had decreased sensation in two fingers was not an impairment that would support a charge that the Americans with Disabilities Act had been violated.[137]

COURT OVERTURNS REJECTION OF APPLICANT WHO HAD CANCER

An appellate court ruled against the Phoenix Fire Department's rejection of an applicant who once had testicular cancer. The issue on appeal was whether an applicant remains "handicapped" after an illness has ended.[138]

TIPS FOR COMPLYING WITH THE ADA

- Examine job applications to make sure they do not ask questions about disabilities. (The application can ask about the applicant's ability to perform a function of the job.)
- Similarly, in job interviews, don't ask about the applicant's disability. Ask about the applicant's ability to perform the "essential functions" of the job.
- Review your job qualification criteria, screening procedures and aptitude tests to make sure they are relevant to the "essential functions" of the job and are not discriminatory.
- Write your job descriptions to distinguish between essential and nonessential aspects of a job.
- Examine your policies regarding medical examinations. Make sure they relate to job suitability. All applicants must take the same examination. The medical examination cannot be given until a conditional job offer has been made to the applicant. The information sought in the examination must be job-related.
- Conduct an audit of your physical facilities to attempt to make them handicapped accessible. (Organizations representing the disabled will assist you in this effort.)
- Keep a file of all accommodations which you make for applicants, employees and the public. If you turn someone down who requests an accommodation, document your reasons and be prepared to defend the "undue hardship" which the accommodation would have caused or the "business necessity" of your actions or the safety conditions which would have been jeopardized.

- Think of imaginative ways in which you can serve the needs of those with disabilities.
- Conduct sensitivity training for your employees, especially supervisors, including problems regarding those with AIDS and mental illnesses.
- Keep good records regarding all your training programs related to ADA.
- Make sure your worker's compensation rules jibe with the ADA.

REMEDIES

To initiate a claim under the ADA, the disabled person must file a charge of discrimination with the EEOC within 180 days of the alleged discrimination, or 300 days if there is also a state agency which processes discrimination claims.

The EEOC, in processing the complaint, usually follows four steps:

1) Notification of the employer about the charge and requesting a response.

2) Review of the complaint and response. More information may be requested.

3) The parties may be asked to appear at a fact-finding conference to obtain additional information or to explore the prospects for a settlement.

4) After the EEOC completes its review, it issues a Letter of Determination, advising the parties of its findings. If the EEOC finds no cause to believe discrimination occurred, it will send the complainant a "Right-To-Sue" letter. The complainant then has the right to initiate a lawsuit within 90 days after receiving the letter.

Successful plaintiffs can receive back pay, reinstatement, restored benefits, attorneys fees and court costs. In the case of intentional discrimination, punitive damages can also be awarded.

SPECIAL HELP AVAILABLE RE ADA

The federal government has set up ten regional Disability and Technical Assistance Centers to help people comply with the Americans with Disabilities Act. Call 1-800-949-4232. The Department of Justice also has a special number to answer ADA inquiries: 202-514-0301. Help is also available from the Job Accommodation Network, 1-800-526-7234 and Project Access, 708-390-8700.

Some suggested references include:

"Americans with Disabilities Act of 1990," Equal Employment Opportunity Commission, Federal Register, Vol. 56, Number 144 of 29 Code of Federal Regulations 1630 (July 26, 1991. U.S. Government Printing Office, Washington, D.C.

"ADA Training Manual for Managers and Supervisors," Commerce Clearinghouse Law Editors, Chicago, Illinois, 1992, Pages 13-14.

"Employers Guide to the ADA," James Frierson, Bureau of National Affairs, Washington, D.C., 1992.

"The Americans with Disabilities Act, What Supervisors Need to Know," By Joan Ackerstein, Business One, Irwin/Mirror Press, Burr Ridge, Illinois.

CHAPTER THIRTEEN

DRUG TESTING

Drug testing of employees is one of the most controversial legal issues of the 80's and 90's. Those who favor widespread drug testing argue that drug-impaired workers may cause accidents, safety risks, accidents, faulty products, and increased risk of liability for their employers. Drug testing is an effective method for coping with these risks.

Under the doctrine of *respondeat superior*, employers may be liable for the torts which their drug-impaired employees commit during the scope of their employment. Negligence in hiring or retaining drug-using employees may also be the basis for suits. The Occupational Safety and Health Act requires employers to maintain a safe workplace. It is not possible to do this, some claim, unless they have the right to identify drug users.

Conflicting rights permeate all the litigation involving drug testing. The government has a legitimate interest in making sure that firefighters and police officers are free from drugs for safety reasons, and the employees have a legitimate interest in protecting their privacy and have the constitutional right to be free from unreasonable searches and seizures. In each case the court tries to balance these conflicting interests.

Opponents of extensive drug testing claim that it is unduly intrusive, an invasion of the employee's privacy. They claim it violates the Fourth Amendment's protection against unreasonable searches and seizures, the Fifth Amendment's privilege against self-incrimination, the Fourteenth Amendment's guarantee of due process, the Equal Protection clause and the right of privacy.

The court cases are also in conflict on the question of drug testing. Some courts have rejected the Fifth Amendment argument saying that drug testing does not constitute self-incrimination, but filling out forms which ask the question, "Do you use drugs?" might be.

With respect to the Fourteenth Amendment regarding due process and equal protection, the question is: "Does the employee have a property right in the job?" At-will employees do not. Others probably do. However, some states recognize an implied covenant of good faith and fair dealing even in at-will contracts.

Although the U.S. Constitution is silent regarding the right of privacy, the U.S. Supreme Court in the *Griswold v. Connecticut* case recognized such a right of the citizen to be free from unwarranted governmental intrusion.[140] Suits have challenged drug testing on the issues of wrongful discharge, breach of contract and invasion of privacy.

The most frequent issue raised to challenge urinalysis drug testing is the U.S. Constitution's Fourth Amendment which states:

"The right of the people to be secure in their persons, houses, papers and effects against unreasonable searches and seizures, shall not be violated, and no warrant shall issue , but upon probable cause, supported by oath or affirmation and particularly describing the place to be searched and the person or things to be seized."

The courts have made clear that urine testing is a "search" because it invades privacy expectations. The next question is to determine if the search is "reasonable."

The courts recognize that drug testing demands prompt action so it is not practical to require a warrant before conducting the test. The courts have also held that, even if employees consent to a search which is illegal, it may still be unconstitutional. However, consent to the test in advance negates privacy expectations. The courts have also said that urine tests must be conducted in a "reasonable" way. The valid interests of the employer must be balanced against the intrusions into the employee's privacy.

In the *Schmerber v. California* and the *Rochin v. California* cases (342 US 165) the Supreme Court said that minor bodily intrusions will be tolerated, but all circumstances surrounding the testing will be weighed.[141a] Does the interest in disclosure outweigh the intrusion?

The courts are divided on whether or not a public employer can test police or correctional support personnel, firefighters, dispatchers or EMS employees. For example, *Anabel v. Ford* held that urinalysis is unconstitutional. *Jones v. McKenzie* held that it is constitutional but it requires probable cause, not mere suspicion.[141b] (A drug-sniffing dog can provide probable cause.) *City of Palm Beach v. Bauman* held that it is constitutional and only requires reasonable suspicion, not probable cause.

The questions to ask with respect to drug-testing programs always are: "Is it reasonable?" "Is the employee 'regulated'." (Police, employees on trains, and jockeys would fall into the "regulated" category.) "Is there good, reasonable, individualized suspicion that a particular employee uses drugs?" "Is the program a random, fair testing process?"

Generally, when there is a reasonable suspicion that an employee is using drugs, s/he can be requested to submit a urine sample for testing. What constitutes a reasonable suspicion? When s/he is seen using drugs or with drug paraphernalia, when the employee is acting erratically, is involved in unexplainable accidents, when s/he shows drastic mood swings etc.

REASONABLENESS OF TEST MUST BE JUDGED FROM ALL CIRCUMSTANCES

In *Skinner v. Railway Labor Executives Association*, the U.S. Supreme Court by a 7 to 2 vote held that tests given in reliance on federal authority involve enough government action to trigger Fourth Amendment concerns; that drug tests do constitute a "search"; that neither a warrant nor probable cause is required for testing; and the reasonableness of the tests will be determined by all the circumstances.[142]

The court declined to hold (as other courts have done) that particularized suspicion is needed to ensure that tests will detect current impairment. The Supreme Court said where privacy interests are minimal and important governmental interests would be harmed by such a requirement, a search may be reasonable even without particularized suspicion. The Supreme Court said sometimes the government need justifies suspicionless searches. Also relevant, the court said, is the degree of interference with liberty.

ERRATIC BEHAVIOR OF FIREFIGHTER PROVIDES REASONABLE SUSPICION

A New Orleans firefighter was granted permission to take a short leave from duty, supposedly to drive his sister home from work. What he really did was drive a friend from work to pick up her child at a babysitter's house. He was observed driving erratically by two police officers and, when they pulled him over for reckless driving, he was abusive and belligerent. He fought and cursed the officers. They telephoned his supervisor. They expressed the opinion that he was on drugs because of the way he was acting. Back at the fire station, the supervisor and his superior gave the firefighter a drug test. He tested positive for cocaine use and was suspended pending an investigation. At a hearing the following month he was told why he had been suspended and that they believed they had had reasonable suspicion to give him the drug test. At that hearing and a second one before the Department's Board of Internal Affairs, he was given an opportunity to respond to the charges.

He was fired for using cocaine while on short-term leave from duty. He appealed to the city's Civil Service Commission which upheld his firing. He appealed to the courts, arguing that the drug test was an illegal search and that he had been denied due process of law.

The court upheld the termination, stating that the department had a reasonable suspicion that he was using illegal drugs based on the police officers' report about his unusual behavior when they stopped him, so the drug test was proper. The court said the information was reliable because it came from the police officers and the information they furnished was enough to justify the reasonableness of the testing.

In response to the firefighter's claim that he had not been afforded notice or a proper hearing, the court said he had been given adequate written and oral notice of the cause of his suspension and firing. He was given a pre-termination hearing at which he had an opportunity to contest the evidence. He failed to show any evidence that his drug test was improperly conducted or processed. His termination was upheld.[143]

A police chief could not order a police officer to submit to a urine sample after he had been seen with a fellow officer who was arrested for drug sales. A federal appeals court upheld the officer's damage suit for wrongful search.[144]

RANDOM TESTING FOR U.S. CUSTOMS EMPLOYEES APPROVED

In *National Treasury Employees Union v. von Raab* (109 US 1384) the High Court upheld testing rules of the U.S. Custom Service. The court said that where potential harm is substantial, the need to prevent it may justify reasonable searches designed to achieve that goal.[145]

In *National Treasury Employees Union v. Hallett*, a federal judge also upheld random testing of broad categories of U. S. Customs Service employees, including non-sworn inspection assistants and clerks, messengers, paralegal, chemists and student trainees. The Court noted that "the government's interest in assuring the integrity of its work force would not justify drug testing for every federal employee." It is proper when the employer can show a nexus between "the nature of the employee's duty and the nature of the feared violation."[146]

Drug screening is lawful only when the demand is reasonable. Although random and periodic testing of public safety employees is usually held to be reasonable, when a certain person is targeted for the urinalysis test, there must be an individualized suspicion of that particular person.

Prison employees claimed that a departmental policy calling for searches of vehicles and their persons, violated their Fourth Amendment right against unreasonable searches and seizures and their right of privacy. The lower court granted the preliminary injunction against enforcement of the policy and it was affirmed by the U.S. Court of Appeals.

The appellate court held that urinalysis testing when conducted within the institution could be made uniform by systematic random selection, or on the basis of reasonable suspicion. The court pointed out that, while urinalysis has been deemed to be a search and seizure under the Fourth Amendment, the department's need to search must be weighed against the individual's right to privacy.

RANDOM TESTS FOR COURIERS AND MAIL CLERKS AT U.S. HHS DENIED

A U.S. District Court judge ruled that random drug tests for couriers and mail clerks employed by the U.S. Department of Health and Human Services could not be given because the employees' privacy interests outweighed the government's interest in safety. He distinguished the case from a prior case which upheld random tests for some HHS employees on the basis of whether the drivers carried passengers. The judge said, "It would be unreasonable and hence unconstitutional to subject these motor vehicle operators to random drug testing."

NEW JERSEY COURT UPHOLDS RIGHT TO TEST

A New Jersey appellate court in the case of *Local 194A v. Bridge Commission* upheld the right of a public employer to periodically test for drug use when the employees' duties affect public safety. The court held that adoption of such a test is a managerial prerogative and the issue does not have to be submitted to collective bargaining before testing. The court said, however, that the procedures to be used in the testing do constitute a subject for mandatory collective bargaining.[147]

U.S. COURT OF APPEALS VOIDS DRUG TESTING FOR FIREFIGHTERS

In *Lovvorn v. City of Chattanooga*, the question asked was, "Was there a 'reasonable suspicion' that the person did drugs?" It requires a particularized suspicion. This case voided a random drug testing plan for fire fighters. Although it acknowledged the legitimate interest of the city in drug-free firefighters, the court noted that urine samples were given under observation which raised concern about employees' privacy. It then said that the permissibility of random drug testing depends, not on whether an industry is heavily regulated, but on the nature of the industry and the harm that would likely result to society if mandatory tests were forbidden. Finding the likelihood of enormous societal losses because of an impaired firefighter to be relatively low, the court held that, for a test of firefighters to be reasonable, there must be either evidence of a department drug problem or suspicion of an individual. Other courts might logically conclude that a drug-impaired firefighter does, in fact, constitute a serious safety risk to fellow employees and the general public and could result in "enormous societal losses" to borrow the Sixth Circuit's words.[148]

MICHIGAN FEDERAL COURT GIVES VIEWS ON TESTING

A U.S. District Court in Michigan held that where the governmental interest at stake is great or the public danger is grave, reasonableness within the meaning of the Fourth Amendment may require less factual justification for the testing. On the other hand, where the court perceives that no grave danger nor fundamental public interest is at stake in routine enforcement of drug laws, the government's rights must be tested against the U.S. Constitution's firmest protections of an individual against exercise of official power.[149]

FLORIDA FEDERAL COURT ENJOINS TESTING AS PART OF ANNUAL PHYSICAL

A U.S. District Court in Florida enjoined a drug-screening program as part of an annual physical examination given to firefighters. The judge acknowledged the importance of mandatory drug testing for police officers, but he refused to allow suspicionless testing of firefighters. He admitted that "firefighters are engaged in hazardous work involving the public safety" but they "do not carry firearms and are not required to use deadly force in the regular course of their duties." He distinguished the case from the *Skinner* case referred to above on the basis that the railroad employees in the *Skinner* case had a history of causing accidents "resulting in numerous fatalities and millions of dollars in property damage" and in the firefighter case there had been no allegation that there was a drug problem in the fire department. He said that balancing the interests of employee privacy with the department's concern for public safety, the tests were not justified, even though the firefighters provide urine specimens as part of their annual physical examinations. The samples had not previously been tested to ascertain drug use.

The Court said that firefighters "have a legitimate interest in keeping their personal life shielded from the government's prying eye, especially when the activity revealed is frowned upon by a large segment of the community and may constitute a crime." He said the city failed to "demonstrate a compelling interest that outweighs these privacy concerns."[150]

MASSACHUSETTS COURT REJECTS CADET'S CLAIM OF PRIVACY INVASION

The Massachusetts Supreme Court unanimously rejected the claim of a fire police cadet who felt his privacy was invaded when a superior told other Boston police cadets that he had been fired for failing a drug test. The court said that the department had a "legitimate interest in deterring drug use by police cadets" and the disclosure of his firing might deter others from using drugs.[151]

NEW YORK COURT UPHOLDS FIRING POLICE OFFICER WHO TESTED POSITIVE FOR MARIJUANA

A New York appellate court upheld the firing of a police officer who had tested positive for marijuana use. The court held that the officer's "severe mood swings" and a tendency toward violence and delusions, provided a reasonable suspicion that his behavior was the result of drug use, justifying the urinalysis by the police department.[152]

CLEVELAND FIRES CADETS; COURT SAYS TESTS WERE UNREASONABLE, BUT...

The Cleveland, Ohio, chief of police received a tip that police cadets were using drugs. He ordered a surprise drug test and several cadets tested positive for marijuana. Their resignations were requested. They filed suit in U.S. District Court under Section 1983 of the Civil Rights Act. The court denied the defendant's liability, even though it said the tests were an "unreasonable search because there were no regulations governing the manner in which the urinalysis was conducted." The appellate court affirmed and the U.S. Supreme Court denied certiorari. The court said that the tests were not administered in conformity with a formalized testing directive, but the samples were double tested using the EMIT (enzyme immunoassay) and TLC (thin layer chromatography) and were confirmed with the RIA (radioactive immunoassay) test. The GC/MS (gas chromatography/mass spectrometry) test, the most reliable confirmatory test, was not employed prior to the resignations, but it was used after disciplinary action had been imposed and it confirmed the accuracy of the prior results.

The court said, while the tests used were not as reliable as the GC/MS test, they were not so unreliable as to be irrational and did not shock the conscience of the court or offend notions of liberty. The court said the plaintiffs did not have a due process right to have the GC/MS confirmatory tests used before the police department demanded their resignations.

NEW TECHNOLOGY MAY SOLVE PRIVACY CONCERNS IN DRUG TESTING

Developing technology might make it easier for the issue of drug testing to be resolved. The concern in every drug-testing case is to balance the employees' right to privacy against the employer's legitimate interest in testing. The U.S. Supreme Court has said minor intrusion will be upheld. A company in Cambridge, Massachusetts, (Psychemedics Corp.) uses hair samples to test for drugs. This test tells whether the employee has used drugs during the past three months while the urine test only reveals whether drugs have been used within days prior to the test. In sampling hair, the intrusion is minimal and would undoubtedly be upheld by federal courts. Hair testing for drugs, however, costs more than double what a urine test costs. Furthermore, some claim that hair testing is unreliable and is biased against blacks. Opponents also claim the hair will pick up drug residue from the air.

The American Civil Liberties Union, which opposes all drug testing, argues that employers have no right to know if employees have used drugs three months previously.

The hair-testing process is based on the principle that, whenever drugs are consumed, traces enter the blood stream and reach the hair. The more drugs consumed, the more will be found in the hair. About 50 hairs, the width of a pencil point, are cut from hair on an employee's scalp (to get the most recent three months of growth) and are shipped to a laboratory in Culver City, California, where they are washed and liquefied and tested for specific drugs.

POINTS TO CONSIDER IN DRUG TESTING

It is difficult to say with certainty what the law is with respect to drug testing since there are so many divergent views by the courts. Some points elicited from various cases, however, are instructive for fire departments:

- Public and some private employment testing involves a search as contemplated by the U.S. Constitution.
- Neither a warrant nor probable cause is needed to test, but the test must be "reasonable." Reasonableness is determined by weighing its intrusion on privacy against valid interests in testing.
- If an industry affects the public welfare or has a history of intense state regulation or a drug problem, random testing is legal.
- The fact that tests cannot measure impairment and are not 100 percent accurate is no basis for automatically voiding them. The more protection afforded respecting the employee's privacy, the more likely the tests will be upheld.
- Neither advance notice not employee consent validates a test, but they increase the likelihood that they will pass legal muster.

FIRE DEPARTMENTS SHOULD HAVE FIRM, CLEAR POLICY ON DRUG TESTING

Every fire department should have a firm, clear policy on drug testing and adhere to an established random testing program. In testing a specific individual, make sure you have an individualized, particularized suspicion of that person. Always protect the integrity of the specimen and use confirmatory tests.

Where facts reasonably support a drug use suspicion, an employee can be required to submit a urine sample for analysis.[153]

CHAPTER FOURTEEN

AGE DISCRIMINATION

Employers frequently prefer younger workers, especially since it usually costs less to pay them. Can older workers be dismissed in favor of younger workers? The answer is generally "No."

The federal Age Discrimination in Employment Act (ADEA), (29 U.S.C. 621), of 1967 which is enforced by the Equal Employment Opportunity Commission, protects workers 40 and older from discrimination in all aspects of employment: hiring, firing, pay, promotion and other conditions of employment. For example recruitment advertisements can't say "Recent graduates wanted." It also forbids retribution against complainants.

The law is applicable to all employers engaged in business affecting interstate commerce with 20 or more employees, labor unions, state and local government agencies and employment agencies.

However, if age is a bona fide job qualification — in other words, if there is a good reason to consider age — or if the employment action is based on a good reason other than age such as misconduct, the law does not apply.

MASSACHUSETTS MANDATORY RETIREMENT AT 50 UPHELD

The Massachusetts State Police required its officers to retire at age 50 to insure that its force was physically fit. An officer in excellent health and physical condition was forced to retire on his 50th birthday. He sued the State Police, arguing that this mandatory retirement age denied him equal protection of the law as guaranteed by the U. S. Constitution's Fourteenth Amendment.

The U.S. Supreme Court rejected his argument. The Court said the action by the State legislature in establishing the mandatory retirement age is presumed to have been done for valid reasons. The Court accepted the fact that physical agility generally declines with age and it, therefore, was legitimate to establish some age for retirement. The Supreme Court upheld the Massachusetts law.[154]

COURT SAYS EMPLOYER MUST JUSTIFY AGE CUTOFF

A U.S. District Court in Boston examined the Massachusetts statute which consolidated Motor Vehicle Enforcement Officers and Metro District Police with the State Police. Employees of the former agencies retired at 65; the State Police retired at 50. The consolidated force was to retire at 55. Members of the former forces, who could previously have stayed in their jobs until they were 65, sued under the ADEA. The Court granted a preliminary injunction against enforcement of the law's mandatory retirement provisions. The State argued — unsuccessfully — that a 1984 decision which approved retirement at 50 was binding. The Court said a subsequent U.S. case, *Western Air Lines v. Criswell*, rejected the "reasonable standard" on which the 1984 case rested in favor of a "reasonable necessity" standard in 1985.[155]

The Employer must show either:

a)   all or substantially all persons over 55 would be unable to perform their jobs safely and efficiently, or

b)   it is impossible or highly impractical to deal with older employees on an individualized basis.

The Court concluded that the plaintiffs had demonstrated a likelihood of success on the merits and were entitled to injunctive relief.[156]

GEORGIA STATE POLICE MANDATORY RETIREMENT AT 55 ALLOWED

The U.S. Court of Appeals upheld a 55-year-old mandatory retirement age for State police in Georgia. The Court said the statute was not a subterfuge to evade application of the ADEA.[157]

STATE'S AGE RULE MUST MEET "REASONABLE" FEDERAL STANDARD

A Wyoming game warden was involuntarily retired at age 55. He sought help from the EEOC which then sued the State of Wyoming.

When the case went before the U. S. Supreme Court, the court held that the Tenth Amendment did not immunize Wyoming from extension of the ADEA. The court said the State was free to pursue its goals so long as it observed a few procedural requirements in the process. The state could fulfill its goal of assuring the physical fitness of its employees by conducting individualized fitness examinations or by demonstrating that the ages set out in its mandatory retirement policy constituted bona fide occupational qualifications (BFOQ's) for particular positions. Justice Brennan said that the ADEA did not completely override Wyoming's "discretion to achieve its goals in the way it thinks best but merely tested the States's methods against a "reasonable federal standard."[158]

CONGRESS' TEMPORARY PUBLIC SAFETY EXEMPTION HAS NOW EXPIRED

In response to requests from numerous state and local governments, Congress granted a temporary exemption for public safety personnel from provisions of the ADEA while the EEOC studied the matter.

Amendments to the ADEA enacted in 1986 created certain exemptions from its coverage that ran from January 1, 1987 to December 31, 1993, allowing state and local governments to discharge or refuse to hire firefighters, prison guards, and law enforcement officers because of age. During the interim the EEOC was to propose guidelines on the use of physical and mental fitness tests for these types of employees. The study showed there was no justification for the exemption.

The exemption has now expired and, although the U.S. House of Representatives passed a bill in 1994 which would have made the exemption permanent, it died in the Senate. When a Congress ends, all pending legislation aborts and the legislative process starts over again in both Houses. At this writing, neither House of the U.S. Congress has acted on the legislation which would allow these exemptions for public safety employees. So, at this point fire departments and other public safety agencies are subject to all provisions of the ADEA as any other employer.

ADEA PROCEDURES

The ADEA has a complex procedure for filing a claim. The complainant must first file with the EEOC or other appropriate state agency. Unlike Title VII, it does not require a right-to-sue letter from the EEOC. The EEOC can also sue and, if it does, a suit by the offended party is blocked.

The ADEA has a Statute of Limitations of two years for non willful violations and three years for willful violations.

POSSIBLE DEFENSES

The following are possible defenses to an ADEA suit:

1) Seniority
2) Bona fide age criteria
3) Good cause
4) Reasonable factors other than age
5) Bona fide occupational qualifications.

Among the remedies for successful plaintiffs under the ADEA are back pay, liquidated damages (if it was a willful violation) hiring, reinstatement and promotion. Usually no punitive damages are available or compensatory damages for pain and suffering.

Although the federal law only relates to those 40 and over, some states have laws or regulations forbidding age discrimination in general.

CHAPTER FIFTEEN

SEXUAL HARASSMENT

S urveys in 1981 and 1988 by the U.S. Merit Systems Protection Board showed that 42 percent of the women in the federal government and 14 to 15 percent of the men had experienced some form of sexual harassment. A 1990 study in the military disclosed that two out of every three women questioned said that they had been sexually harassed. Seventeen percent of the military men reported they had been harassed by males or females.

In 1990 *Women in the Fire Service* conducted a survey of women firefighters. The following are some of the examples of fire service sexual harassment the study revealed:

"An officer put moves on me. I rejected him and was treated badly."

"There's nothing I can prove. Obscene things put in my bed, locker, gear. I feel more aggravated, let down, lonely, like I'll never be totally accepted."

"Comments like: 'It's much better when you work in a t-shirt. We can see your [breasts] better."

"Continuous! Posters on walls, asking for dates, requests for sex, hugging, touching, leering, urinating in front of me, being screamed at that I don't belong here...."

"I am harassed all the time. Most of the men are not well-educated and have never related to a woman other than in sexual terms. They do not know how to treat a woman as an equal."

Fire departments should take the problem of sexual harassment very seriously because huge recoveries are possible in sexual harassment suits. Baker & McKenzie, the world's largest law firm with 1700 lawyers, was hit with a whopping sexual harassment verdict. A secretary, who only worked at the firm for three months in 1991, alleged that one of the lawyers had lunged at her breasts, grabbed her hips and made sexually suggestive comments to her. Seven other women made similar allegations at the trial. A six-woman, six-man jury awarded the secretary $50,000 in general damages and $7.1 million in punitive damages, $6.9 to be paid by the law firm and $225,000 by the offending lawyer. (A San Francisco County Superior Court later concluded that the award was excessive and cut it to $3.5 million.)

SEXUAL HARASSMENT DEFINED

What is sexual harassment?

It is defined as:

"Unwelcome sexual advances, requests for sexual favors, and other verbal or physical conduct of a sexual nature constitute sexual harassment when:

1) Submission to such conduct is made either explicitly or implicitly a term or condition of an individual's employment,

2) Submission to or rejection of such conduct by an individual is used as the basis for employment decisions affecting such individual, or

3) Such conduct has the purpose or effect of unreasonably interfering with an individual's work performance or creating an intimidating, hostile or offensive working environment."[159]

As we have seen, Title VII of the 1964 Civil Rights Act prohibits employment discrimination based on sex, race, color, religion or national origin. It covers hiring, firing, compensation and the terms and conditions of employment.

The Equal Employment Opportunity Commission enforces the Act. Both the EEOC and the courts have interpreted the Act as prohibiting sexual harassment as a form of sex discrimination.

Prior to 1991, sexual harassment complaints could only be heard by a judge without a jury and the successful sexual harassment victims generally could only recover back pay, lost wages etc. In 1991 the Act was amended to allow the victim the option of having a jury trial, and made it easier to sue. It also offered more incentives for bringing a sexual harassment complaint. This has contributed to the proliferation of these suits. These amendments also provided increased monetary awards and allowed punitive as well as compensatory damages in addition to recovery of back pay, attorneys fees and reinstatement. Compensatory damages can now cover pain and suffering.

Under this 1991 Civil Rights Act, an employer with fewer than 100 workers can be liable for compensatory and punitive damages up to $50,000. If the company has 100 to 201 employees, the amount is $100,000; $200,000 if it employs 201 to 500 and $300,000 if it has more than 500 workers.

Sexual harassment claims fall into two categories: quid pro quo" (Latin for "this for that") and "hostile environment." These two types of cases were defined in the landmark case of *Meritor Savings Bank v. Vinson.*[160]

QUID PRO QUO

Quid pro quo: In a typical case, a supervisor says to an employee, "If you don't go to bed with me, ... you're fired" or "... you won't get that promotion;" or "You're not driving that engine unless you give me a kiss;" or "I'll be on the interview board for your promotion. If you expect me to give you a good rating, you'd better sleep with me;" or "If you'll go out with me tomorrow night, I'll make sure you get all the training you need to make your probation." In these cases an employee's supervisor uses his/her authority to hire, fire, promote or discipline the employee in order to gain sexual favors. Fearing for his/her job, the employee agrees to have sex with the boss. After s/he breaks off the relationship, s/he is fired or demoted and s/he files suit against the supervisor and the employer alleging sexual harassment.

It is not an adequate defense to say that the employee submitted to sex voluntarily. The question for the jury to decide is: "Was the relationship 'welcome'?" The alleged victim determines whether or not the conduct is "welcome." The courts have begun to apply a "reasonable person," or in some jurisdictions, a "reasonable woman" standard to determine whether the alleged victim's claims are valid.

Sexual harassment can include repeated derogatory, abusive language, dirty jokes, or suggestive comments, comments about the person's anatomy, innuendoes, sexual pictures in the workplace, etc.

HARASSMENT NEED NOT BE SEXUAL

The harassment does not have to be overtly sexual in nature. It could involve screaming at someone or berating subordinates or damaging personal belongings because of the victim's gender.[161]

A supervisor shouts to a woman: "Don't you dumb women know ANYTHING?"

This type of treatment over an extended period of time would constitute sexual harassment, even though it is not "sexual" in nature.

A jury in Alexandria, Virginia, awarded an employee $675,000 after finding that she was a victim of sexual discrimination by a supervisor who routinely insulted her.

EMPLOYER MAY BE LIABLE FOR ACTS OF SUPERVISORS

It is usually very clear as to what quid pro quo cases are, but they are usually difficult to prove. There usually are only two individuals who know what really happened and they often give conflicting stories. One makes the accusation and the other denies it.

Under the doctrine of *respondeat superior*, employers will usually be liable in quid pro quo cases involving supervisors. An example of this is found in the *Heelen v. Johns-Manville Corp.* (1978) case where an employer was liable for sexual harassment when a supervisor demanded that an employee share his motel room with him on a business trip and fired her when she refused.

Quid pro quo cases can only be perpetrated by someone who has supervisory control over another employee. The theory — under the doctrine of *respondeat superior* — is that the supervisor exercises authority over the employee as the agent of the employer. Employers may he held responsible for the sexual harassment, even when the employer had no knowledge of the harassment and the objectionable conduct was not only unauthorized but was specifically forbidden by company policy.

In the *Meritor* case, however, the Supreme Court rejected the idea that employers are strictly liable for the actions of their supervisors. The court failed to give guidelines, but said courts should use the legal principles of the law related to agency for guidance. Courts usually hold that employers are liable where they knew, or should have known, of the improper conduct of subordinates and did not take immediate steps to remedy the situation.

HOSTILE ENVIRONMENT

In the hostile environment cases, the employee is not claiming the loss of any economic benefits, the victim claims that the conduct of one or more individuals has created an intimidating, abusive or hostile work environment.

These cases include conversations of a sexual nature, sexual comments about a person's body, or sexual relations, the display of sexually explicit material such as magazines featuring nudes or sex acts, frequent teasing, taunting, off-color jokes, photographs or posters of scantily clad persons on the wall, crude and vulgar comments degrading to women, inappropriate and repeated unwelcome touching of a person of the opposite sex, a supervisor who openly and frequently criticizes employees of one sex, but not those of the other sex etc. Comments about clothing may even be prohibited. For example: "That tight sweater really makes your boobs stand out. Whew!" That is obviously an inappropriate comment. All these activities could constitute a basis for hostile environment cases. One isolated dirty joke will not constitute a hostile environment, but if the jokes continue frequently and become so severe and pervasive that they affect employment conditions, they may, by considering the totality of the circumstances, be determined to constitute a hostile environment.

Hostile environment cases are more difficult to prove. In these cases there is no discriminatory effect on wages, job assignments or other work benefits. A complainant must prove that unwelcome, gender-based conduct was so severe or pervasive that it unreasonably interfered with the person's work performance by creating an intimidating, hostile or offensive working environment.

It should be kept in mind that these hostile environment cases may be committed by co-workers as well as by supervisors. In the *Zabkowicz v. West Bend Co.* case (1986), the company was held liable for sexual harassment where co-workers directed obscene remarks and gestures at the plaintiff and posted drawings which depicted her engaged in sexual activities.[162]

THE MERITOR CASE

In the *Meritor Savings Bank v. Vinson* case, which was decided by the U.S. Supreme Court in 1986, the standards for sexual harassment were set forth.[160]

Ms. Vinson claimed that she was constantly being subjected to sexual harassment by her supervisor. Shortly after she had completed her probationary period as a teller, he began sexually harassing her and demanding sexual favors. She said she initially resisted, but eventually gave in for fear of losing her job. She said that over the next several years she had sexual intercourse with him on numerous occasions, that he fondled her in front of other employees, had exposed himself to her in the bank and had forced sex on her on several occasions. She said she did not report these incidents because she feared reprisal. The supervisor denied the allegations and suggested instead that Ms. Vinson was merely a disgruntled employee who made up the stories in response to a work dispute.

The U.S. District Court held that she was not a victim of sexual harassment, that it was a voluntary relationship and was not related to her continued employment. The court ruled that the bank would not have been liable for the supervisor's behavior in any event because it had no notice of same. The court took note of the fact that the bank had a policy against discrimination and a procedure by which an employee could file a grievance. The court noted that she had declined to follow this procedure.

The U.S. Court of Appeals reversed the decision of the lower court, holding that whether or not the relationship was voluntary was irrelevant to the question of sexual harassment as a condition of employment, that the key question is: "Was it 'welcome'?" Moreover, the court held that an employer is absolutely liable for sexual harassment committed by a supervisor under the doctrine of *respondeat superior* and the issue of actual or constructive notice to an employer is irrelevant.

The U.S. Supreme Court agreed to hear the case and focused on three questions: 1) Do unwelcome sexual advances that create a hostile or offensive environment constitute discrimination based on gender and therefore rise to a violation of Title VII of the Civil Rights Act? 2) Does the "voluntariness" of a sexual relationship obviate any finding of sexual harassment? and 3) Is an employer strictly or absolutely liable for sexual harassment by its supervisory personnel, even though the employer has no knowledge and could not reasonably have known of such activities?

The Court had no trouble resolving the first question. The Court unanimously felt that sexual harassment is covered by Title VII of the Civil Rights Act. Furthermore, the Court agreed that a plaintiff need not show economic or tangible injury to bring such an action. Employees have the right to work in an environment free from the kind of hostility or abuse created by insults, jokes, or other forms of degradation based on race, sex, national origin or religion. The Court found that, when such abuse is so severe and continual as to change the victim's working environment, it will constitute a Title VII violation.

Regarding the second question, the High Court said that the "voluntary" aspect of a sexual relationship would not necessarily preclude a finding of sexual harassment. Instead, the Court said the central question should be whether or not the conduct was "welcome."

With respect to the bank's liability, the court was not clear. It looked at agency law, but rejected the absolute liability idea and further noted that absence of notice was not an automatic preclusion of liability. The Court said employers will not avoid liability simply by asserting that it had a grievance procedure and an anti-discrimination policy in effect at the time of the alleged sexual harassment and that the harassed employee failed to use the procedure. In this case, her first line of complaint was the offending supervisor!

At a minimum, such a policy must clearly and specifically address sexual harassment as prohibited behavior and must demonstrate a sincere commitment on the part of the employer to eradicate such behavior. Employers should develop policies and practices that not only prohibit discriminatory behavior, but encourage victims of such discrimination to seek remedies.

THE OFFENSIVE CONDUCT MUST BE FREQUENT OR SEVERE

The conduct must be sufficiently frequent or severe to alter the conditions of the victim's employment and create an abusive working environment. A lower court, after establishing that there was conduct which created a hostile environment, then must determine if the employer is liable for it. If the employer knew about it and did nothing to try to stop it or if the conduct was so pervasive or severe that the employer's knowledge of it can be inferred — in other words the employer SHOULD have known about it — but did nothing to stop it, the employer will be held liable.

The offending employee, of course, will always be liable if the case is proved, but victims prefer to sue the employer under the "deep pocket rule" — the employer usually has more money than the perpetrator.

Usually the offensive incidents must have been repetitious. A single sexually offensive remark generally is not sufficient to establish harassment. A polite request for a date that is rejected, standing alone, will not constitute sexual harassment. The courts will look at the totality of the circumstances. But, if it is serious enough, one incident may constitute sexual harassment. For example, a supervisor physically fondling an employee would be considered serious.

EMPLOYER HELD LIABLE FOR NOT DEALING WITH FALSE RUMORS OF AFFAIR

The harassment does not have to involve physical conduct. Words alone might create the hostile environment. In *Jew v. University of Iowa* (1990), the court ruled that the University's medical school was liable for a hostile work environment because it failed to stop false rumors which had been circulated by the plaintiff's colleagues that she had engaged in sex with her department head to gain favor with him. The court ordered her promoted to a full professorship and compensated her for back pay and attorneys' fees.[163]

conducive to creation of an atmosphere of hostility did in fact occur and, if so, the employer must attempt to dispel workplace hostility by taking prompt remedial steps.[168]

APPROPRIATE ACTION SHOULD BE TAKEN

Employers should use oral and written warnings, transfer the victim or harasser and be alert to future activities of the offending employee because knowledge of past activities would invariably make the employer liable for subsequent harassment misconduct.

D.C. FIRE DEPARTMENT WINS/LOSES FONDLING CASES

The District of Columbia Fire Department was sued for a emergency medical technician's fondling of a woman who was being transported to the hospital in a city ambulance. The city won that case on the basis that it had no prior knowledge of the EMS technician's propensities. A second suit was filed by another woman on the same charge against the same EMT and the city lost the suit because it had prior notice of the man's behavior and did nothing to correct the situation.

SCHOOLS EVEN HAVE SEXUAL HARASSMENT PROBLEMS

There have been a flurry of lawsuits and federal sexual harassment complaints for activities ranging from school boys' lists of promiscuous girls to obscene name calling on school buses. Many school systems are developing sexual harassment policies forbidding "leering" and "spreading sexual gossip."

CHARGE LEADS TO TEACHER'S SUICIDE

A popular math teacher in the Fairfax County, Virginia, school system had a very friendly relationship with his students and they treated each other with jovial familiarity. He was very obese and had a medical condition which made his upper body painfully sensitive to touch. A female student, joking with him in a friendly way, began poking her finger into his chest. He urged her repeatedly to stop, telling her that she was hurting him. She continued and commented,

"You shouldn't have such a big chest." He countered spontaneously, "And YOU shouldn't have such a SMALL chest." She told her mother and a sexual harassment complaint was filed against him. After a hearing before the school board, he became very depressed, believing that he would lose the case. He committed suicide.

SCHOOL MAKES $20,000 SETTLEMENT

In another complaint against a California school system, a girl won a $20,000 out-of-court settlement because eighth grade boys regularly made crude references to her breasts.

In Texas an 11-year-old fifth grade girl filed a suit against an 11-year-old boy for calling her a whore and rubbing her leg with his penis.

A 7-year-old girl in Eden Prairie, Minnesota, won a sexual harassment suit because the school failed to take "timely and effective responsive action" when boys on the school bus called her a "bitch" and graphically told her to perform sex with her father.

COUNTY DEFENDS HARASSMENT CHARGE TO DISCOURAGE FUTURE CLAIMS

A Montgomery County, Maryland, County Councilman was sued by a female former employee who claimed he had required her to have sex with him as a condition for keeping her job. He denied the charges. After a celebrated trial, the county won the case. The county spent over $1 million in defending the suit when the suit probably could have been settled out of court for substantially less. Why did the county choose to absorb such vast legal expenses rather than settle the case for a fraction of that amount? Because settling the case would have given others motivation to sue the county on spurious sexual harassment charges on the assumption that, regardless of the merits of their case, the county would "buy them off" with an out-of-court settlement.

VICTIMS MAY ALSO SUE IN STATE COURT

It should be kept in mind that sexual harassment cases, as violations of the Federal Civil Rights Act, are federal cases, but the victim may also be able to sue in state courts under state anti-discrimination statutes or under general tort law for intentional infliction of emotional distress, assault, battery, invasion of privacy, defamation, etc.

VOLUNTEER COMPANIES MAY ALSO BE LIABLE

In 1992 the Rising Sun (MD) Fire Company reinstated a female volunteer who had complained to the Maryland Commission on Human Relations that she was suspended from her job after she charged a co-worker with sexually assaulting her in 1990. Before a hearing on the complaint was held, the Fire Company reached a settlement with the woman whereby she was reinstated and the company agreed to adopt a policy against sexual harassment. She also filed a multimillion dollar suit against the Fire Company for discharging her.

The Germantown (MD) Volunteer Fire Department Chief was removed as Chief by the Montgomery County Fire and Rescue Commission because of his history of racial slurs and sexual harassment charges. The Chief was being prosecuted in State court on criminal charges of sexual contact and battery. The charges grew out of a complaint by a female emergency medical technician volunteer at the Germantown Station who alleged that three times in 1993 at the station he had improperly touched her. In one incident, she said, the chief "wrapped his arms around me and wrapped his right leg around my waist." In another she alleged he had "pressed his body" against hers, and in the third she said he "grabbed my face and puckered his lips as he pulled my face toward his."

The Chief went into State court in an effort to get re-instated as Chief, but the court dismissed the case because he had filed his case in court instead of exhausting his administrative remedies first. He should have asked the Commission to re-consider its decision before going to court. Because of this, the State judge said the court lacked jurisdiction to hear the case. Criminal charges were also filed against him. In 1988 the same man had been forced to resign from the Hyattstown, Maryland, Volunteer Fire Department for making a racial slur against a black firefighter.

Sexual harassment cases may be handled administratively. In 1994 a Hyattstown (MD) Volunteer Fire Department sergeant was stripped of his rank and suspended for 15 days for sexual harassment of two women members of the company. He was found guilty of "actions unbecoming an officer" by a trial board after the charges leveled against him by the women were aired at a hearing. The board found that he had violated policies and procedures of the personnel regulations.

TO BE ACTIONABLE, HARASSMENT MUST BE SEVERE OR PERVASIVE

A white female fire inspector/code enforcement officer for the City of Swanee, Kansas sued the city under Title VII of the federal Civil Rights Act, claiming she was subjected to sexual harassment and a hostile working environment. She claimed that the fire chief called her a dumb blonde, her supervisor had falsely told another employee she was having a sexual affair with someone in the fire department and the city manager said she was a bitter woman. She claimed that these comments "affected her attitude toward work and made her feel she wasn't respected and wasn't going to get anywhere while these people were her superiors." The city contended that the three allegedly sexist comments were not sufficiently severe or pervasive to support a hostile work environment claim. She also contended that she was treated differently because of her gender. She said she was not allowed to attend training seminars, was disciplined for accompanying the deputy fire chief at his request to a fire, was occasionally required to perform secretarial work while the secretaries were absent, was subjected to different discipline, her requests were acted on more slowly than those of male firefighters, she was denied maternity leave, and was removed from police protection by the fire chief during an arson investigation.

One officer testified that she was treated differently from other paid firefighters, but not because of her gender.

The city moved for summary judgment.

The court said for sexual harassment to be actionable, it has to be so severe or pervasive it must alter the "conditions of the victim's employment and create an abusive working environment." To prove a basic case of sexual harassment, the court said, it must be shown that: plaintiff belongs to a protected class, must have been subjected to unwelcome sexual harassment based on sex, the harassment must have affected a "term, condition or privilege" or employment, the conduct must be persistent and routine rather than isolated. The court said the conduct need not be "clearly sexual in nature" It is sufficient that the treatment is unequal and would not have occurred except for the victim's gender.

The court did not find in the comments complained of any support for her claim because they could not be categorized as having been made because of her gender. The court ruled that the evidence was not enough to support a claim of a hostile work environment.[169a]

MEN MAY ALSO BE VICTIMS OF SEXUAL HARASSMENT

Women are not the only victims of sexual harassment. Sexual harassment applies to both sexes. In 1993 nine percent of the sexual harassment complaints received by the EEOC — almost 12,000 cases — were filed by men. If it is difficult for women to file a sexual harassment complaint, it is far more difficult for a man because the tendency is to laugh it off and not take it seriously.

At Avon Products a woman employee kept pestering a male co-worker, asking for dates, calling him at home and leaving harassing notes. It got so bad that he couldn't function on the job and was very nervous because of her aggressive behavior so he filed a complaint.

In Fairfax County, Virginia, a woman deputy was under investigation in 1992 for fondling two male deputies. The woman, a six-year veteran of the force, allegedly "touched or attempted to touch the groin area of the deputies." The two men did not make the complaint. The action was observed by someone else who reported the matter, but when the male deputies were confronted with the information they confirmed what had happened.

MEN SEXUALLY HARASSING MEN USUALLY NOT COVERED BY CIVIL RIGHTS ACT

There are also many cases of men sexually harassing men. With more and more gay men "coming out of the closet," this trend will undoubtedly accelerate. In the *Wright v. Methodist Youth Services, Inc.* case, a former employee claimed he was fired for rejecting his male supervisor's sexual advances. The court found that, although Title VII usually involves men harassing women, "the obverse of that coin — an alleged demand of a male employee that would not be directed to a female" — similarly constitutes illegal sexual discrimination.[169b]

However, claims of sexual harassment based on an individual's sexual orientation have also been held to not be covered by Title VII. In *Dillon v. Frank*, the sixth federal appellate circuit, while sympathizing with the plaintiff's situation that he "was taunted, ostracized and physically beaten because he was a homosexual" and that his co-workers' actions were "cruel," this harassment was not based on gender and, therefore, was outside the purview of Title VII.[170a]

It is quite possible, however, that state statutes might make this type of discrimination illegal.

OUTRAGEOUS CHARGES THROWN OUT OF COURT

A woman dispatcher at a New Jersey police department who had been reprimanded for tardiness and absenteeism complained that her sergeant had solicited her for sado-masochistic sex, exposed himself to her and masturbated in front of her. The Chief of Police promptly responded to the complaint. He interviewed 18 women who had worked for the same sergeant and found no corroboration whatsoever of the charges or of any past similar behavior. An independent arbitrator, a woman, was hired to investigate the complaint. After a thorough investigation, she concluded that the alleged sexual harassment did not take place. The alleged victim filed suit in US. District Court and her suit was dismissed.[170b]

CHIEF IS FIRED AFTER INSURANCE CO. SETTLES CLAIM

The Macedonia (Ohio) City Council dismissed its fire chief when it learned that an insurance company had paid $100,000 to settle a sexual harassment claim by a female paramedic. She claimed her job as Macedonia Fire Department EMS coordinator was eliminated after she complained to the mayor about sexual harassment by the chief. She claimed that, despite her protests, the chief had hugged, kissed and grabbed her, repeatedly asked her for dates and made sexually explicit and offensive remarks. The mayor had suspended the Fire Chief following a U.S. Equal Employment Opportunity Commission report which backed the claims of the paramedic.

COURSE AND VULGAR LANGUAGE NOT NECESSARILY HARASSMENT

The Supreme Court of Maine ruled that course and vulgar language used by police officers and firefighters was offensive and unprofessional, but before it will be recognized as a cause of action for sexual harassment, it must be shown that it was directed at someone because of that person's gender. The court said, "In order for conduct to create a hostile work environment it must be so severe and pervasive that it alters the conditions of employment and creates an abusive working environment." In this case sexually suggestive and offensive or vulgar jokes and comments were made by both male and female co-workers. There was no evidence that the remarks were directed at the plaintiff or that they would not have occurred except for her gender. The employer received a directed judgment in its favor.[171]

EMPLOYER, AWARE OF PREVIOUS SEXUAL MISCONDUCT, LOSES $908,000 SUIT

A law student who was working as a summer intern for the Arizona Prosecuting Attorneys Advisory Council claimed the Council's former executive director had sexually assaulted her at a prosecutors' conference. As the result of which, she claimed she suffers from post-traumatic stress disorder and rape trauma syndrome. It was shown that the perpetrator had been hired by the Prosecuting Attorneys Council even though they knew that he had been fired from a previous job for allegations of sexual misconduct with underage informants. The intern received an award against the former executive director for $1,480,000. The employer, under the doctrine of *respondeat superior* had to pay $908,000 in compensatory damages.[172]

HAVING POSED NUDE FOR MAGAZINES DOESN'T BLOCK WOMAN'S SUIT

In the case of *Burns v. McGregor*, the court held that the fact that a woman employee posed nude for magazines does not prevent her from bringing a sexual harassment complaint against her supervisor. She claimed she resigned her job because the company's male owner persistently demanded oral sex. He denied wrongdoing and pointed out that her claim that the advances were unwelcome was inconsistent with her behavior as a nude model. The plaintiff had posed nude for two national motorcycle magazines. One photo showed a full, frontal, nude view revealing a pelvic tattoo. Two other photographs highlighted jewelry attached to her nipples.

The trial court held that a person who would pose nude in a national magazine could not be offended by the behavior she complained of. The appellate court reversed this finding, stating that the trial court must view the behavior complained of in the totality of the circumstances.

The appellate court said there was no evidence that the plaintiff had solicited the conduct complained of (which took place after the nude pictures had been published). The court said, however, that evidence regarding the plaintiff's sexually provocative speech or dress and her serving as a nude model is relevant in determining whether or not the sexual advances were welcome.[173]

FIREFIGHTER HAS CONSTITUTIONAL RIGHT TO READ *PLAYBOY* IN FIREHOUSE

The Los Angeles County Fire Department, in revising its sexual harassment policy, forbade the reading of sexually explicit magazines such as *Playboy*, *Penthouse*, etc. in the fire stations. A challenge to the ban was filed by a captain who claimed the prohibition abridged his First Amendment rights. The Fire Department argued that the regulation was required by federal law because the sexually explicit periodicals would create a hostile work environment for female firefighters. A federal judge ruled against the Fire Department and said that the firefighter's constitutional right to read *Playboy* outweighed the Fire Department's policy on sexual harassment. The judge said the Department had failed to prove that the firefighter's quiet reading of the magazine during hours of relaxation created a sexually hostile work environment.

Two female firefighters had testified that the presence of *Playboy* offended them and that male firefighters often made abusive remarks to them while looking at the magazine.

In spite of this case, fire departments should be cautious in allowing "girlie" posters and magazines featuring nudes and sexually explicit material to be openly displayed at the fire station. Notwithstanding the court's ruling in the Los Angeles case, if employees insist on bringing such material into the station house, they should be required to keep it in their lockers out of sight when not being read.[174]

REASONABLE PERSON TEST VS. REASONABLE WOMAN TEST

In the *Rabidue v. Osceola Refining Co.* the Sixth Circuit applied the "reasonable person" test and found that crude language and sexually oriented posters would not interfere with a reasonable person's work conditions "when considered in the context of a society that condones and publicly features and commercially exploits open displays of written and pictorial erotica at the newsstands, on prime-time television, at the cinema and in other public places."[175]

In the Third Circuit, however, the U.S. Court of Appeals held otherwise. In the case of *Andrews v. City of Philadelphia* (1990) a female employee of the Philadelphia Police Department alleged that abusive language, the display of pornographic pictures, anonymous phone calls and the destruction of property by fellow employees and supervisors created a hostile environment. The Court said that men and women may have different perceptions about such conduct and, while men might find obscenity and pornography innocent and harmless, it is highly possible that women may feel otherwise. The court said that the trial court, therefore, should look at the actions "to see if they produce a work environment hostile and offensive to women of reasonable sensibilities."[176] The court's logic in adopting this gender-based standard of scrutiny was to eliminate stereotypes and harassment which would likely persist in traditionally male-dominated workplaces such as the Philadelphia Police Department or, it could be added, most fire departments.

In the *Ellison v. Brady* (1991) case, the Ninth Circuit U.S. Court of Appeals also adopted the "reasonable woman" test rather than the "reasonable person" test. Ellison, a female IRS Agent working in California, complained about two "bizarre" love letters she had received from a fellow worker. IRS management counseled the man to leave her alone and later transferred him to another office. He filed a grievance over the transfer and the IRS agreed to return him to his original office if he promised not to bother Ellison again. She filed a sexual harassment complaint. The IRS and the EEOC rejected it. She sued. The U.S. District Court agreed with the IRS and the EEOC, stating that the man's conduct was "isolated and genuinely trivial" and dismissed Ellison's case.

On appeal, however, the Ninth Circuit U.S. Court of Appeals sent it back to the lower court with instructions to consider the alleged conduct from the perspective of a "reasonable woman", not a "reasonable person" because it would be male-biased and tend to ignore the experiences of women. The court said that sexual harassment in the workplace is a major problem and adopting the victim's perspective ensures that the courts will not "sustain ingrained notions of reasonable behavior fashioned by offenders," in other words, men. The court pointed out that Congress did not enact Title VII to codify existing sexual prejudices and that, hopefully over time, both men and women will learn what conduct offends persons of the other gender.[177]

There was a strong dissent in that case, objecting to the gender-based test. This dissent may prove to be the majority view when the U.S. Supreme Court addresses that issue. The dissent, however, recognized that a workplace where men are predominantly employed may be a different situation.

MALE-DOMINATED WORKPLACE

The *Robinson v. Jacksonville Shipyards, Inc.* (1991) case addressed this male-dominated workplace issue. Ms. Robinson, a welder, sued the company for sexual harassment based on the hostile environment argument, putting into evidence pictures of nude and partially clothed women depicted in positions exposing their genitals and engaging in various explicit sex acts. These types of pictures had been posted extensively throughout the shipyard. Management of the shipyard, not only approved of these posters, but often had their own similar pictures. Part of the company's defense was to show the "social context" such as that accepted by the Sixth Circuit U.S. Court of Appeals in the *Rabidue* case. The *Rabidue* decision said that Title VII

FIRE DEPARTMENTS SHOULD NOT OVER-REACT

Fire departments should be careful not to over-react to the current tide of sexual harassment cases. The complaints should definitely be taken seriously and investigated, but all instances of abusive language or gender-based comments, jokes or teasing may not necessarily rise to the level of actionable sexual harassment. It becomes discriminatory only when it is so severe or pervasive that it would adversely affect the working conditions of a "reasonable person" or a "reasonable woman," depending on the jurisdiction in which the case is filed.

SUGGESTED POLICY ON SEXUAL HARASSMENT FOR FIRE DEPARTMENTS:

The Central City Fire Department is committed to maintaining a work environment which is free from inappropriate and disrespectful conduct and communication of a sexual nature or which might make the workplace uncomfortable to someone because of his or her gender.

In an effort to avoid even the appearance of impropriety, this policy against sexual harassment has been adopted and will be vigorously enforced. Fire service personnel violating the letter or spirit of this policy will be appropriately disciplined.

The Central City Fire Department strongly opposes sexual harassment in any form. Sexual harassment in the workplaces under the jurisdiction of this Department by an officer, employee or non employee will not be tolerated.

This policy will be equitably enforced whether the victim is male or female.

Sexual harassment is not only contrary to the policy of this Department, but is also a violation of Title VII of the Civil Rights Act as well as the applicable laws of this state.

Sexual harassment is behavior of a sexual nature which is not welcome by the intended recipient, is personally offensive, erodes morale or interferes with the work performance or effectiveness of personnel. Unwelcome sexual advances, requests for sexual favors or other oral or physical conduct of a sexual nature constitutes harassment when:

— submission to such conduct is made explicitly or implicitly a term or condition of an individual's employment.

— submission to or rejection of such conduct by an individual is used as a basis for an employment or promotional decision affecting that person and /or

— *Such conduct has the purpose or effect of unreasonably interfering with a person's work performance or creating an intimidating, hostile or offensive working environment.*

Sexual or so-called "dirty" jokes will not be permitted which might embarrass or ridicule an employee because of his/her gender.

If comments or conduct of a sexual nature are unwelcome by a person, they may constitute harassment. The Department will not accept as an excuse for such objectionable behavior that the offender was "only joking" or that "I didn't think the other employee would object."

Employees are prohibited from displaying on Fire Department property any sexually oriented posters or other pictures which show nudes or explicit sexual activity. Magazines or books which feature such material shall not be openly on display on Fire Department property. If an employee insists on bringing such material onto the Fire Department property, when it is not being read, it should be kept in that employee's locker and should not be shown to or shared with other employees.

No employee is to be treated differently than any other employee because of his/her gender. This policy specifically prohibits criticizing, berating or ridiculing an employee because of his/her gender.

All officers are responsible for implementation of the Fire Department's policy against sexual harassment and other discriminatory practices. They should ensure that all employees under their supervision clearly understand the Department's policies in this regard and what the penalties will be for infringing those policies. Supervisors should take or assist in taking prompt, appropriate corrective action when necessary to ensure compliance with this policy.

Any employee who feels that s/he has been victimized by violation of this policy should promptly register a complaint with the Fire Chief or his/ her designated representative. All employees are urged to use this complaint procedure if they believe they have been subjected to discrimination or harassment or have knowledge of same in the Department.

Registering such a complaint in good faith will under no circumstances be grounds for disciplinary action. For such a complainant to be disciplined, or in any other way disadvantaged for filing the complaint, not only violates the Department's policy, but is also a violation of law.

The above sample policy is offered only as a suggested guide. Each Fire Department should adopt its own policies geared to its own unique circumstances. As in all legal matters, you should have access to a competent attorney to guide you through the complicated rocks and shoals of sexual harassment.

The following suggestions are offered by Women in the Fire Service:

STOPPING SEXUAL HARASSMENT: THE FIRE CHIEF'S ROLE

Fire chiefs of career, volunteer, and paid-on-call departments who wish to minimize problems of sexual harassment should consider the following steps:

- *Adopt written policies prohibiting sexual discrimination and sexual harassment, and a workable, confidential, step-by-step procedure for the filing of complaints.*
- *Provide training for all personnel — firefighters, officers and support staff — on multi-cultural issues, including sexual harassment. Anti-women behavior is part of the dominant culture of many firehouses and should be addressed as such. These sessions should facilitate discussion, and not simply consist of reading the department's policy and considering everyone to have been "trained."*
- *Remember that "what you permit, you promote." Do not ignore harassment. To do so sends the message that you are in agreement with the harassing behavior or discriminatory attitudes. Do not place all of the burden for reporting and correcting the problem on the harassed individual or group. Each stage of prejudiced behavior encourages the next; extreme behavior develops when more subtle behavior is permitted to continue.*
- *Support — do not discourage — people who bring complaints of harassment to your attention. Handle complaints confidentially and in a timely manner. An open-door policy goes a long way towards resolving problems at the lowest level, long before the chief is faced with a polarized situation or an expensive lawsuit.*
- *Prevent retaliation against those who have filed complaints, and against any witnesses who support their allegations. If the charge is found to have merit, discipline the harasser appropriately. Don't "solve the problem" by transferring the target to a new station against his/her will.*
- *Ensure personal privacy for everyone in station accommodations. Lack of privacy increases the tensions and resentment that lead to harassment.*
- *Keep in mind that just because you do not hear about sexual harassment occurring does not mean that it is not there. Use exit interviews (interviews with people who leave the job) as a reality check. A good manager will want to understand the reasons behind an employee's request for a change in status, whether it is an application for transfer or a notice of resignation.*

- *Be a role model. Avoid patronizing behavior or tokenism toward any group; be aware of language, policies, images, or situations that have questionable gender or racial connotations (e.g. the term "non-white") and correct any stereotyping that occurs. Be aware of, and correct, behavior that reflects your own prejudices.*

THE STATION OFFICER'S ROLE

All station officers or acting officers, whatever their rank, are responsible for maintaining a harassment-free work environment in their stations. Departmental procedures may assign the officer a specific role or responsibility in the harassment complaint procedure, but even in departments that have no clear anti-harassment policy, first-line supervisors are in a crucial position with respect to preventing sexual harassment.

- *Stop any behavior occurring in your station or crew that is clearly harassment. In order to do this, you must be able to identify harassment when it occurs and to deal with it fairly, effectively and quietly. If the training provided by your department on these issues is inadequate, use your own initiative to get additional training and information.*
- *Educate your crew on what does and does not constitute harassment. This should be a routine part of departmental training, but if your department does not offer such training, or if the training has been inadequate, consider developing and implementing a comprehensive anti-harassment training program for the people you supervise. Every fire department employee should have a working knowledge of sexual harassment as well as of the department's policies and procedures.*
- *If a member of your crew brings a complaint of sexual harassment to you, be supportive. Do not downplay the reported incident or attempt to make excuses for the behavior. Follow through on the complaint swiftly, impartially and without discussing it with those whom it does not concern. Targets of harassment will not feel free to use the system if they fear that the whole station will immediately know of their complaint. Nor is it fair to someone who has been accused of harassment for the accusation to be made known, particularly if it should turn out to be unfounded. Only those directly involved should know of the complaint; those called as witnesses should be cautioned not to discuss the matter with others.*
- *Prevent any kind of retaliation against the person who has filed the complaint. Harassment targets often believe that filing a complaint would do no good and would probably only make matters worse. If this kind of attitude prevails in your fire station, harassment will not be reported and will only escalate, creating a nightmare for the*

officer as well as for the target(s).

- Do not overreact to harassment, but do not fail to act. Discipline should follow the department's established guidelines, and should reflect whether the offense is the individual's first or just the latest in a long series. It must also be appropriate for the severity of the behavior. A firefighter guilty of attempted sexual assault would logically be disciplined more severely than one who brought copies of pornographic magazines to work.

- Creating a harassment-free work environment includes stopping harassment by those who are not in the workplace. Be alert to discriminatory or harassing behavior by visitors to the stations: members of the public, friends of firefighters, equipment salespeople or repairers and others. Handling these situations will require tact and diplomacy, but any person who behaves in that way must be made aware that you do not support their behavior or attitude, and that they will not be welcome in the station if the behavior continues.

IF A CO-WORKER IS BEING SEXUALLY HARASSED

Many workers who would never harass anyone themselves are guilty of tolerating harassment that occurs in their workplace. Some onlookers may mistake the target's silence for acceptance; others may not want to get involved or may fear losing the camaraderie of their co-workers by being a "spoilsport." Fair-minded and supportive co-workers, however, will refuse to condone harassing behavior in the workplace. Your support of the target, as a friend or a witness, can be crucial to her becoming aware that the behavior need not be tolerated.

- If you observe behavior at work that seems questionable, don't assume it's welcome just because the recipient doesn't complain. Talk with her privately, letting her know that you felt the behavior might be inappropriate. Especially if she is a new recruit or brand new in the station, remind her that she doesn't have to accept unwelcome sexual behavior, whether it involves comments, physical contact, pornographic posters or videotapes. Emphasize to her that she has your support in getting the behavior to stop.

- If the behavior is clearly harassment but the target is unwilling to take steps to stop it, talk with the harasser in private. Let him know that you find the behavior inappropriate and that it's not okay with you. If your own work environment is being made uncomfortable by the sexual behavior of others, you have the right to ask that the behavior stop, or even to file a sexual harassment complaint yourself.

- *If you think your own behavior towards a co-worker, present or past, might be or might have been harassment, stop the behavior. To determine where to draw the line between friendly behavior or joking and harassment, consider:*
  - *Is it an equal exchange?*
  - *Would you do or say the same thing to your mother or father? To another person of your gender?*
  - *Would you want to see coverage of your behavior on the six o'clock news?*

  Ask the co-worker about it privately. Let him or her know that the behavior will not continue if it's a problem, and that you can still be friends. If, after you've asked about the behavior, you're still not sure about it, stop the behavior anyway. There are hundreds of different ways for women and men to interact enjoyably and professionally in the workplace; don't risk harming a positive work relationship by behaving in ways that make either of you uncomfortable.

IF YOU ARE BEING SEXUALLY HARASSED

Fire department employees who are the targets of sexual harassment should:

- *Know the department's policies prohibiting sex discrimination and sexual harassment, and the procedures for filing a complaint.*
- *Politely and firmly tell the harasser to stop, in front of witnesses if possible. If you are unable to confront the harasser directly, write a letter and give it to the harasser in the presence of a witness. Keep a copy of the letter in a safe place, not at work. Or speak to your supervisor, to Personnel or to an EEO officer.*
- *If the harasser repeats the conduct, inform your supervisor immediately and follow it up with a note or letter to the supervisor. Again, keep a copy in a safe place, not at work. It is important that you give notice of this offensive behavior to your employer and that you have a record that you did so. (Courts will generally only hold an employer liable for sexual harassment where it can be shown that the employer knew, or should have known, about the conduct.)*
- *Document all incidents in a diary with time, place, names of witnesses, what was said or done, and an exact account of your response and any physical or emotional stress you experienced. Keep this log in a safe place, not at work.*
- *Talk with other women on the job or who have left the job, especially those who have worked with the harasser in the past; they may also have been targets. Do not be surprised or discouraged if other women do not support your decision to fight or report the harassment.*
- *Keep records of all positive evaluations, promotions, etc., in the event that your complaint results in retaliation against you by your*

employer, officer or other firefighters.

- *Contact your union, labor organization or employee group for assistance. If you work under a contract, be familiar with its anti-harassment clauses and your rights under the by-laws of your union.*

- *Contact women's organizations for support. You are not the first woman to be victimized by this kind of behavior. Some chapters of the National Organization of Women (NOW) sponsor support groups for victims of harassment, and many chapters of 9 to 5 can give you support as well. Other women's professional and trade groups, such as Women in the Fire Service, may be able to offer advice and resources.*

- *Be aware that sexual harassment, and the decision to take action against it, are very stressful. Take advantage of any employer-sponsored employee assistance programs. Try not to internalize guilt. Any incidents of sexual harassment that happened to you were not your fault, the harasser did not "mean well", and he was not doing it because he likes you. Sexual harassment is not something you have to tolerate because you chose to enter a non-traditional field; you do not have to adapt to the ways of your co-workers "at any cost." If you have friends and co-workers who can offer support, including corroboration of incidents, depend on them. Be aware, however, that the general public is largely uneducated and often unsympathetic on this issue.*

- *If informal or internal remedies fail to stop the harassment, if you lack confidence in the internal remedies (for example, if the person in charge of investigating your complaint is the one who is harassing you, or has threatened retaliation), or if you are retaliated against for complaining about the harassment, you should consult an attorney. Seek out a lawyer who has experience in handling employment discrimination complaints. Try to obtain private counsel, even though local, state or federal human rights agencies may assign a staff attorney to handle your formal complaint. The women's bar association or working women's advocacy groups may be able to refer you to an experienced attorney.*

- *File a formal complaint with the federal Equal Employment Opportunity Commission or your state or local Fair Employment Practices agency. There are specific time limits for filing such complaints, depending on whether the harassment is continuing or retaliation occurred and whether your state or locality has a FEP agency. Title VII requires that you file with the EEOC within 180 days of the last act of discrimination. In areas with state or local FEP agencies, that time period is extended by 90 days while you attempt to resolve the problem internally. You can always drop the charge if matters are resolved to your satisfaction.*

TIPS ON HANDLING SEXUAL HARASSMENT COMPLAINTS

— Have a policy prohibiting sexual harassment. Draft it with care, taking into consideration special fire service problems and situations. Make sure everyone knows the policy. Put it in the employee handbook, bring it up at training sessions, make sure everyone knows what sexual harassment is and how it will be dealt with by the department. Have a clear avenue open to employees for registering complaints. Don't have the first line of grievance be the person's immediate supervisor who might be the party perpetrating the sexual harassment.

— Take the complaint seriously. Don't minimize or ridicule it. This is extremely important, even if you don't personally think much of the complaint or the complainant. If you don't take the claim seriously and investigate it, the Department may face substantial liability.

— Be compassionate, yet professional, in listening to the complaint. Filing such a complaint can be emotionally traumatic for the victim. Show your concern and be sympathetic, but maintain your objectivity. Don't become unduly influenced by the complainant's emotions.

— Don't prejudge the case strictly on the basis of the complaint. Be professional in striking an appropriate balance. It is not appropriate at this point to pass judgment. It is also not appropriate to make gratuitous comments such as, "If you didn't wear those tight-fitting mini-skirts and sweaters, maybe this would not have happened." Concentrate on getting the facts and don't be judgmental.

— Conduct all interviews in a private room.

— Explain to the complainant how the investigation will be conducted.

— Find out what happened. Get specifics, not vague generalities. Information you may wish to obtain during the first interview:

    Name and position of the person complaining.

    Identity of the person or persons who allegedly did the harassing and their positions in the Department.

As many facts as you can obtain to find out exactly what happened, even though this might be embarrassing or painful for the complainant. It's important to get the whole story.

How often did the objectionable conduct occur and on what dates?

Where did the alleged incidents take place?

What was the complainant's reaction to the conduct?

Did the complainant tell anyone else about the incident? When, who, where, what and under what circumstances?

Does the accused have control over the person's compensation, promotions, working conditions, shift assignments etc.?

Has the accused made any threats or offered any inducements to the complainant with respect to the conduct or the complaint?

Does the complainant know of any other possible victims of the same behavior?

Try to find out from the complainant what effect the alleged harassment has had on him/her.

Urge the complainant to let you know if there are any reoccurrences of the objectionable behavior or any threats or retaliation. Inform him/her that you will speak with the accused to make sure that the conduct is not repeated.

Get the names of any possible witnesses.

Ask the alleged victim how s/he thinks the matter should be handled. Sometimes the complainant only wishes to have the wrongdoers cautioned to alter their behavior and does not want to pursue the matter beyond correcting the problem. It is important to find this out early.

Try to assess the complainant's credibility, keeping in mind the "reasonable person" or "reasonable woman" tests discussed above.

If appropriate, take a signed statement from the complainant. Sometimes people subsequently change their story as to what happened, what the complaint was about or what transpired during the interview. The signed statement sets the record straight.

Preserve your notes from the interview.

— Following the initial interview, investigate the complaint promptly, aggressively and fairly. The complainant should be assured that a thorough investigation will be conducted promptly. Consider having an outside, objective person investigate the complaint. Have the person or persons conducting the investigation be of the same sex as the complainant.

— Create a confidential file on the matter.

— Don't promise the complainant complete confidentiality, because you will have to talk to the accused and possible witnesses, but don't discuss the matter with other employees in such a way that the complainant is likely to be embarrassed. It is important that you talk to the accused person or persons so the objectionable activity will cease.

— After a complaint is received, review the department's policy and procedures regarding sexual harassment to see if they are adequate.

— Suggestions for the investigator:

> The person doing the investigation ideally should be someone from outside the department or at least someone outside the particular station house where the alleged harassment occurred.

> The investigator should interview the alleged perpetrator, but make clear that no assessment has been made at that point as to the truthfulness of the allegations.

> Identify the complainant and state the basis of the complaint, giving the accused an opportunity to respond to, explain or deny the charges.

> Ask for the names of witness who can corroborate his/her position.

> Assess his/her credibility. Consider taking a signed statement.

> Preserve the notes of the interview.

> Interview all witnesses.

> Distinguish between firsthand knowledge and hearsay, but include both in your report. Assess the credibility of the witnesses, including any bias for or against the complainant or the accused.

Again, consider taking a signed statement from witnesses, but at least preserve your notes.

In evaluating the matter consider the "reasonable woman" and "reasonable person" tests and the difference between "welcome" and "voluntary" sexual conduct.

Write a thorough, objective report. It should describe the exact allegations in the complaint and the response of the accused. Note any documents reviewed and report all interviews conducted.

If appropriate, set forth in the report the investigator's opinion as to what happened, as to whether or not the sexual harassment occurred, but make it clear that this is the investigator's opinion and explain logically how it was arrived at. (Some department policies might call for the investigator to only set forth the facts and not provide his/her own opinions about the matter, and leave these conclusions to be drawn by the decision-maker.)

Make recommendations regarding corrective action, changes in policies or procedures, etc., which are reasonably calculated to prevent further harassment.

Submit the report on a confidential basis to the decision-making official, ideally the Fire Chief.

— The decision maker should not just rubber stamp the report but view it carefully to find inadequacies, unfairness, deficiencies etc. Ask follow-up questions of the investigator. Conduct his/her own interviews if this is felt to be appropriate.

— Document all actions.

— Follow up with the complainant and the accused as to the investigator's findings and what future action will be taken.

— Keep the Department's legal counsel advised every step of the way. Seek and follow his/her advice.

— While sexual harassment is very serious, don't over-react and assume the accused is guilty merely because of the allegations. Don't feel that some type of punishment must be meted out, regardless of what the investigation shows. If the individual is falsely accused and punished, s/he might file a defamation suit or wrongful discharge suit. If the accused is in a protected class him/herself because of age, race, creed, national origin etc. the accused may bring a suit claiming the termination for sexual harassment was merely a pretext to cover an ulterior discriminatory motive. The law does not require sacrificial lambs. If a good faith investigation is conducted and the allegations are not substantiated, don't discipline the accused.

— Sometimes it will not be possible to find out what actually happened and the decision maker is faced with two conflicting views and no corroborating witnesses for either party. In such a case the chief may wish to separate the employees to prevent possible future problems. However, sometimes this itself will result in a grievance being filed by the transferred employee.

CHAPTER SIXTEEN

LEGISLATION

W hen you discover during a legal audit, or in studying
after-action reports, or in some other way you learn of a
conflict in the laws under which the fire department
operates, or that the immunities afforded to the department and its
personnel are inadequate, or that the department cannot realistically
perform the burdensome mandates imposed on it by law, or that the
department lacks the authority to do something it deems essential for
performance of its duties, or that you find other flaws, confusion,
inadequacies or other difficulties with the existing law, you should
take steps to have the law changed.

There is a tendency for many people to consider the laws immutable, but laws can be changed and they can be repealed. They
frequently are. The law is a dynamic, not static process. It changes
constantly either through legislation or court decisions. The fire
department should harness this process to improve its legal situation.

HOW THE LEGISLATIVE SYSTEM WORKS

Whether the law falls under the jurisdiction of the city council, the
county commissioners, the State legislature or the U.S. Congress, provisions are made for amendments to the law. Legislators will usually be
very receptive to well-reasoned requests for changes in the law.

It is important for the good manager to understand how laws are made and how they are changed.

CHECKS AND BALANCES

Our legal, judicial and legislative systems are complex, diverse and independent of each other. They are all related in their activities and yet they are independent. The constitutions (state and federal) mandate that the executive, judicial and legislative branches of government be independent so that one branch does not obtain too much power. This is our system of checks and balances. The entire system is based on this premise.

While legislative bodies at the federal, state and local levels operate somewhat similarly, there are distinctions. It is important to learn these distinctions in your state and on your local level.

Federal laws are passed by the U.S. Congress. They must pass both the Senate and the House. They are signed into law by the President. State laws are passed by the state legislature which consist of a lower chamber, the House, and an upper chamber, the Senate. (Nebraska has only one body in its State legislature.) Both bodies must pass a proposal and it must be signed by the Governor/President before it becomes law. The exception for both the Congress and state legislatures, of course, is when the Governor or the President vetoes a bill and the legislative body overrides that veto to enact the law without the executive's signature.

On the local level there are a vast variety of governmental structures. Some have separate executive and legislative branches. Some have a single body such as the Board of Aldermen or County Commissioners, etc., which operates as both the executive and legislative arms of the local government.

The legislative process is very slow. It is important to know the intricacies of each specific system, particularly the one in which you are seeking your legislative changes.

Politics complicates the process, so it behooves you to understand, not only the mechanics of the legislative process, but the political system with which legislation is inextricably entwined.

PREPARE

Once you decide you wish to seek changes in the laws affecting you, a program should be developed to achieve your objectives.

Part of your action plan should be to gather information and research material to justify the changes you seek. This is an appropriate opportunity to tap your network of fire service friends throughout the country who might have valuable ammunition for you in your legislative efforts.

TIPS FOR INFLUENCING LEGISLATION

Once your issue is identified, research it thoroughly so you know all aspects of it, not only reasons why it should be passed, but also reasons why some people will think it should NOT pass. Knowing the anticipated arguments against the bill, will help you to muster arguments to overcome them.

Organize a plan to present the issue in a manner which will gain it the most support. Make sure you cover all the points which should be covered. Make sure you have answers to all the questions which the legislators might ask you.

The department's position on the issue should be clarified. If there is opposition within the fire department to the idea or elsewhere in the fire service, do your best to neutralize this opposition.

FOLLOW THE PROCESS CLOSELY

Plan an action program for promoting the legislative idea. Monitor it closely as it works its way through the legislative process. If a legislator tells you s/he will support it, don't assume that s/he will actually vote for it. It is not necessarily an intentional deception on the legislator's part (although politicians do have a tendency to tell constituents what they want to hear), but there may be legitimate reasons for that legislator to change his/her mind. Acquisition of other information, or persuasive arguments by opponents, or changes in the bill before it comes to a vote — all might cause the legislator to change his/her mind. Stay in close touch with legislators who pledge support to keep them on board. Ask them to help persuade their uncommitted colleagues. Try to get them deeply involved as internal promoters of your bill. The more involved they are, the more committed they will become to your proposal.

EVALUATE AND COMPROMISE WHERE IT MIGHT HELP

Periodically during the legislative process, evaluate the prospects for passage. If they are dim, consider what might be done to make the bill more palatable to the public and the legislators, or what changes might be made to the bill to make it more effective or more acceptable. A vitally important point to always keep in mind is that compromise is the lubrication which makes the legislative machinery work.

PERSONAL CONTACTS WITH LEGISLATORS

If your visit with the legislator is at the Capitol, make an appointment. Staff members play a vital role in the legislator's job and have unusual influence on what s/he does. Consequently, you should not become discouraged if you do not see the member him/herself, but are referred to an aide.

If the legislator is "on the floor" (that is, in the legislative chamber) you may try to see him/her in the lobby outside for a brief meeting. (This is where the term "lobbyist" comes from). However, be very conscious of not wasting his/her time. State your case clearly and succinctly and offer to get additional information if s/he desires, or to answer any questions about your proposal which s/he might have in the future.

If you have an appointment, be on time and be ready to present your case because the legislator may have a parade of constituents lined up to see him/her after your appointment.

Your contacts with the legislators can be oral or in writing.

Tell your story in your own words, but in nontechnical language which is easily understood. S/he probably won't understand fire service jargon.

Respect the legislator's position if it is contrary to yours and don't argue. Explain, but don't argue.

Always follow up your visit with a letter, thanking the legislator for seeing you. This also affords you a chance to reiterate your points in a form which gives a clear, permanent summary of your position.

Monitor progress on the proposal as it works its way through the complex legislative process.

While the legislator is at home in his/her legislative district, you might arrange a meeting with him/her at a restaurant, his/her office, a golf course, or wherever it might seem appropriate. You might also approach him/her at a banquet where s/he is the guest speaker, or chat with him/her at a civic or fraternal gathering.

You might invite the legislator to attend local events such as fire prevention week activities. Ask him/her to make a speech, cut a ribbon, present an award or give him/her an award.

Invite the legislator to tour your facilities to meet your employees. Politicians always appreciate an opportunity to meet constituents.

All of these things give the legislator good exposure to his/her constituents and give you good exposure to the legislator. The more dealings you have with legislators and develop a friendly rapport with them the better your chances will be to get a favorable hearing of your ideas. Friends always get preferential access. That's human nature.

TIPS FOR COMMUNICATING IN WRITING

- Write on your personal or fire department stationery.
- Type the letter and sign your name.
- Give a complete return address on the letter and on the envelope. Add your telephone number and fax number.
- Identify clearly the subject about which you are writing.
- Explain how the issue affects you, the fire department or, most importantly, the legislator's own constituents.
- Avoid stereotypical phrases and sentences.
- Avoid form letters
- Be reasonable.
- Ask the legislator's position. Be understanding if it turns out to be different from your own. Persuade, don't argue. Be respectful and considerate.
- Consider your timing. In politics, timing is EVERYTHING!
- Remember to send "thank you" letters for a personal visit, for a favorable vote in committee or on the floor, and even for the legislator's listening to you and considering your request.

- Know your subject matter thoroughly, know how the legislative system works, know the key legislators who are important to that particular matter and know who your allies and adversaries on this issue are, both inside and outside the legislature.
- Find out who a legislator's important political supporters are. If you can convince them, you will go a long way toward convincing the legislator.

PREPARE TO FACE OPPOSITION

You may face strong opposition to your idea, but don't give up. Figure out how to overcome it. It might take you several legislative sessions to achieve your goal, but if your cause is just, you will probably eventually prevail.

You might find the legislative challenge stimulating. When your idea is enacted into law, you will definitely find it extremely gratifying.

TESTIFYING BEFORE A LEGISLATIVE HEARING

When an individual is called upon to be a witness at a legislative hearing certain things should be kept in mind. The testimony should consist of the facts. Statements should be brief but complete, and directly related to the subject matter of the hearing.

A prescribed time period will be allotted for the hearing, and a number of other witnesses will be scheduled to be heard. The legislators will not be favorably impressed if you infringe on these time constraints with long, rambling testimony.

When someone is an expert in a particular matter, it is sometimes very difficult to be succinct. There is a tendency to explain more than the hearers are interested in hearing and more than they need to hear.

When you take a position the legislators will want to know the basis for your statement. As part of your preparation, you may wish to seek guidance from others. For example, if it is a code change or code adoption, model code organizations, other code officials, or other professional groups might help you by giving you arguments and case histories to support your position.

Before testifying, you should conduct sufficient research to be prepared to answer questions which expand on your position. You should thoroughly understand the problem and be able to define it and explain how the proposed solution will solve it. The testimony should be seasoned with experiences from other communities or your own with respect to the problem.

When questions are asked, answer in a clear, concise manner, giving as much detail as is necessary, but no more. The more things you say, the more material opponents will have to ridicule or discredit your testimony.

Know your audience. Direct your testimony to that audience and not to the spectators in the hearing room. Don't use technical jargon. Your testimony should be easily understood by those to whom you are directing it.

Your overall demeanor, including your nonverbal communication, can impede or enhance your testimony. If you are nervous, shifty and do not use good eye contact, your listeners may perceive that your nervousness is because you are being less than honest in your testimony.

HELP THOSE WHO HELP YOU

Cooperation is a two-way street. Try to help those who help you. Help them to get favorable publicity, Make sure fire personnel know of their good work. Help them during their re-election campaigns. (Make sure there is no prohibition in your local or state laws against your political participation before you get involved in elections.)

GETTING A BILL PASSED IS A MANAGEMENT PROBLEM

The legislators are ordinary people just like you. Don't be awed by them or by the legislative process. Approach it as you would any other management problem. Study the problems and develop strategies for overcoming them.

CHAPTER SEVENTEEN

FIRE PREVENTION

The objective of a fire prevention program is to maintain a fire safe community, not to prosecute as many people as possible for violations of the fire code.

If a significant number of people are uncooperative, fire department representatives might consider meeting building owners and managers or other landlord groups to resolve any difficulties. Education and public relations are as important in a fire prevention program as the legal aspects.

Fire prevention includes day-to-day code enforcement and compliance. It includes issuing permits, licenses, waivers etc. as well as keeping abreast of new developments in technology and techniques to make sure the fire prevention program and the fire code are as effective as they can be.

Inspectors should carry some means of identification. Inspections should be conducted at reasonable hours.

INSPECTION PRIORITIES

Priorities should be established for various types of inspections. Consider the following sequence of priorities:

- **Those required by law**. If your state law says that all nursing homes must be inspected every six months, you should have in place a system to make sure that this is done. Your failure to do so would increase your liability exposure if a fire occurs and you failed to inspect on the time schedule mandated by law. This would be considered negligence per se. All the plaintiff would have to prove in order to show your negligence is that you failed to comply with the statute's provisions regarding inspections. Sometimes the code will specify certain occupancies as "life hazard uses" and require more frequent inspections. Know what these requirements are and set up a system for complying.

- **Known violations**. Failure to correct a dangerous situation inevitably leads to liability. This is a long-standing common law principle. Failure to follow up on cited code violations falls into this category and is one of the main sources of liability for fire departments. Repeated inspections can help to make a case against a scofflaw who fails to correct fire code violations.
  In the case of *Adams v. Alaska*, the State fire marshal's office had no obligation to inspect the Gold Rush Hotel, but when requested to do so, an inspector was sent to the hotel. He found a number of serious safety violations. The owners were told they would be contacted by the fire marshal's office about needed corrections. The inspector reported the violations to his supervisor and suggested that someone visit the hotel as soon as possible. The hotel owners were never contacted again and no one visited the hotel. A fire occurred which killed several hotel guests. It was claimed that the violations discovered during the inspection caused the deaths. The Court held the fire marshal's office liable for failing to follow up on the known violations.[181] Even though the fire marshal's office had no duty to inspect the hotel in the first place, once the inspection was undertaken, the common law legal duty of a volunteer to carry out the inspection in a non-negligent manner was imposed.

- **Complaints**. Failure to follow up on complaints — especially those notifying the department of a dangerous situation — will have the same result as failing to follow up on discovered code violations. When the department is on notice that a dangerous condition exists, it should aggressively take steps to correct that hazardous situation. Reports of imminent danger must be responded to immediately.

How should you handle anonymous complaints? In the opinion of the author, you should follow up on them vigorously. As observed above, the main objective of the fire code is to maintain a fire safe community, not to prosecute offenders, so you should not be concerned about having the complainant appear as a witness against the violator. That being the case, what difference does it make if the complainant is anonymous? An employee might know of a very dangerous fire hazard at his/her workplace, but fear retribution by the boss if s/he complains about it. In the employee's mind, the solution is to notify the fire department of the hazard anonymously. From a liability point of view, the fire department cannot afford to ignore such a notification of a dangerous situation.

FOLLOW UP

It is essential that procedures be established to follow up on a violation once it is discovered. Failure to follow up to make sure the violation is corrected could result in liability of the department for negligence. As stated above, courts go out of their way to impose liability when a known violation is allowed to go uncorrected and someone is injured as a result.

Aside from legal considerations, a sound follow up system is good management. If the public gets the impression that fire code violations will be aggressively enforced, voluntary compliance will increase.

CERTIFIED INSPECTORS

If your state law requires that fire inspectors be certified, they must be certified. Inspection by a non-certified inspector would not be valid. If the law requires that certain types of structures be inspected according to a certain time schedule, an inspection by a non-certified inspector does not meet the requirements of the law. This would increase your chances of liability.

GENERAL RULE: ENTRY REQUIRES A WARRANT

Obviously inspectors must have access to the buildings they inspect. The Fourth Amendment to the U.S. Constitution protects citizens against unreasonable searches and seizures. Both dwellings and commercial buildings are protected under this provision. A routine fire inspection is a "search" under the law. Violation of the Fourth Amendment not only will cause the evidence seized to be inadmissible, but it might also lead to a civil rights suit.

The general rule is that inspectors may not enter a building without a warrant. However, there are a number of exceptions to this requirement:

EXCEPTIONS TO WARRANT REQUIREMENT

- **Consent**. If the owner (or someone with appropriate authority) gives permission, the inspector doesn't need a warrant. Remember that the permission can be withdrawn at any time and, if it is, the inspector must leave. To continue the inspection, a warrant would be needed. If consent is denied, a warrant must be obtained. The only "probable cause" required is to show that the inspection is part of an administrative program being carried out pursuant to law.
- **Exigent circumstances**. During an emergency, warrant requirements are dispensed with. Obviously, firefighters do not need a warrant to enter a burning building to suppress the fire. Cause-and-origin investigations may also be conducted following suppression without a warrant.
- **Public areas**. If an area is open to the public, the owner has no expectation of privacy, so a warrant to enter that area would not be necessary. The fire department would, therefore, have the right to inspect the sales area of a department store without a warrant, but not the personnel office or the warehouse because those areas are not open to the public.

FIRE DEPARTMENT CAN ACT IMMEDIATELY TO AVERT DANGER

The law recognizes that it might not always be practical to follow normal operating procedures. When the situation is life-threatening or might cause irreparable damage if allowed to continue, and it is not practical to get court approval in advance, immediate steps can be taken.

An overcrowded night club must be brought within the safe occupancy level immediately. There is no time to give the owner a hearing or a right to appeal from the inspector's order before it is carried out.

A hazardous construction project being carried on without a permit or approved plans justifies issuance of an immediate stop-work order, or a hazard — a danger of explosion, or a building about to collapse —requires that immediate steps be taken to avert an imminent danger.

The constitutional protections, which normally would apply, are dispensed with. However, as soon as the danger is averted, the owner should be afforded an opportunity to challenge the government's decision. Where appropriate, the fire department should go to court after the action has been taken and the crisis has passed to get a permanent solution approved by the court.

FOURTH AMENDMENT IS BROAD

In 1967 the Supreme Court of the United States held that, if the owner does not give consent, government inspectors must have a warrant to inspect a building.[182] At the same time the Supreme Court held that this protection applies to commercial buildings as well as to residences.[183]

The Court has said that the Fourth Amendment extends beyond the "typical entry into a private dwelling by a law enforcement officer in search of the fruits or instrumentalities of a crime... The officials may be health, fire or building inspectors. Their purpose may be to locate and abate a suspected public nuisance, or to simply perform a routine periodic inspection. The privacy that is invaded may be sheltered by the walls of the warehouse or other commercial establishments not open to the public." The Court said the test is essentially an objective one: whether the expectation of privacy is one which society is prepared to recognize as "reasonable." If there are reasonable expectations of privacy, the inspector must get a warrant.

Some courts have held that an application for a permit or a license is implied consent to enter private property to conduct the required inspection. Find out what the status is in your jurisdiction regarding this exception.

ADMINISTRATIVE WARRANTS

When the Supreme Court held that inspectors could not enter private property without a warrant even though it was for a routine, non criminal "search," it faced a dilemma. It recognized that these routine inspections are essential for the public safety but how could these inspectors get a warrant when they could not show "probable cause" that a crime had been committed or that they would find evidence of a crime during their "search?" The Court created (in the case of *Marshall v. Barlow*), a new type of warrant — an administrative warrant.[184]

Administrative warrants are much easier to obtain than criminal warrants. The inspector only has to show that the inspection is part of a rational, organized program under which inspections are conducted in accordance with law. The statute which gives the authority to conduct the inspection also gives the authority to obtain an administrative warrant.

Having an established procedure and effective forms for obtaining administrative warrants, worked out in cooperation with your attorney and the chief judge of the appropriate court will save you time and trouble, as well as help reduce mistakes and potential liability problems.

CIVIL v. CRIMINAL

In some jurisdictions a fire code violator is given a citation similar to a traffic ticket which reduces the paperwork involved in the process.

Some jurisdictions allow the option of processing a fire code violation under either criminal or civil law. You should determine whether your jurisdiction falls into this category. If it does not, you might wish to consider seeking legislation to authorize these options.

Having the choice is a distinct advantage. Proceeding under criminal law, of course, presents the possibility that a serious offender could go to jail. There is also a greater stigma attached to a criminal conviction. However, in a criminal case, the guilt of the accused must be proved "beyond a reasonable doubt" while in a civil suit the standard of proof is only by a "preponderance of the evidence," a much easier task.

In a civil case, a formal complaint and warrant are not required and the due process considerations are less stringent.

Judges might also be more receptive to bringing someone to answer civil charges for a code violation rather than a criminal charge, particularly if the judge is one who doesn't take fire code violations seriously. (If the government's prosecuting attorney or the judges have this frame of mind, the fire department should engage in an educational effort to convince them that fire code violations are matters of life and death, and the department will only request prosecution of those violators who have persistently refused to bring their properties into compliance with the code.)

Civil procedures also allow a wider variety of tools to use against the violator.

INJUNCTION COULD BE OBTAINED

For example, the department could obtain an injunction, an order of the court compelling someone to do something or to refrain from doing something. Failure to obey the court's order could result in the person being cited for contempt. Contempt is a summary offense for which the person can be summarily jailed without a trial and can be confined there until the judge releases him/her which would normally be when his property is brought into compliance.

MASTER/RECEIVER COULD BE APPOINTED

The Court could appoint a master or receiver who would assume control of the property, make the necessary repairs and bill the owners. If the bill is not paid, it becomes a lien on the property. The master or receiver could also set up an escrow fund, which is useful if the property has rental income. The money collected from rents would go, not to the landlord, but to the court or to the master or receiver appointed by the court. The owner is denied his income until the repairs are made.

CITY COULD MAKE REPAIRS, BILL OWNER

Under another technique, the local government, after getting authority from the court, makes the repairs to the property and is reimbursed from either the income stream of the property or through a lien on the property. Failure to satisfy the lien can result in a forced sale of the property.

EVERY DAY COULD BE A NEW VIOLATION

Sometimes a property owner will pay a fine, but not make the repairs, considering it a small cost of doing business. One way that some communities cope with this problem is by considering every day the property is in violation a separate violation so, instead of one fine being imposed, a fine is levied for every day the property is not in compliance with the code.

CONSUMER PROTECTION LAWS COULD BE USED

Some jurisdictions have used consumer protection laws against violators. The theory is that, when a rental property owner advertises a property for rent, there is an implied warranty that it is fit for human occupancy. If it violates the code, it is not fit for human occupancy, so this constitutes false advertising which is prohibited under the consumer protection laws.

TAX DEDUCTIONS COULD BE STOPPED

Some states deprive the property owner of tax deductions for the operating expenses of a property as long as it is in violation of the code.

SOME COURTS SENTENCE SLUMLORDS TO LIVE IN THEIR BUILDINGS

Some courts have ordered slumlords to live in their substandard property until the violations are corrected.

These types of creative solutions to cope with fire code violations might make your fire prevention program more effective and more efficient. If you do not have the authority legally to do these things, you might wish to seek legislation to enable you to use these techniques.

By understanding the department's legal responsibilities and setting up effective code enforcement procedures, you can make the legal system work for you instead of against you.

CHAPTER EIGHTEEN

FIRE CODES

A code is a compilation of laws, rules and standards pertaining to the same subject matter arranged systematically for easy reference and use. It tells you when and where to do something or to refrain from doing something.

Your jurisdiction probably has a Building Code, Plumbing Code, Electrical Code, Fire Code, etc.

STANDARDS

The code will usually include references to standards for use in evaluating compliance with the requirements of the Code. A standard is a benchmark set up by the enforcement authority against which products, techniques and systems are measured. It describes the requirements and specifications which must be met. It tells you what to do and how to do it, or how not to do it.

For example, your fire code might require automatic sprinklers in certain types of buildings. It might explain how to install the system and list specifications for the equipment. The legislative body in enacting it would indicate when and where the sprinkler systems must be installed. This is the function of the code.

MAKE SURE OF YOUR AUTHORITY

In your fire prevention program one of the steps that should be taken is to study existing codes to make sure they provide the necessary authority and cover all the possible hazards which can be anticipated. If the code is inadequate, steps should be taken to initiate legislative action to remedy it.

PURPOSE OF CODE

The purpose of a community's fire prevention code is to provide a reasonable level of protection for life and property from the threat of a fire, explosion or similar emergency. It sets forth minimum requirements regarding general fire safety provisions, means of entrance and exit, fire-impeding systems and devices, requirements regarding occupancies, use and maintenance of equipment and processes, how and where explosives, hazardous materials and combustible or flammable liquids are to be stored etc. It also specifies requirements for the installation and servicing of fire protection equipment.

BUILDING CODE vs. FIRE CODE

A building code establishes the minimum requirements for the design and construction of buildings to ensure public safety, health and welfare.

Generally, building codes pertain to requirements BEFORE and DURING construction of the building and things which must be done in connection with the construction. Fire codes provide the authority and criteria for periodic inspections of buildings AFTER they have been built. Frequently, conflicts will exist between the building code and the fire code. In these instances, clarification should be sought from the legislative body serving that jurisdiction.

An important part of fire prevention is the plan review of proposed buildings, in concert with other agencies of the government, to ensure that the buildings will not contain any undue hazards which could be prevented. Sometimes various agencies, each looking at these plans from their own perspective, will disagree. For example, the health department might insist that a half inch of ventilation be left at the bottom of lavatory doors to facilitate air flow for sanitary purposes while the fire department might insist that the doors be airtight to serve as a fire block. In these cases, the chief policy maker for the jurisdiction or the legislative body should resolve the conflict.

PERFORMANCE CODES/SPECIFICATION CODES

Codes can be divided into performance codes and specification codes. A specification code spells out in detail the material to be used, the method of assembly etc. The performance code details the objective to be met and establishes the criteria for determining whether or not the objective has been met.

The code provides only minimum requirements.

OTHER TOOLS

Some of the other tools available to the fire department in code compliance are: persuasion, permits, inspections, third-party inspections, show-cause hearings, administrative orders, citations, stop-work orders, court injunctions, temporary restraining orders, criminal prosecution, etc.

AUTHORITY COMES FROM THE STATE

State statutes usually provide authority to local governments to enact fire codes. Provisions in the law empower the local government to mandate correction of hazardous conditions or prohibit those operations which might constitute fire hazards which cannot be adequately regulated to assure safety.

MINI-MAXI CODES

Some states have adopted mini-maxi codes. This means that their code contains both minimum and maximum requirements. Although enforcement of these codes is a local responsibility, the authority for amending the code is usually retained at the state level. A mini-maxi code prevents local governments from providing both less protection than the state code calls for and more protection. It prevents local governments from exceeding the maximum requirements because this might place an undue burden on those who must comply with the code. Architects, builders and other contractors usually favor the mini-maxi codes because they establish uniformity throughout the state and make it easier for them to operate on a statewide basis. The state is often concerned that excessively stringent fire codes might discourage potential employers from locating in the state.

SOME CODE PROVISIONS

The authority to regulate storage and usage of combustibles, explosives, fireworks, electrical installations in buildings etc, is usually specifically provided for in the code.

Some important criminal law provisions in the code include arson, explosives, criminal damage to property through explosion or fire, fraud against insurance companies, etc.

MODEL CODE ORGANIZATIONS

Various organizations prepare model codes on which building and fire prevention codes are based. Some of these organizations are: the American Insurance Association, the Building Officials and Code Administrators International, Inc. (BOCA), the International Conference of Building Officials (ICBO), the National Fire Protection Association (NFPA) and the Southern Building Code Congress International, Inc. (SBCC).

It is the function of these organizations to pool experiences and technical information and compile a body of suggested regulations, materials and systems to help ensure health and safety.

It is time-consuming, expensive and unnecessary for a community to start from scratch to write its own code. It is much more appropriate to adopt a model code or parts of various model codes. These model codes establish sound performance requirements based on the technical expertise of the entire infrastructure of the model code organization — committees, consultants and staff. Model codes have been professionally written with the help of expert technicians and lawyers. The suggested provisions have probably already been tested for reasonableness and enforceability and the code provides uniformity. These codes also offer flexibility with a convenient and economical revision process.

These model code organizations conduct research into new products and technology and this work is reflected in their suggested codes. They also periodically adopt changes to their codes to reflect new developments.

The community might wish to consider adopting both a building and a fire code prepared by the same model code organization to maintain uniformity, thereby helping to avoid conflicts. One model code might refer to a rooming house while a different model code might refer to the same type of structure as a multi-family residence. By adopting building and fire codes from the same organization, this confusion will be avoided.

CODE MUST BE ENACTED

These model codes have no legal authority in themselves. They are merely recommendations. In order for one of these codes to have the force of law in a community, the government of that community must formally adopt it through ordinance. A government might adopt the model code of one of these organizations and thereafter refer to it as that organization's code. The government does not always enact the model code wholesale. It may modify the suggested model code to conform to local conditions or problems or its own perception of what its code should provide. In adopting codes, it is perfectly appropriate for a government to pick and choose provisions from various suggested model codes.

Not only does a model code have no legal efficacy in your jurisdiction until your government adopts it, but even if your jurisdiction has adopted that model code, when that code-writing organization adopts a change to its model code, this change does not go into effect in your jurisdiction automatically. It must be enacted.

Once a fire code is adopted by your jurisdiction, the fire department is charged with the duty for enforcing it.

ADOPTING THE CODE

The legislative body can adopt the model code by reference. In other words, the ordinance which enacts the code cites the specific edition of a particular model code and adopts it as its own. The ordinance must be very specific in citing a particular edition of the code being adopted.

Another method for enacting the code is by transcription. In other words, the model code is actually copied into the ordinance in its entirety. The transcription method is more expensive, but it has the advantage of setting forth the provisions verbatim in the community's own set of laws, making it readily available to the general public. It is simpler and less expensive to adopt the model code by reference unless extensive modifications of it are being enacted. If the code is to be adopted by reference with modifications, those modifications must, of course, be specifically and clearly set forth.

In addition to providing the code itself, model code organizations will also provide model ordinances for enacting the code.

A copy of the code must be available for public inspection both before and after its adoption. Having it available in all fire stations and libraries might be appropriate.

PENALTIES

The adopting ordinance must contain a charging clause which makes it illegal to violate the code's provisions. It must also contain a penalty for each specific violation. Simply because there is a technical requirement in the code does not necessarily mean a penalty is provided automatically. The code must specifically spell out the penalties.

The powers of a court to enforce the violations must also be provided in the law. Different courts have different powers and different jurisdictional authority. Bringing the enforcement action in the wrong court could result in it being dismissed. Following the proper procedures is an essential aspect of effective enforcement.

POLITICS

The process surrounding the adoption of a code or amending it is political. Manufacturers of certain products will lobby the legislative body and the fire department to obtain favorable consideration for their type of product in the code. For example, jurisdictions where cast iron soil pipe is manufactured would probably be under intense pressure to prohibit the use of plastic pipe as an acceptable alternative to cast iron. The foundry which makes the cast iron pipe provides employment in that community, so it is probable that cast iron pipe would be specified in its code rather than allow an alternative type of pipe which is manufactured elsewhere, notwithstanding what the model code recommends. The fire department should be familiar with these political realities as it approaches the legislative or code amendment process.

TECHNICAL AND LEGAL EXPERTISE

Code writing requires technical as well as legal expertise. If these skills are not available within the fire department, consultants are available to assist. The fire department might have the technical capacity and the city or county attorney's office might provide the legal expertise. Help is also available from a variety of professional organizations, including those model code groups listed above. From whatever source, technical people should be available to interpret the developing technology in the context of code requirements while others use their legal skills to interpret the code and propose amendments in proper legal form.

CODE REQUIRES CONSTANT ATTENTION

Constant review and revision of existing laws and regulations are necessary to keep the fire prevention program focused on fire problems facing the community while reaping the benefits of new technology, innovation and change. Someone in the fire department should have the responsibility for staying abreast of developments in new products, technology and model code revisions to anticipate new problems and plan a strategy for coping with them.

Real life experiences might also suggest appropriate amendments to the code. Certain regulations may prove to be impractical, unreasonable or unenforceable. After each fire, after-action reports should be studied to ascertain whether or not the fire could have been prevented if the code had contained certain provisions.

A legislative audit might reveal many confusing provisions, inconsistencies or conflicts in the fire code and other statutes. When such a discovery is made, legislation should be sought to amend the statute. The person in the fire department responsible for keeping abreast of new developments in codes should recommend future amendments to the fire code in response to these new developments.

GETTING CODE PROVISIONS ADOPTED

When the fire department wishes to change the code, the economic impact of the change should be assessed. Those responsible for promoting adoption of the legislation should anticipate where the allies and opponents will come from. For example, builders might oppose certain restrictions which would increase the cost of construction, thereby making the builders less competitive with existing buildings or other communities. It is much easier to succeed in having the legislation enacted if there is little opposition to it and considerable support for it.

The plan to adopt or change the code should be coupled with a program of public education, including public hearings at which the need for the change is explained and the public is afforded an opportunity to express its views pro and con. Your program should include involving citizens and various interest groups in the process to enlist their support or to neutralize their opposition.

Public hearings also enable you to gather information from the public in a formal way regarding the proposal and methods for improving its effectiveness. Public education also leads to more voluntary compliance with the regulations and, therefore, a more effective fire prevention program.

BE REASONABLE, NOT ARBITRARY

In developing and enforcing the code, the objective should be to achieve acceptable levels of risk which the community has determined it is willing to accept.

Unreasonable or arbitrary enforcement of overly stringent requirements may impede the economic vitality of the community by discouraging job-creating companies from locating there. These provisions might also be unconstitutional.

FORMS/PROCEDURES

When the inspector discovers a code violation, the process toward assuring correction should begin. The good manager will develop standardized forms and procedures to make sure that the process not only begins right, but is properly conducted throughout.

NOTICE IS IMPORTANT

The due process clause of the U.S. Constitution requires that violations be cited in sufficient detail so that the violator will have adequate notice so s/he will be able to prepare a defense if s/he desires to do so. Furthermore, if the violator does not know exactly what is wrong, s/he may have difficulty correcting the problem.

Even when the violation is corrected in the presence of the inspector, a record of it should be made to prevent the violator from later claiming that the violation was not cited or that s/he did not know what the violation was. It is possible that, on a subsequent inspection, the same violation might be discovered again so it is important to have a record of the first violation.

GOOD RECORDS ARE ESSENTIAL

Most fire code violations are never prosecuted. Rightly so, because the object of the fire code is to obtain compliance with the code to maintain a fire safe community. But, if the case ever does go to trial, good records will be essential. They also might be helpful in eliminating or minimizing liability. Lack of documentation always makes it difficult to prove something.

The specific procedures for notification required by statute in your jurisdiction should be followed or the charges against the violator might be dismissed.

The notice should state the exact section of the code, statute or regulation which has been violated. Failure to do so could result in dismissal of the case.

SETTING DEADLINE MAY EXPOSE DEPARTMENT TO LIABILITY

While some lawyers for local governments disagree, the author recommends that the violator not be given a specific date by which the code violation must be corrected. The citation should call for it to be corrected "as soon as possible." By giving a specific date, the fire department is in the position of knowing of the dangerous condition and allowing it to exist for the period of time between its discovery and the date when it is supposed to be corrected. Procrastination is part of the human condition and the violator will invariably wait until the deadline is about to expire before beginning to make the correction. This could increase the department's liability if a fire occurs during the grace period. Courts follow a long-standing legal principle that, if someone knows of a dangerous condition and does not correct it, liability will be imposed if an injury results. It is preferable to put the onus on the violator to correct the problem "as soon as possible." Some corrections will obviously require more time than others and the "as soon as possible" requirement will be measured against the "reasonable and prudent person" standard. Make the violator decide what a reasonable time is rather than defining it for him and putting the fire department in the position of knowing about a violation and allowing it to exist for a period of time.

APPEALS

The system must allow for appeals to an independent body and the procedure for filing an appeal should be spelled out in the code itself as well as the powers, duties, membership and procedures of the appellate body. Failure to allow an appeal is a violation of the due process clause of the U.S. Constitution. It is good policy to advise the applicant that s/he has the right to appeal from your decision.

The appellate process must be impartial and fair. It must allow the appealing party a full opportunity to present his/her side of the case. People are entitled to have the decision based on the merits of the case, not on friendship, politics or personal considerations.

VARIANCES

Your code enforcement program should include a sound system for considering variances (also called "waivers" and "equivalences"). By variance we mean a deviation from the specific provisions of the code. If you deviate from the law, you must have the authority to do so. The variance is a safety valve which allows reasonable judgment to control when the literal word of the code is not appropriate.

Most codes allow for "alternative methods and materials." The question to keep uppermost in mind when considering variances is whether or not what is being proposed offers an equivalent measure of fire safety as the specific provision in the code. If the answer is "yes", grant it. If not, deny it. The substituted solution should meet the intent of the code if not its specific provisions. Put your decision in writing so there can be no question later about what is being allowed or rejected and why. Have valid, technical reasons to support your decision, regardless of what it is.

Treat all parties with similar problems the same way. Otherwise your decision will be deemed arbitrary, capricious and unfair and a violation of due process.

Suppose a builder rehabilitating an historic building proposes installing a smoke-and-heat detection system, fire-retardant wall and floor coverings and sprinklers instead of enclosing an architecturally significant staircase. In considering his request, the questions to be asked are: "Do I have the legal authority to accept the substitution?" and "Am I satisfied that the substitute will provide an equivalent level of safety?" If the answer to both questions is "yes," the alternative solution should be accepted. If the answer to either or both is "no," the matter should be rejected. You are entitled to demand proper documentation if you are not sure a proposed solution will provide an equivalent level of safety. You might also want to require the applicant to pay for tests by independent laboratories before you make your decision.

Code variances should not be considered on a hit-or-miss basis. Well thought-out policies and procedures should govern the manner in which they will be considered.

Whatever you decide in a variance case, you should have valid, technically supportable reasons for your decision. In all cases, you should explain the reasons for your decision.

All requests and approvals should be in writing to avoid disputes later over what was agreed to.

Sometimes when you have doubt about your interpretation of the code, or whether or not a waiver of one of its provisions is proper, it might be judicious to defer the decision and refer it to the Board of Appeals (or whatever the appellate body is called in your jurisdiction.) This is in keeping with due process requirements and the Board of Appeals, as a policy-making body with discretion, will undoubtedly have greater immunity than you have.

However, the initial responsibility for evaluating the waiver request lies with the fire department. The Board of Appeals will expect the fire department to have evaluated the proposal thoroughly and professionally before arriving at its decision.

CONNECTICUT COURT UPHOLDS ADMINISTRATIVE WARRANT

In December, 1987 the fire marshal for East Hartford, Connecticut inspected three rooming houses under an administrative warrant after the owner refused him access. When the fire marshal went to inspect the buildings, he could not find the owner so he entered by force. He found several violations and so advised the owner by letter.

The fire marshal later reinspected the premises and, finding the violations still uncorrected, he arrested the owner and charged him with failure to abate fire safety code violations.

The owner was convicted and sentenced to 30 days in jail and fined $2500. The owner appealed.

The appellate court affirmed the conviction, rejecting the owner's argument that the statute allowing the fire marshal to inspect his property was unconstitutional because it allowed an administrative search warrant without probable cause.

The court said that probable cause to issue a warrant to inspect for safety code violations "exists if reasonable legislative or administrative standards for conducting an area inspection are satisfied." Those reasonable standards, the court said, include the periodic annual inspections such as those authorized by this statute. The statute was held to be constitutional and the conviction by the trial court was affirmed.[185] (The U.S. Supreme Court has previously made it clear that probable cause, which is required for a criminal warrant, is not required for an administrative warrant.)

CONSTITUTION DOES NOT BAR HEALTH AND SAFETY LAWS

In 1970 a man obtained a building permit in Boulder, Colorado to enlarge and convert a building into commercial rental space and apartments. The Boulder Fire Department, following inspections in 1986 and 1987, issued an order listing 10 violations of the safety code which were then in effect. The owner was given a deadline of July 1, 1987, to make the corrections and was warned that failure to comply was unlawful.

The owner was also notified that he had the right to appeal to a hearing officer which he did. Following the hearing, he was ordered by the hearing examiner to correct four of the 10 violations.

He appealed the hearing officer's decision to district court. The district court affirmed the hearing officer's decision and the owner appealed again, contending that the building permit issued in 1970 and his reliance on it created a vested right. He claimed that he was constitutionally protected from impairment by a subsequently enacted code which imposed more stringent safety requirements than those which existed when the permit was issued.

The court said that, while a building permit might create a vested right if the permit holder takes steps to rely on it, this does not insulate him from complying with changes in the law which have been enacted subsequently under the state's police power for protection of the public.

The court held further that the U.S. Constitution does not prevent a city from enacting and enforcing ordinances to protect the health and safety of the community.

The owner also claimed that the fire department's order constituted a taking of his property without just compensation in violation of the Fifth and Fourteenth amendments. The court said a taking only results if the government prevents all economically viable use of the property and this was not the case here.

The owner also claimed that he had presented evidence showing that his buildings had some features which provided more protection than the code required and some which provided equivalent protection to the code's requirements. The court said that the hearing officer had viewed this evidence and concluded that there were four violations, so the court will not re-assess the evidence. The decision of the district court was upheld.[186]

LACK OF SMOKE DETECTORS CAUSES DEATHS, BUT CONVICTION OVERTURNED

In a 1987 Connecticut case, a fire resulted in the death of five occupants of a three-story apartment building. There were no smoke detectors in the apartments in violation of the code. One resident testified that, if he had been alerted to the fire earlier, he could have warned the other occupants and saved their lives.

The owner was convicted of eight counts of criminally negligent homicide and three counts of failure to install smoke detectors in violation of statutes and the fire safety code. He appealed on the ground that the fire safety code was unconstitutional because it was vague and violated the due process clause and that he had not grossly deviated from the standard of care of a reasonable and prudent person.

On appeal the conviction for negligent homicide was overturned and remanded to the lower court to render a judgment of not guilty. The appeals court held that the trial court erred in denying the owner's motion for acquittal. The appellate court said he could not be convicted of criminally negligent homicide for violating the statutes and the fire code which required him to install smoke detectors.[187]

TREAT EVERYONE THE SAME WAY

It is worth repeating: all parties with similar situations should be treated the same way. Otherwise the decision could be overturned on appeal and could give rise to a civil suit for acting in an arbitrary and capricious manner. The U.S. Constitution's due process clause requires that all your proceedings and consideration of all requests be fair.

CHAPTER NINETEEN

MEDIA AND THE FIRE SERVICE

T he fire department's Public Information Officer should be particularly concerned with the First Amendment of the U.S. Constitution. The media broadly interpret the First Amendment as giving them virtually unlimited access and unlimited rights. The courts do not agree with them.

The First Amendment reads:

"Congress shall make no law respecting an establishment of religion or prohibiting the free exercise thereof, or abridging the freedom of speech or of the press or the right of the people peaceably to assemble and to petition government for a redress of grievances."

FREEDOM OF THE PRESS NOT AN UNFETTERED RIGHT

While the First Amendment says that Congress shall make no law abridging freedom of speech or the press, this is not a completely unfettered right. It does, however, give news media as well as individuals, the right to speak, write, and criticize government and government officials.

Although free speech and the press are paramount in a free society, there are circumstances when other values take priority and win out in a conflict over rights. There is an on-going struggle in our society trying to balance freedom of speech and press against the need of government to keep some things confidential.

In the *Campbell v. Seabury Press* case, the court said, "The First Amendment mandates a constitutional privilege... First, is the privilege to publish or broadcast facts, events and information relating to public figures. Second, is the privilege to publish or broadcast news or other matters of public interest... The privilege extends to information concerning interesting phases of human activity and embraces all issues about which information is needed or appropriate so that individuals may cope with the exigencies of their period."[188]

Query: What is news? It is more than what the media refers to as "hard news" and includes human interest features as well.

PUBLIC'S RIGHT TO KNOW vs. NEED FOR CONFIDENTIALITY

There is always a need for the department's PIO to balance the public's right to know against withholding information for the public good.

There is often an inherent adversarial relationship between the media and government. Under our system of government, the media have a legitimate responsibility to serve as the people's watchdogs on government. They should not be lap dogs and they should not be attack dogs.

Government and the media will never agree on what should and should not be made available to the media. The media want access to everything, every place and everyone. They claim that the First Amendment encompasses a right to gather government information as well as to publish it.

COURTS DON'T RECOGNIZE MEDIA'S RIGHT OF ACCESS

The courts have not acknowledged this "right of access" under the First Amendment except access to public records, criminal court trials etc.

In *Worthy v. Herter*, the U.S. Court of Appeals said, "Freedom of the press bears restrictions... Merely because a newsman has the right to travel does not mean he can go anywhere he wishes. He cannot attend conferences of the Supreme Court or meetings of the President's cabinet or executive sessions of the committees of Congress."[189]

In *Pell v. Procurier*, U.S. Supreme Court Justice Potter Stewart wrote: "The Constitution does not...require government to accord the press special access to information not shared by members of the public generally."[190]

MANY LAWS RESTRICT FIRST AMENDMENT RIGHTS

There are numerous restrictions on freedom of speech and freedom of the press. The individual's right to a good reputation limits freedom of speech and press through libel laws. Copyright, privacy, obscenity, and other laws also restrict First Amendment rights. Laws regulating business, industry and trade also circumscribe the activities of the media which are, after all, for-profit businesses.

Aside from constitutional considerations, some state statutes give the media special privileges which are not mandated by the First Amendment. Know what they are in your state.

PIO NEEDS TO KNOW LIBEL, PRIVACY, ETC.

Those working in public information should have a working knowledge of libel and slander, invasion of privacy, copyright and other legal aspects of communications.

DEFAMATION

Defamation is a false, derogatory communication to a third person which tends to expose a person to hatred contempt or ridicule, lower them in the esteem or others, cause them to be shunned, or injures them in their business or occupation. (Find out how your state defines defamation.)

Libel is printed, written or broadcast information.

Slander is oral defamation and, therefore, of limited reach.

Court decisions have created protection from libel for the media if the person libeled is a public official or a public figure. In those cases in order to recover, the plaintiff must prove "actual malice" on the part of the media.

"Actual malice" is defined as "with knowledge that it was false or with reckless disregard as to whether it was false or not."

There are five essentials for defamation:

1) Publication
2) Identification
3) Defamation
4) Fault
5) Injury

Libel laws vary from state to state and from time to time.

PRIVACY

Balanced against the Freedom of Information Acts are the Privacy Acts and the law of privacy. Privacy is the right to be let alone, the right of a person to be free from unwarranted publicity. The term "right of privacy" is a generic term encompassing various rights recognized to be inherent in the concept of ordered liberty and such right prevents governmental interference in intimate personal relationships or activities. The right of privacy gives individuals the freedom to make fundamental choices involving themselves and their family and their relationships with others.

In the 1901 case of *Roberson v. Rochester Folding Box Co.* the lower court gave $15,000 to a little girl whose picture was used by a flour company on its packages without her permission. The Court of Appeals reversed, stating there was no precedent established for a right of privacy.[191]

In 1902 the New York legislature created one. It enacted a statute which made it a misdemeanor and a tort to use the name, portrait or picture of any person for advertising or trade purposes without that person's consent.

Two years later the Supreme Court of Georgia judicially recognized the tort of privacy.

LAW OF PRIVACY IS STILL EVOLVING

The law of privacy is relatively new and is still evolving. Some of its provisions appear in various state statutes, some in court decisions. Privacy is not specifically mentioned in the U.S. Constitution.

Some of the matters covered in these state privacy statutes involve:

— The right be let alone
— Right to privacy in family matters
— Protection against intrusions into physical solitude
— Prohibiting publication of private facts
— Forbidding putting a person in a false light
— Prohibiting the appropriation of a person's name or likeness or persona for commercial or trade purposes.

Nearly every state now recognizes in some measure the right of privacy.

The law of privacy is not uniform. There is conflict from state to state and from jurisdiction to jurisdiction.

Some courts have held that public figures have given up to some extent their right of privacy, that persons who seek publicity — actors, explorers, politicians — make themselves "news" and part with some of their privacy.

When courts and legislatures address the law of privacy, they attempt to balance the right to privacy against the public's right to know.

COPYRIGHT LAW

Copyright is the right to control or profit from literary, artistic or intellectual property.

You cannot copyright an idea, only the literary manner or style of expression.

How can you tell if something is copyrighted? The copyright notice which consists of the word "copyright" or the abbreviation for the word or "C" in a circle, the year of copyright and the name of the owner of the copyright should appear on the work.

Essentials for securing a copyright:

1) Putting the notice on the work (see above)
2) Filling out the application and filing it with the Register of Copyrights, Library of Congress
3) Depositing two copies of the work with the Register of Copyrights
4) Paying a fee.

The Federal statute gives five years to register the work.

The copyright law provides exclusive protection for the life of the creator plus 50 years.

PHOTOCOPYING

Some fire departments might be unknowingly infringing the copyright law in copying newsletters. The copyright law prohibits such photocopying of copyrighted material.

Suppose you come across an excellent article in one of the fire service magazines which you think someone else in the department might find useful. You make a photocopy of it and send it to that person. Have you violated the copyright law?

No, because your photocopying the article is a "fair use" of the article. If you made extensive copies for distribution, the answer would be different.

Another example: the fire department subscribes to a newsletter which provides current, valuable information about events and developments in the fire service. You think several others in the fire department should read it. You could route the copy of the newsletter to them, but that will take too long and invariably some recipients would delay reading it and hold onto it for so long that it would no longer be current when the others received it. The newsletter is very expensive and the budget can't stand subscribing for all of the people you think should read it, so you photocopy enough copies to route it to them. Have you violated the copyright law?

Yes. That is not fair use. You have denied income to the newsletter publisher.

Under the Copyright Act, the owner of copyrighted material is given the right to control duplication and distribution of that material. Anyone who copies the copyrighted material without the copyright owner's consent infringes on that copyright, unless the copying falls within one of the Act's exceptions.

Fair use is the only exception that you might be able to take advantage of for photocopying something. The law implies the consent of the copyright owner to a fair use of his material for the advancement of science and art.

What is "fair use" is a question of fact.

In addition to copyright infringement, the tort of unfair competition might also be available for relief of the copyright owner.

FOUR FACTORS WEIGHED IN DETERMINING "FAIR USE"

The courts consider four major factors in determining whether copying constitutes fair use.

1) Whether the use is for such purposes as reviews, criticism, comment, news reporting, teaching (including multiple copies for classroom use), scholarship, research, etc.

2) Nature of the material copied. Making one copy of one page from a newspaper, magazine or journal will ordinarily be a fair use of the work, while copying even one page of material such as test answer forms or workbooks or other material which is intended to be consumed when used, will usually be considered infringement of the copyright for that material.

3) How substantial the copying is. Courts look primarily at the size of the copied material in relation to the size of the work as a whole. They also consider the importance of the content copied in relation to the whole work. Copying one chapter from a book or an article from a journal may be fair use, but a single page from a newsletter might be infringement.

4) Effect on potential value of the copyrighted work. If the infringer sells the copied material or otherwise restricts the copyright owner's ability to obtain income from his copyrighted work, this will not be considered fair use. Regularly copying issues of a newsletter denies income to the copyright owner and is, therefore, prohibited. Making a copy or several copies of an article of special interest in one edition of that newsletter would probably be considered fair use.

Suppose you are preparing a training manual and there is an excellent article from a magazine which you would like to include. How should you proceed? Contact the publisher or owner of the copyright and seek permission, preferably in writing, to use the article. If you take a portion of the article and use it in your manual and put quotation marks on the copied material and give full credit to the copyright owner, that will usually be considered fair use. But, if you use the exact same material and do not use quotation marks and do not give credit to the copyright owner that will not be considered fair use.

PENALTIES

Congress in enacting the copyright law allowed for statutory damages against infringers ranging from $100 to $50,000 per work infringed, together with court costs and attorney's fees. The amount of damages is up to the discretion of the court, but the recovery for the copyright owner can be very large. Let's take that hypothetical case of copying the newsletter and sending copies to others in the fire department. If the copyright owner can prove you duplicated and distributed an entire year's newsletters, the damages could amount to $2.6 million ($50,000 multiplied by 52) plus court costs and attorney's fees.

There can also be criminal penalties for wilful infringement of copyrights for commercial purposes.

How could the copyright owner ever find out that you breached the copyright? Usually a disgruntled employee will inform him/her and serve as a witness at the trial.

FREEDOM OF INFORMATION ACT

PIO's should be familiar with Freedom of Information Acts and Sunshine Laws.

The Federal Freedom of Information Act applies to documents and computer data.

There are provisions for reasonable charges for copying, searching etc.

In addition to the Federal government's Freedom of Information Act, many state and local governments have adopted their own versions of the Freedom of Information Act. If your jurisdiction has such a law, make sure you know what it provides, because the most frequent users of this statutory access to government files and records will be the news media.

As a matter of interest, the following information is exempted from access under the Federal Freedom of Information Act:

— National Defense or foreign policy matters

— An agency's internal personnel rules and practices
— Information which other statutes prohibit disclosing
— Trade secrets and commercial, financial or confidential information
— Interagency or intra-agency memoranda or letters;
— Personnel, medical and similar files, the disclosure of which would clearly be an unwarranted invasion of personal privacy
— Investigatory files compiled for law enforcement purposes
— Material in reports of agencies responsible for regulating or supervising financial institutions
— Geological or geophysical information with data (including maps) concerning wells.

State and local Freedom of Information Acts may have similar exemptions.

RELEASING INFORMATION TO INSURANCE COMPANIES

Privacy protection might expose the fire department to liability problems for the release of information regarding fire investigations to insurance companies. If an insurance company denies payment of insurance proceeds to an individual because s/he has been investigated for arson, an action in libel or slander might be brought by the subject of the investigation against the fire department, claiming s/he was defamed by the department's release of information about an investigation when s/he was charged with no crime.

To insulate itself against such liability, the fire department should make sure that the investigator seeking access to those files produces a waiver of privacy signed by the subject of the investigation. The insurance company might make signing such a waiver a condition of its policy: if the person won't sign, the company won't pay. The fire department should not put itself in the middle on this. Just as presentation of a waiver should be insisted on when someone asks to see medical or personnel records, the same requirement should be applied to insurance investigators.

Some states have passed statutes giving the fire department immunity against liability for exchanging information with insurance companies. If your state has such a statute, the requirements should be followed exactly. If your state does not have one, you might consider putting this on your legislative agenda.

Not only should the fire department insist on seeing a waiver of privacy rights before making information available to investigators, but a photocopy of the signed waiver should be placed in the appropriate file in case it is ever needed to prove that release of the information was authorized.

SUNSHINE LAWS

Sunshine laws require that government meetings be open to the public, not only to the press. The press has no special right to be there beyond what the public has.

There are usually provisions for executive sessions under certain circumstances. Most statutes require minutes or transcripts of those meetings which are closed.

If your state or local government has a Sunshine Law, is there an exemption for meetings held by emergency personnel during a major incident? If not, the news media will have free access to your command center and all your deliberations there. This could lead to a confrontation between fire department executives and the media. This will put the PIO in the unenviable position of arbiter. This potential controversy should be resolved long before the conflict occurs. If you don't like the idea of the media (and the public) having access to your command center, convince the legislature to amend the Sunshine Law to provide an exemption for emergencies.

SHIELD LAWS AND CONTEMPT OF COURT

In general, news people are required to testify in court and reveal their news sources. Many have been cited for contempt of court and sent to jail for refusing to do so.

Many states have adopted "Shield Laws" which protect news people from having to reveal their news sources. (This privilege is usually not available in libel cases.) This privilege is NOT given to the media under the First Amendment, but by a specific state statute.

VIRGINIA COURT ORDERS ABC TO REVEAL SOURCE

In January, 1995, a state judge in Virginia ordered ABC to reveal the identity of a confidential tobacco industry source in connection with Philip Morris' $10 billion libel suit against the television net work. The source, called "Deep Cough," was featured in a "Day One" report which alleged that Philip Morris and other tobacco companies add nicotine to cigarettes to addict smokers. "Deep Cough" appeared on camera in unrecognizable silhouette with a disguised voice. Lawyers for the tobacco company argued that learning the identity of the source was crucial to proving malice. The judge ordered ABC to reveal the identifies of "Deep Cough" and four other sources. (*Washington Post*, January 28, 1995.

LIABILITY FOR NOT WARNING OF RISKS

From a liability perspective, what responsibility do you have to warn the media regarding risks? Will the doctrine of "assumption of risk" relieve you of liability? These are important questions you should discuss with your government's attorneys.

Some years ago a man pulled a van up to the Washington Monument in the Nation's Capital and claimed that the van was filled with explosives. He said, if his demands were not met, he would blow up the Washington Monument. The police warned the media that the only safe area for them, if the explosion occurred, was beyond a certain perimeter. Within a closer perimeter, there would be injuries and within a still closer perimeter there would be almost certain death. The media supervisors gave their employees the option of going closer or staying back in the area of safety. Many took the risk and went into the dangerous areas, gambling — correctly in this case — that the van was not filled with explosives. The police department's warning would help to reduce liability in the event of injury.

WHEN TO KEEP YOUR MOUTH SHUT

The PIO's job is to provide information and answers to the media, but when there is a possibility of a law suit, the PIO should be very circumspect as to what is said.

Some examples:

In the Chicago underground flooding, an employee announced that he had called the dangers to the attention of the city and they were not corrected. This admission impedes the government's ability to defend itself.

Several years ago in Los Angeles County a bridge collapsed. An employee of the county was interviewed on television saying, "I told them that bridge was unsafe. I wrote a memo advising them the bridge was unsafe and I'm sure it's still in county files." Again, the county would be hard pressed to effectively defend lawsuits.

In the 1994 Atlanta sink hole case, someone told the media that an engineering study had shown that the danger of sink holes existed in that area. Having notice of the dangerous condition will invariably lead to liability.

The Salt Lake County, Utah, PIO responded to a fiery auto crash on a winding, county road in which there were a number of injuries. When asked by a reporter to speculate on the cause, the PIO commented that that road had always been a dangerous road. His statement was a red flag for a lawsuit. A county employee (and therefore the county) knew it was a dangerous road. Since the county had notice, why was it not corrected? This is negligence.

Three teenagers were crushed to death at a rock concert at a Salt Lake County-owned arena. Sheriff's deputies speculated to reporters that the kids might have been using drugs. This infuriated the parents who brought suit to clear their children's names as well as for damages. The case was settled out-of-court.

These types of comments make it virtually impossible for the government to defend lawsuits effectively. It is very important for government employees to be very guarded as to what they say in such situations.

ACCESS TO FIRE AREAS

One area of frequent conflict between the media and PIO is restricting access to dangerous areas in the interests of safety.

Some states, including California and Wisconsin, have statutes which allow the media access to disaster sites. These laws, thus far, relate only to public property, but most fires occur on private property. What then? Some courts have addressed this issue.

FLORIDA COURTS ALLOW MEDIA TO ENTER PRIVATE PROPERTY AFTER FIRE

One afternoon in 1972 17-year-old Cindy Fletcher was alone in her Jacksonville, Florida, home when a fire of undetermined origin did severe damage to the house and killed Cindy. When the fire marshal and a police sergeant arrived at the house to investigate, they invited news media to join them. This was their standard practice.

The fire marshal wanted a clear picture of the "silhouette" left on the floor after the body had been removed to show that the body was already on the floor before the fire heat damaged the room.

The Fire Marshal took a Polaroid photo of the outline, but the picture was not clear and he had no more film. He asked a photographer from the Florida Times-Union to take the silhouette picture. The photographer took the picture and it was made a part of the official investigative files of the police and fire departments. It was also published in a Times-Union story about the fire.

Cindy's mother first learned of the fire and her daughter's death from the newspaper story.

She sued the newspaper alleging three things: 1) trespass; 2) invasion of privacy and 3) intentional infliction of emotional distress.

The trial court dismissed one count and granted summary judgment in favor of the newspaper on the others. The trial judge said, "The question raised is whether the trespass ... was consented to by the doctrine of common custom and usage. The law is well settled in Florida that there is no unlawful trespass when peaceable entry is made, without objection, under common custom and usage."

Numerous affidavits had been filed with the court by news media including the Chicago Tribune, ABC-TV News, AP, Miami Herald, UPI, Milwaukee Journal and the Washington Post, stating that "common custom and usage permitted the news media to enter the scene of a disaster."

Mrs. Fletcher appealed from the trial court to the Florida District Court of Appeals, First District. This appellate court held that she should have been able to go to trial on the issue of trespass. The newspaper appealed. The Florida Supreme Court ruled that no actionable trespass or invasion of privacy occurred. The court felt that the entry was by "implied consent".

(Query: How can a dead person give consent?)

The Court said, "It is not questioned that this tragic fire and death were being investigated by the fire department and the sheriff's office and that arson was suspected. The fire was a disaster of great public interest and it is clear that the photographer and the members of the news media entered the burned home at the invitation of the investigating officers.

"Implied consent would, of course, vanish if one were informed not to enter at that time by the owner or possessor or by their direction. But here there was not only no objection to the entry, but there was an invitation to enter by the officers investigating the fire."

(Queries: What gives the investigators the authority to invite someone into someone else's home? Since the owner was dead, how could she withdraw her consent?)

The court held there was no trespass by the news media.

MEDIA HAS NO RIGHT TO ENTER PRIVATE PROPERTY IN OKLAHOMA

However, in *Oklahoma v. Bernstein*, reporters were found guilty of trespass for following protesters onto the property of a nuclear power plant construction site. Despite extensive government support, guaranteed loans, close regulation by the Nuclear Regulatory Commission, etc., the court held it was private property. The protests were newsworthy, but the reporters had no right to trespass on private property.[192]

NEW YORK COURT SAYS OWNER CAN BAR MEDIA FROM HOME

Two television stations accepted invitations to accompany a Humane Society investigator when he served a warrant which authorized him to seize animals which might be found confined in an overcrowded area or in an unhealthy situation or not being properly cared for.

When the investigator served the warrant, the TV camera crew went into the home with him. One of the owners asked the TV people to stay outside the home. They entered anyway and the pictures and story were broadcast on the evening news on WROC-TV and WOKR-TV.

The New York Supreme Court for Monroe County held that the First Amendment right to gather news does not allow members of the media to commit crimes and torts in the course of newsgathering. Reporters are not above the law.

The resident of the house had told the TV crew to stay out of her house, but they did not.

The court distinguished this case from the Florida Cindy Fletcher case, saying this was not a "disaster of great public interest."

"...[O]ne may not create an implied consent by asserting that it exists and without evidence to support it... The entry was made in disregard of plaintiff's express instructions to stay out."

ACCESS TO INFORMATION RE HAZARDOUS MATERIALS

Some statutes and federal law require that certain information about the storage and handling of hazardous materials be made part of the public record and be available to the public. Some information such as trade secrets are exempt.

GOVERNMENT CAN'T MUZZLE EMPLOYEES

Frequently, the government will attempt to put restrictions on employees dealing with the media or in making public comments contrary to fire department policy.

Can an employee be punished for complaining to the media or does this violate the First Amendment's free speech guarantee?

It depends.

In 1983 the U.S. Supreme Court addressed this issue in the case of *Connick v. Myers.* The Court held that, if the speech in question is about a purely internal office matter, the employee can be disciplined, but if the speech touches on a "matter of public concern," it is protected by the First Amendment.[193] The employee's constitutional rights must be weighed against the government's interests. Striking the right balance between these conflicting interests usually gives the courts difficulty, and they do not all agree as to what is permitted and what is prohibited.

In one case, a policeman publicly embarrassed his department several times, complaining that a private security force in a subdivision of the jurisdiction was getting preferential treatment. The officer was given a disciplinary transfer. He brought a suit under Section 1983 of the Civil Rights Act. The lower court granted a summary judgment for the city government.

The U.S. Court of Appeals for the Fifth Circuit, however, reversed the summary judgment. The court explained that the officer's complaints touched on a "matter of public concern" within the meaning of the U.S. Supreme Court's *Connick* decision so they are protected under the First Amendment and he cannot be punished for them. The court said his criticism of what he felt was special treatment for the security force working in the subdivision addressed matters of public concern rather than purely private concerns. The court said that his remarks were not merely criticism directed at public officials, and they did not merely constitute complaints over internal police department affairs. He had claimed, in effect, that a particular segment of the community was being singled out for special treatment, and his outspokenness reflected a view that this favoritism was unfair. The Court said the issue of whether one segment of the community was receiving favored treatment from the police department was a matter of public interest and it is essential that public employees be able to speak out freely without fear of retaliation about such matters.[194]

In another case a clerk in a constable's office, upon hearing that President Reagan had been shot, said to a co-worker, "If they go for him again, I hope they get him." She was fired for the comment. When she filed suit, the U.S. District Court held for the government.

The U.S. Court of Appeals for the Fifth Circuit reversed. The Court said, "We are persuaded that the government's interest in maintaining an efficient office does not outweigh the First Amendment interest in protecting [the employee's] freedom of expression. Her comment was made to a co-worker who, with the benefit of the remark's full context, was not offended. [The constable] specifically stated that he did not fire her because of any disruption caused by her comment, and the evidence did not show that the remark threatened the future efficiency of the office."

The court said the government's strongest argument was that a law enforcement agency should not have to employ someone who advocates breaking the law, but the court said in this case the employee's "duties were so utterly ministerial and her potential for undermining the office's mission so trivial" that the government's interest did not outweigh her free speech rights.[195]

A federal Court in the District of Columbia ruled in 1990 that the D.C. Fire Department's requirement that an employee obtain approval before giving statements to the media was unconstitutional.[196]

In 1991 a federal court struck down a city ordinance which prohibited employees making speeches or granting interviews with the media.[197]

A New York court also struck down a Sheriff's Department rule which prohibited deputies from giving interviews to the media.[198]

The Pierce County, Washington, Fire Marshal maintained that he was fired by the County Executive because he opposed passage and enforcement of ordinances which he felt were unlawful and dangerous because they conflicted with fire flow and private road standards, threatened life and property and exposed the county to potential liability. He expressed his views to the County Council and other officials as well as to the general public. The County Executive claimed that he had fired him because of the Fire Marshal's managerial shortcomings.

The Fire Marshal sued the county and the County Executive, alleging violation of his rights under the First and Fourteenth Amendments. The defendants moved for summary judgment on the basis that they were immune. When this motion was denied, the County Executive appealed.

The court said that state and local officials are only entitled to immunity if they did not violate clearly established statutory or constitutional rights. The defendants would only be entitled to summary judgment on the basis of immunity if there was not sufficient evidence to create a genuine issue in dispute.

The court said that the communications for which the Fire Marshal was allegedly fired were matters of public concern and were protected by the First Amendment. Firing an employee in retaliation for speech which is protected by the U.S. Constitution would be a violation of clearly established law about which a reasonable person should have known.

The County Executive contended that his interest as an executive in promoting the efficiency of the government outweighs the Fire Marshal's First Amendment rights and, therefore, the County Executive should be immune.

The Court found that the Fire Marshal had offered substantial proof that he had been fired in retaliation for his speaking out against the ordinances. Since the County Executive's motives were in factual dispute, the matter could not be resolved without a trial, so the denial of the motion for summary judgment on the basis of immunity was proper. The case should be brought to trial.[199]

A U.S. District Court in Washington, D.C. enjoined the disciplinary transfer of a fire captain and an order that he submit to a psychiatric evaluation. The fire chief ordered that a cartoon be removed from fire houses. A fire captain, serving as an acting battalion chief, objected to the order and protested to the chief, using the prescribed chain-of-command channel. His complaint was considered impertinent and an assistant chief transferred him to a non-command position in the Property Section as punishment and he was declared ineligible for overtime and was instructed to submit to a psychiatric evaluation.

The captain filed suit, alleging deprivation of his constitutional rights. The court held that his protest "was clearly on a subject of public concern" and he was, therefore, entitled to a preliminary injunction.[200]

A former firefighter argued that his First Amendment rights were violated when he was fired for making public comments about the operations of the fire department. The trial court found that the city did not have to tolerate such disruptive behavior and ruled against him. The U.S. Supreme Court denied certiorari, in other words, declined to review the case, which leaves the decision undisturbed.

OHIO COURT UPHOLDS FIREFIGHTER'S RIGHT TO SPEAK AT PUBLIC MEETINGS

An Ohio firefighter attempted to make a statement at a public meeting of the city council, but was told by the city manager that, if he wanted to speak about fire-related issues, he had to go through the chain of command. The firefighter protested and insisted that he be allowed to address the council on a fire-related issue. He was later suspended without pay in violation of employment rules and a charter provision which required that the city manager handle all communications with employees.

The firefighter filed suit in federal court.

The court said, "When the people through its [sic] government create a public forum for discussion of issues and the dissemination of public opinions, that government may not limit the forum to discussion by some and not to all of those who wish to speak."

"Specifically, here the government may not prohibit government employees from speaking when the prohibition is based solely on the speaker's status as an employee...."

"...[T]he government... may not selectively prohibit a citizen from speaking at a public meeting based upon the content of the speech or upon the speaker."

The U.S. District Court cited *Local 2106, IAFF, v. City of Rock Hill*, where a South Carolina State law insulated the city council from any dealings with employees who were supervised by the city manager. In that case the U.S. Court of Appeals said the South Carolina law was a "content-based limitation on free speech" and was "an impermissible classification under the equal protection clause."[201]

In the Ohio case at bar, the judge held that the suspension of the firefighter was in retaliation for his attempted exercise of his First Amendment right of free speech, but the firefighter received only nominal damages because he could not prove any compensable loss.[202]

The key point on which these cases turn is whether or not the public comments are about mere routine, internal operational matters or whether they are of public concern. In the latter case, they cannot be prohibited and in the former they can.

CHAPTER TWENTY

HEALTH AND SAFETY

OSHA

The Occupational Safety and Health Act of 1970 was the first federal statute regulating workplace safety. The federal jurisdiction is obtained if the company affects interstate commerce which virtually all enterprises do.

In general, OSHA does not apply to local and state fire personnel. States adopt and operate their own health and safety standards. If those standards are at least as high as those which the federal government has adopted, the states are entitled to receive federal funding on a 50% to 50% basis. Sometimes the state's health and safety rules and regulations are more stringent than those of the federal government. As has been stressed repeatedly in this book, it is important to know your state's specific statutes. In the 23 states where these agreements with the federal government have been made, OSHA's industrial fire safety standards for private companies are applicable to state and local governments.

In the absence of the cooperative agreement between the states and the federal government, OSHA's rules and regulations have no applicability to state and local fire departments.

In 1994, legislation was introduced in Congress which would have placed state and local governments under OSHA's regulation. There was great concern in the fire service about how this would have substantially increased liability for fire chiefs, incident commanders and others in supervisory roles. Volunteers might also have been covered.

Under the provisions of the bill, called the Comprehensive Occupational Safety and Health Reform Act (COSHRA), the jurisdiction of OSHA would have been expanded to include all public sector employees, including fire departments. The bill proposed to increase the potential fines for willful violations of OSHA's standards that result in a fatality up to $250,000 for an individual and up to $500,000 for an organization. If the violation of OHSA standards resulted in a death, criminal penalties would be increased from the six months to 10 years for the first offense and 20 years for the second offense. If a willful violation resulted in bodily injury, a fine of $250,000 and 5 years in prison could be imposed on the individual and the employer would be prohibited from paying the fine on behalf of the offender. Supervisors, line officers and other management officials who violated the OSHA, or who directed someone else to do so, could be held criminally liable if death or serious bodily injury resulted.

At this writing, this OSHA "reform" appears to have no chance of being enacted by the new Republican-controlled Congress. However, fire service personnel should keep a wary eye on future developments.

OSHA attempts to protect safety and health of employees in two ways:

1) It imposes a general duty on covered employers to prevent workplace hazards which might cause death or injury to workers and

2) It authorizes the Secretary of Labor to establish health and safety standards with which employers must comply.

Pursuant to the latter authority, the Secretary of Labor has set maximum levels of exposure for certain hazardous substances.

The law requires employers to report within 48 hours on-the-job deaths and injuries which require hospitalization. Employers must also keep a log of work-related deaths, injuries and illnesses.

The Department of Labor's Occupational Safety and Health Administration enforces the Act. Inspectors may inspect places of employment at any reasonable time, but if the employer does not consent to the inspection, the inspector must obtain a warrant.

Sometimes these inspections occur in response to a complaint by workers, but they also may be on a random, spot check basis by OSHA itself.

Union representatives, if any, as well as the employer have the right to accompany the OSHA inspector as s/he checks the facility.

If the employer requests the inspection and violations are found, no penalties are imposed unless the employer fails to correct the deficiencies which were discovered by the inspector.

Citations can carry fines and even imprisonment. Appeals from a citation are heard by an administrative law judge (hearing examiner) of the Occupational Safety and Health Review Commission. The U.S. Courts of Appeal hear appeals from the decisions of the Commission.

The OSHA statute allows the states to develop and enforce their own health and safety standards.

There may be a violation of OSHA even if no accident occurs. The intent of the law is to set new standards of workplace safety.

Even if the employer might have a good defense, such as contributory negligence, to an employee's suit, this would be irrelevant to an action by OSHA.

OSHA HAS REGULATIONS RE BLOODBORNE PATHOGENS

OSHA has issued regulations regarding exposure to bloodborne pathogens (29 CFR 1910.1030) and has adopted the position of the Centers for Disease Control which assumes every direct contact with certain human body fluids is infectious and that all individuals who come in contact with these body fluids should be protected as though the fluids were infected.

Employers are required to provide appropriate protective equipment and or clothing, including hypoallergenic gloves, gowns, laboratory coats, face shields, masks, eye protection, mouthpieces, resuscitation bags, pocket masks, ventilation devices etc. Employers must also provide for decontamination, cleaning, replacement and disposal of exposed personal protective clothing and equipment. Training is required and the training materials and program must be appropriate for the educational level of the employees.

The fire department should make an exposure determination and

establish an exposure control plan which identifies all tasks, procedures and positions in which contagious exposures might occur. The
employer also must establish a written infection-control plan, including the exposure determination and the schedule and methods for
reducing the risks of exposure.

CONGRESS PASSES LAW ON INFECTIOUS DISEASE NOTIFICATION

In 1991 Congress passed a law (PL 101-381) regarding infectious
disease notification. Under this law, emergency responders can learn
if they had been exposed to an infectious disease while they were
providing emergency medical care. This will enable them to obtain
medical treatment and take precautions to prevent further transmission of the disease. The Public Health Officer for each state must
appoint a Designated Officer for each employer of emergency responders. This Designated Officer has the responsibility to oversee
the notifications required by the Act. Hospitals are required to notify
this Designated Officer of any patient who has been transported by
the department's EMS or ambulances who is found to have an
infectious disease.

Firefighters who think that they might have been exposed to an
infectious disease must report the incident to the department's
Designated Officer who, in turn, will determine the likelihood of the
exposure. If s/he feels an exposure could have occurred, s/he contacts the hospital where the patient had been transported and provides all the details of the incident. If the hospital agrees that exposure is possible, hospital personnel will notify the Designated Officer
within 48 hours if the patient has an infectious disease which could
have been transmitted during the contact with fire department
personnel.

Needless to say, all medical information involved in this process
should be kept strictly confidential.

States that have their own notification programs which are at
least as stringent as the federal requirements may apply for a waiver
from the federal regulations. If a state does not apply for a waiver,
the federal program will be applicable in that state.

NEW TB CASES SURGING

Tuberculosis (TB), long felt to have been conquered by medical science, is making a surging comeback and to make matters more serious, the new strains are drug-resistant.

OSHA enforces a Tuberculosis Safety Standard (29 CFR 1910.134) which requires a minimum level of respiratory protection to prevent the transmission of TB where a person who is known or suspected of having the disease is being housed or transported. As part of its industrial hygiene inspections in high risk workplaces, OSHA checks for occupational exposure. The high risk workplaces include health care settings, emergency medical services, correctional institutions, homeless shelters, nursing homes, drug-treatment centers and home health care providers.

The OSHA standards and Centers for Disease Control guidelines call for employees to be provided and be required to wear NIOSH - approved, high efficiency particulate air respirators to the minimum level of respiratory protection.

The respiratory protection program is mandated for all employees who transport a person who is known or suspected of having TB.

Obviously, the possibility of transmission of a disease from a patient to a rescuer is very great because firefighters, emergency medical technicians and paramedics routinely respond to situations where they are directly exposed to blood and other body fluids. Therefore, universal precautions should be taken. This means it should be assumed that everyone these personnel come in contact with is infected. This precaution gives the utmost protection to responders without diminishing patient care.

Employers in violation of the guidelines are subject to fines.

Most state and local governments have adopted their own protection programs which include: identifying infected persons, protective clothing and equipment, training, medical surveillance, protection of employees and confidential record keeping.

Fire departments should develop policies, procedures and training programs to minimize risk of exposure for all personnel. Failure to do so could result in liability for the department.

State workplace safety laws as well as OSHA require that employers maintain a safe workplace and warn employees of hazards.

In this context, it would seem permissible for fire departments to warn employees either formally or informally when someone with whom they are about to come in contact has a contagious disease.

CONGRESS SCRUTINIZES OSHA

At this writing (March, 1995) Congress is taking a close look at alleged abuses by OSHA with a view toward reducing the regulatory burden on employers.

FIRE DEPARTMENT OCCUPATIONAL SAFETY AND HEALTH PROGRAM

The National Fire Protection Association's 1500 Standard on Fire Department Occupational Safety and Health Program requires fire departments to create an official written risk management plan which examines the nature and scope of risks likely to be faced and sets down procedures for each risk. The plan must cover administration, training, vehicle operations, protective clothing and equipment, operations at emergency and non-emergency incidents, and other related activities.

Four elements make up the process of risk management, according to NFPA:

1. Risk identification. In this phase, a department should list all potential problems it has encountered or may encounter, and problems identified in facility or apparatus inspections.
2. Risk evaluation. Each risk identified should be reviewed for its likelihood of occurring and the potential severity and expense if it does occur.
3. Risk control. If possible, the identified risks should be eliminated or avoided altogether. If that is not possible, then appropriate procedures, training and inspection should be implemented to control the risks and reduce the likelihood that they will occur.
4. Risk management monitoring. The steps being taken to control risks should be reviewed periodically and changed if they have not been effective in reducing risks.

CHAPTER TWENTY-ONE

AIDS AND THE FIRE SERVICE

HIV/AIDS CAUSE MASSIVE LEGAL AS WELL AS MEDICAL PROBLEMS

The Human Immunodeficiency Virus (HIV) which causes Acquired Immune Deficiency Syndrome (AIDS) has given rise, not only to a fearsome epidemic, but also has spawned more lawsuits than any other disease in American history. This creates a complex interplay between issues of public health, constitutional rights, personal relationships, the health and lifestyle of individuals, medical testing, public policy, needle exchange by illegal drug users, the health and safety of employees, gay rights, social services, government-financed research, privacy, discrimination and legal liability.

The HIV virus destroys the body's immune system, causing AIDS. Someone may test positive for HIV, but have no symptoms for many years after becoming infected. A person who is actually HIV positive may show negative results in a test, especially in the early stages. It may take someone from one to six months to show signs of the HIV virus after being exposed to it. During this time, the infected person may show negative results on a screening test, but be fully capable of transmitting the disease to others. The average time from HIV infection to developing AIDS is eight to eleven years.

Virtually all HIV-infected persons will eventually develop AIDS and will remain infected for life. There are no known cases of anyone infected with HIV successfully overcoming the disease. Since there is no known cure, HIV-positive persons face almost certain death.

The Centers for Disease Control defines AIDS to include HIV-positive persons with no symptoms who have a CD4 lymphocyte count of 200 or less per cubic millimeter of blood. The CD4 lympho-cyte count is a measure of cells that are important to the body's immune system.

HIV is transmitted through sexual contact — homosexual or het-erosexual — through exchange of contaminated needles, through ex-posure to infected blood or other body fluids or from mother to fetus.

FIRE SERVICE PERSONNEL ARE PARTICULARLY VULNERABLE

Because fire service personnel are particularly vulnerable to exposure to the disease and the far-reaching legal implications of the disease, the issues involved should be given the closest scrutiny by everyone in the fire service. Those who remain ignorant about these problems or choose to ignore them potentially expose themselves to substantial legal damages.

How should the fire department proceed in balancing the duty to afford privacy protection to those infected by HIV and the duty owed to employees to protect them from transmission of the disease? A very important question with vast policy and legal ramifications.

The legal issues, both judicial and legislative, are still evolving and different jurisdictions vary greatly in how they handle the matter. With this point in mind, readers should be aware that changes in statutes or judicial interpretations might negate the information contained herein.

SOME AREAS FOR LIABILITY CONCERN

Liability may arise from unauthorized testing, unauthorized disclosure of HIV test results, discrimination against infected persons, taking some inappropriate administrative action against a person who is HIV positive and for failure to adequately prevent transmis-sion of HIV.

The challenged testing program of the fire department could be nullified by the courts and not only would the department not be able to use the information gained from an impermissible test but the department and its employees could also be held liable for violating the constitutional rights of the person tested.

DILEMMA FOR FIRE DEPARTMENTS

The dilemma for fire departments is that they may face liability for breach of confidentiality in disclosing information, but they also may face liability for failure to disclose the information.

A government agency was held liable for using red stickers to identify HIV-positive prisoners. The court held that this violated the prisoner's right of privacy and may encourage discrimination against those infected.

On the other hand, a government can be found liable for not warning others who are at risk of being infected by someone who is HIV positive. A hospital lost a suit by a security guard and was assessed $1.5 million in damages for failure to warn him that a violent patient he was trying to subdue was HIV-positive.

TESTING PROGRAMS FOR HIV/AIDS SHOULD BE NARROWLY SHAPED

Testing programs should be narrowly shaped to achieve an important governmental interest since taking blood is intrusive on privacy and the results obtained may be highly sensitive. The courts are most likely to prohibit mandatory testing programs unless the government can show that the infected individuals would not be able to perform the functions of their jobs or that they pose a significant risk to others.

Medical examinations must be voluntary, job related or be a business necessity. The fire department would have to demonstrate that being free from the HIV infection is essential to the fire department's business.

The law in many states is that the testing must be voluntary. Mandatory testing is usually only authorized where the risk to others justifies the invasion of the tested person's rights. The legislation usually reflects sympathy for firefighters, police, EMT's and victims of crime who might be exposed to the disease. Mandatory testing might be allowed before the risk of exposure occurs or following an incident in which the health care worker, police officer or firefighter might have been exposed to the disease. A test will usually be allowed where a health care worker can show that s/he was, or might have been, exposed to a patient's blood under conditions in which HIV transmission could have occurred. Usually employee testing is allowed only when being free of the virus is a bona fide occupational qualification.

Even though the state might authorize a testing program, there are usually confidentiality or procedural requirements the breach of which could lead to liability. The existence of a state statutory authorization does not necessarily immunize the department from a federal constitutional challenge.

Most general employee screening tests where the risk of transmission of the disease is nonexistent and the test is not job related or a business necessity would contravene federal and state statutes because the governmental interest would not justify the intrusion on the individual's privacy.

Notwithstanding the District of Columbia firefighter applicant case discussed immediately below, most courts recognize that HIV status is important to police, prison officers, firefighters and emergency medical technicians who may be at increased risk for HIV infection. It could also be argued that public safety and health care workers pose a risk of transmission to others if they are injured in the course of their contact with them. These workers also work in a highly regulated environment in which regular medical examinations are routinely used.

Most states require informed consent for HIV testing. Most states also have statutes governing the confidentiality of AIDS information, but most include exceptions as to when the information may be disclosed to protect third parties who may be at risk. Most states also have statutes prohibiting discrimination against HIV/AIDS-infected persons. The Americans with Disabilities Act specifically includes HIV/AIDS as disabilities which are protected under the Act. Most states also treat HIV/AIDS as a protected handicap.

Governments must balance the risk of invading an individual's right of privacy against the need to conduct the test. Both mandatory and voluntary testing programs can be a source of potential liability. If the test is consented to, there will be less liability risk. However, in devising a testing program, state statutes should be carefully reviewed to insure compliance. Failure to comply with the statutory requirements will invariably lead to liability.

The HIV-related information gathered during the test entails additional liability risk. Certain medical information must be kept confidential and may only be breached for a greater societal interest in disclosure of the information. Even then, the information may not be disseminated more widely than is absolutely necessary to accomplish the legitimate purpose.

Some states allow mandatory testing of a patient in advance of a procedure which could put the health care workers at risk of HIV transmission. Some states allow testing following a possible exposure in which the health care worker, police officer, firefighter or emergency medical technician has probable cause to believe that s/he might have been put at risk of transmission. Firefighters, EMT's, police etc. may get a court order to compel the testing of a person who might be HIV-infected.

It is clearly in the legitimate interests of fire departments to test in order to reduce the risk of transmission to personnel, to ensure that an employee is capable of performing the functions of his/her job without risk to others, to take appropriate public health steps and to provide medical care or benefits.

FIRE DEPARTMENT PERSONNEL SHOULD BE WARNED

Notwithstanding the liability risk for disclosure, fire departments should devise some system for warning their employees when they are to be exposed to HIV/AIDS risk. At the same time, however, care should be taken to restrict disclosure of this information to minimize infringement of the individual patient's privacy rights. Perhaps the department could adopt an internal code to designate HIV-positive persons, but scrupulously avoid having those outside the department learn of this information.

CONSTITUTIONAL CONCERNS

If a person is subjected to testing without his/her consent, s/he may raise objections under the Fourth, Fifth and Fourteenth Amendments of the U.S. Constitution. To successfully defend against these challenges, the government must show that the testing program is reasonable and is designed to achieve an important governmental purpose.

The courts have consistently held that collecting blood and urine is a "search" as covered by the Fourth Amendment's protection against unreasonable searches and seizures. The question to be addressed is: "Is the individual's justifiable expectation of privacy outweighed by the government's purpose in doing the testing?"

The courts will consider all the surrounding factors including the manner in which the test is conducted, whether the occupation is one which is highly regulated and the risks posed by others.

COURTS UPHOLD TESTING OF FOREIGN SERVICE OFFICERS

In the case of *Local 1812, American Federation of Government Employees v. U.S. Department of State*, the Department's policy of testing foreign service applicants and employees for HIV/AIDS was challenged. The union argued that the testing was an unreasonable search and seizure prohibited by the Fourth Amendment. The government argued that because of inadequate medical care at some foreign posts, HIV-infected individuals would not meet the statutory requirement for foreign service officers that they and their dependents were capable of assignment anywhere in the world. The court refused the plaintiff's request for a preliminary injunction against the program holding that the testing program was rational and closely tailored to ascertain the applicants' and employees' fitness for duty in a specialized government agency and the invasion of privacy was minimal because the test only involved an additional examination of a blood sample which the individuals were required to give under a system which had been in place for many years.[203]

A different conclusion was reached in the case of *Glover v. Eastern Nebraska Community Office of Retardation*. In that case a mental retardation agency subjected its employees to testing for HIV antibodies to prevent HIV transmission from employees to patients. The U.S. District Court said that "even if staff members were infected with a chronic infectious disease, the risk to clients is extremely low and approaches zero." The Court said, "Such a theoretical risk does not justify a policy which interferes with the constitutional rights of staff members." The U.S. Court of Appeals affirmed this holding.[204]

TESTING FIREFIGHTERS FOR AIDS UPHELD

A similar testing program was upheld in the case of *Anonymous Fireman v. City of Willoughby*. A firefighter sued the Fire Department, claiming that subjecting him and other members of his squad to suspicionless HIV-testing as part of an annual physical fitness medical checkup infringed his constitutional rights as an unreasonable search and seizure and an invasion of his privacy. The City defended in part by stating that the firefighters' union had waived the plaintiff's rights in a collectively bargained agreement which allowed the medical exams. The city pointed out that firefighters and paramedics can be exposed to contaminated blood, mucus and saliva and, if the employee has skin punctures or lacerations, infection of the HIV virus could result. The U.S. District Court considered whether this violated the Fourth Amendment rights of firefighters and paramedics. The court said, even though the test was conducted on blood drawn for legitimate purposes, the test did constitute a "search" under the Fourth Amendment.[205]

The court also said that a union cannot bargain away the constitutional rights of its members. However, the legality of the test was upheld. The court said the individual privacy interests of employees were slight and were outweighed by the city's compelling interest in protecting the public from the transmission of the disease from firefighters and paramedics. The court upheld the tests because the employee's expectation of privacy was not very great since blood tests are very common and involve little trauma and because of the highly regulated nature of firefighters' employment. The court said firefighters are at high risk "for contracting and/or transmitting the HIV virus" so the mandatory testing for AIDS is a reasonable search, even without suspicion of an individual. The judge said the governmental interest in containing the epidemic justified the testing.

It should be noted that nothing in the judge's decision indicated that the Fire Department could fire an HIV-positive firefighter, but the judge said that, where there is a probability that others might become infected, the infected employee could be "assigned to other non-high-risk work where the chance of transmission is mitigated or eliminated."

The reader is cautioned that the findings enunciated in this *Willoughby* case are not universally accepted by other courts.

DECISIONS OF COURTS ARE CONTRADICTORY

A number of courts have held that employees with AIDS can be denied employment. Other courts have stated they may not be discriminated against in hiring because they do not constitute a threat to the health and safety of other workers.

D.C. FEDERAL COURT ORDERS THAT HIV-POSITIVE APPLICANT BE HIRED

A federal judge in Washington, D.C., in the case of *Doe V. District of Columbia* ordered the D.C. Fire and Emergency Medical Services Department to hire a firefighter who had disclosed that he was infected with AIDS. The firefighter applicant had challenged the Fire Department's withdrawal of a job offer after he voluntarily told the Department that he was HIV-positive.[206]

In a deposition, the fire chief said that the Department's policy against hiring applicants with AIDS was based on "public perception" and that he would "be crazy not to" take into consideration public sentiment about AIDS.

The judge said that an HIV-positive firefighter poses virtually no risk to the public, even if he must perform mouth-to-mouth resuscitation. The court gave the plaintiff $25,000 in damages for pain and suffering due to the city's intentional discrimination, back pay running from the date in February, 1989, when he had been scheduled to begin work up to the day of the judgment, and ordered that he be hired.

The judge accepted the testimony of experts — infection control specialists — that the man would pose "no measurable risk" to his co-workers or the general public. (The case was decided on the basis of the Rehabilitation Act of 1973.)

Subsequently, the man decided he did not want to take the Fire Department job.

COURT ALLOWS CHANGE OF ASSIGNMENT FOR HIV-INFECTED FIREFIGHTER

This District of Columbia case, however, should be compared with *Severino v. North Fort Myers Fire Control District*, where the court said that regular firefighter duties involve risk of HIV transmission so removal of an HIV-positive firefighter from rescue assignments was a reasonable accommodation.[207]

SHERIFF FORCED TO RE-INSTATE VOLUNTEER DEPUTY WITH AIDS

The Sheriff of Weld County, Colorado had to reinstate a volunteer deputy with AIDS who had challenged his firing in federal court and won a jury verdict against the county. The Sheriff said the volunteer deputy's partners refused to ride with him because they were afraid they might become infected with the HIV virus. The Sheriff commented, "You can't train them to be totally accepting of an individual" with AIDS. Perhaps, but you had better TRY to train them because that is something which the courts will look at when assessing liability against the Department.

ESTATE OF EMPLOYEE WITH AIDS GETS $500,000

The New York Human Rights Division ordered an employer to pay $500,000 to the estate of a man who died of AIDS. The employee had alleged that he was fired after he had developed facial lesions caused by Karposi's Sarcoma, a disease which attacks persons with immune deficiencies such as AIDS. The employer claimed it had no knowledge that the employee was sick and claimed the firing was for inadequate job performance. The facial lesions were patently obvious and the employee had previously received favorable job evaluations so the employer's claims lacked credibility as far as the Civil Rights Division was concerned.[208]

The $500,000 award was for back pay, mental anguish and humiliation.

MEDICAL RECORDS MUST BE KEPT CONFIDENTIAL

Fire departments should be cautious as to how they treat medical records. No records should be released except upon a court order or a signed consent form from the subject of the file, waiving the right of privacy in the records. Insurance companies will usually require their insured to sign such a waiver, but fire department personnel should make sure that no records are released in the absence of such a waiver and a photocopy of the signed consent form should be kept in the person's file.

A New York appellate court approved the filing of a suit against a doctor who disclosed to an employer that one of its employees had AIDS. The employee went to the doctor for treatment of a job-incurred ear and sinus infection and during the course of the visit he told the physician that he was HIV-positive. He asked that this information be kept confidential.

Subsequently, the doctor responded to a subpoena in connection with a contested worker's compensation claim and forwarded the man's entire chart to the employer's attorney, revealing the information about his having AIDS.

The employee filed a suit against the doctor, charging invasion of privacy and breach of contract.

The appellate court held that the plaintiff had stated a cause of action for disclosure of confidential information in violation of a statute protecting HIV victims. The employee had authorized the release of medical information in connection with his worker's compensation claim, but disclosure of the AIDS information was extraneous to his claim and, therefore, unwarranted.[209]

Similarly, in *Urbaniak v. Newton*, a California court held in favor of a plaintiff who had brought a worker's compensation case based on head, neck and back injuries sustained while working as a machine operator. At the request of the defendant insurance company, he was examined by a neurologist. During the examination, electrodes that drew blood were fastened to his body. Because he was concerned about exposure to the medical worker, the plaintiff warned the nurse to be careful in handling the electrodes because he was HIV-positive. He asked that this information not be placed in his medical report. The neurologist, however, in writing his report attributed the plaintiff's muscle tension to his concerns about being HIV positive rather than to the injury he received on the job. The report containing this information was given to the defense attorney, the insurance company, the Worker's Compensation Appeals Board and the plaintiff's chiropractor. The court held that the plaintiff's expectation of privacy had not been reduced by his disclosure to protect the safety of health workers, that this disclosure of the information did not diminish his right of privacy. The court said his constitutional right to privacy had been violated by the improper use of the confidential information about his HIV status.[210]

Doe v. Borough of Barrington held that the Fourteenth Amendment protects a person's privacy interest in not having the government disclose that a person is infected with AIDS. The court said that because of the sensitive nature of the information about AIDS the information should be kept confidential and should not be revealed unless the government has a compelling interest in doing so.[211] (Revealing the information to firefighters and EMT's who might become exposed to the disease is, in the author's view, a sufficiently compelling reason for revealing the information to protect employees.)

In the *Barrington* case, the court also held that the plaintiff's constitutional right of privacy was violated because the police department had failed to train its employees about AIDS and to inform them to keep the identity of AIDS carriers confidential.

REPORTING REQUIREMENTS VARY FROM STATE TO STATE

Under the common law, physicians were generally under a mandate to disclose information to prevent the spread of communicable diseases. Requirements for reporting HIV/AIDS information vary greatly from state to state. Some states prohibit reporting that someone is HIV-positive. In one state a physician learned that a patient who worked as a butcher for a food chain was HIV positive. He was prohibited from reporting this information to public health officials even though the risk of the man cutting himself and contaminating meat would seem to be very great. Other states require the reporting of such information to public health officials.

Some states require that the HIV infection be reported only statistically, but others require that the person's name and address be reported. The states also vary greatly in how much protection must be afforded the information reported, provisions for disclosure under particular circumstances and penalties for unauthorized disclosure. Some states provide criminal as well as civil penalties for unauthorized disclosure of HIV/AIDS information.

Collection of information about HIV/AIDS is generally permissible if sufficient safeguards are employed to keep the information confidential. In *Whalen v. Roe* and *Thornburgh v. American College of Obstetricians and Gynecologists* it was held that any information gathered by governments about HIV/AIDS must respect the confidentiality and privacy interests of the patients and the information must be protected from public disclosure.[212] As always, the rights of the individual will be balanced against the needs of the government.

EMPLOYER MAY REDUCE HEALTH INSURANCE OF AIDS-INFECTED EMPLOYEE

Employers may reduce health insurance and other benefits for AIDS-related claims without incurring liability under the Employee Retirement Income Security Act (ERISA) when the employer's reasons for modifying the plan are to contain costs for AIDS expenses, enabling the employer to provide some level of insurance to all its employees.[213]

EMPLOYER CAN ELIMINATE THREATS TO HEALTH AND SAFETY

Neither the Rehabilitation Act nor the ADA protects individuals who pose a threat to the health or safety of him/herself, the public or fellow workers. If, however, an individual does pose a direct threat because of his/her disability, the employer must determine whether a reasonable accommodation would either eliminate the risk or reduce it to an acceptable level. If no accommodation exists which would eliminate or reduce the risk, the employer may refuse to hire an applicant or may discharge an employee who poses such a threat.

For additional information on this subject, readers are referred to "AIDS and Governmental Liability" published by the American Bar Association, which is an excellent, comprehensive guide to legislation, legal issues and liability affecting state and local governments in HIV/AIDS matters.

It behooves fire and emergency medical personnel to know what their particular state provides in these areas and be guided accordingly.

CHAPTER TWENTY-TWO

WOMEN AND THE FIRE SERVICE

I n addition to sexual harassment and grooming issues, there are other legal issues which particularly relate to women in the fire service.

The gradual conversion of the male-dominated fire service into a gender neutral public service is now well under way and many of the early problems have been worked out and there is less friction between the sexes, but there are still some important areas of concern which fire service managers should address to ensure that all remnants of gender bias are eradicated.

For a comprehensive discussion of important issues involving women and the fire service, readers are directed to "The Changing Face of the Fire Service: A Handbook on Women in Firefighting," a publication prepared by *Women in the Fire Service* (P.O. Box 5446, Madison, Wisconsin 53705, 608-233-4768) under a contract with FEMA's U.S. Fire Administration.

ENTRY-LEVEL PHYSICAL TESTING

One area of friction and potential liability is that of entry-level testing. Employment tests are regulated under Title VII of the Civil Rights Act and the EEOC has issued regulations implementing the Act. Any employment selection process such as physical tests can be challenged if they have an adverse or disparate impact on a protected

group. If the test disqualifies women or racial minorities in dispro-
portionate numbers, they may contravene the Act.

According to the EEOC's guidelines, a selection rate for a protected group that is less than 80% of the rate for the predominant
group, will generally be evidence that a test has a disparate impact.
If it can be shown that the test has a disparate impact on a protected
group, the burden is then on the fire department to conduct a study
that will validate the test. The department would have to show that
the test does, in fact, accurately predict which applicants will better
perform on the job. The person or group challenging the test may
show that other tests, which would have a less adverse affect on the
protected class, would serve the employer's applicant-selection needs
just as well.

Consequently, in the development and administration of its
entry-level tests, the fire department should take great care to avoid
discriminating against a protected class.

Fire departments sometimes have had difficulty in showing that
the tasks required in the entry-level test actually measure what they
are intended to measure and have not been able to justify the scoring
used in evaluating the tasks.

Some fire departments have adopted tests that assess the overall
physical fitness of the individual applicant. Scoring may be geared to
the applicant's age, body weight and gender with general fitness
standards for each group. The advantage of this system is that it
provides a comprehensive view of the applicant's overall physical
fitness.

A test must be job-related. In other words, the tasks required in
the testing must be closely related to those tasks which the individual
would actually perform on the job. In order to achieve this, the
department should find out what firefighters do and how they do it.
This analysis should provide a picture of the job and its tasks. The
department then must determine the relative importance of the
various skills which the job requires and the degree of competence
needed with respect to each skill. This analysis is made by interview-
ing workers, supervisors and administrators, studying training
manuals and actually observing the performance of the workers
doing their job. Applicants should not be required to perform tasks

during the tests which firefighters do not perform on the job. Neither should applicants be expected to perform at a higher level of competence than those who are already working in the job.

There is an increasing trend in the fire service to require existing personnel to meet certain physical criteria as a condition for continued employment.

Linking entry-level testing to on-the-job performance leads to the development of a single performance standard for all firefighters within the department such as that reflected in NFPA's revised 1582 "Standard on Medical Requirements for Firefighters."

Fire departments should be aware of advances in medicine and exercise physiology to improve their methods for evaluating specific physical qualities related to the job of firefighter.

Some departments have never used physical performance tests for applicants, but rely instead on the department's training program to weed out those who are incapable of doing the job. This places a heavy burden on the training staff and may be unfair to the applicant who may have given up another job to join the fire department.

The overarching objective of entry-level testing should be to select from the pool of applicants those who, with appropriate training, could become skilled and competent firefighters. Each applicant should be assessed objectively and fairly against a standard so that no one is disadvantaged because of factors not related to job performance.

FEDERAL COURT IN NEW HAMPSHIRE ENJOINS BIASED FIREFIGHTER TESTS

A woman plaintiff who scored number two in a 120-hour testing program for firefighters was not hired. There were eleven women applicants and none of them were able to pass the agility test, while 30 of 35 men passed it. Part of the test consisted of climbing and descending a 100-foot ladder, removing a ladder from a fire truck, placing it and returning it to the truck, completing a 1.5-mile run in 12 minutes, and pulling a 1-inch booster hose 200 feet within 35 seconds.

The judge pointed out that the old NFPA standard allowed 13 minutes to complete the 1.5-mile run, rather than 12, which the plaintiff was able to achieve. It was also noted that both the run and the hose pull were deleted from the 1992 NFPA standards. The court said these tests were gender-biased and the defendants were unable to prove that they constituted a bona fide occupational qualification of the job.

The city attempted to justify the tests by pointing out that they were being used by four other fire departments in nearby Rhode Island. The judge said there was no evidence that those four Rhode Island cities had conducted proper job analyses or validation studies of the tests. The judge said, "Follow the leader is not an acceptable means of test validation."

The court issued an injunction against further use of the agility test and obstacle course and ordered that the plaintiff be hired as an entry-level firefighter.[214]

PROVING THE VALIDITY OF TESTS

There are three ways to prove the validity of a test:

Criterion validity — the test accurately predicts or is significantly correlated with the actual work proficiency of employees.

Content validity — the test duplicates or is representative of the job's actual duties.

Construct validity — cites the mental and psychological traits required for successful job performance.

Only the first two are relevant to the fire service because it has been determined that construct validity is not applicable to job duties requiring physical abilities.

These validity studies are usually conducted by industrial psychologists or other experts.

MATERNITY LEAVE

Fire department management should clarify what the department's policy is regarding maternity leave.

The Pregnancy Discrimination Act of 1978 (PDA), an amendment to Title VII of the Civil Rights Act, broadened the definition of discrimination to include bias based on pregnancy and childbirth. It reads in part as follows:

"Women affected by pregnancy, childbirth or related medical conditions shall be treated the same for all employment-related purposes, including

the receipt of benefits under fringe benefit programs, as other persons
not so affected but similar in their ability or inability to work...."

The law is applicable to all employers who have fifteen or more employees and to employment agencies and labor unions.

The purpose of the PDA is to prevent arbitrary and discriminatory treatment of pregnant women employees. The PDA guarantees that they will have access to the benefits available to other workers. An employer's health insurance plans, for example, cannot exclude coverage for pregnancy and childbirth and an employer cannot refuse to hire or promote a woman because she is pregnant. If disabled workers are given light duty, pregnant women must be offered light duty. A pregnant woman cannot be arbitrarily fired nor can she be required to take extended leave unless it is medically necessary. The PDA also allows women to continue working well into their pregnancy and to return to their jobs as soon as they are physically able.

EMPLOYERS CANNOT FORCE PREGNANT WOMEN TO AVOID HAZARDOUS DUTIES

Some employers, including fire departments, adopted policies designed to protect the fetus from dangerous exposure to hazardous environments. Some substances or conditions in the workplace could have an adverse affect on unborn babies. These fetal-protection policies required pregnant women to leave a certain type of job, such as fire suppression, which might expose her and her unborn child to environmental or chemical conditions which might jeopardize their health.

These policies were based on concern for the fetus as well as concern for the possible liability consequences if an unborn baby's health was adversely affected by the mother's workplace exposure to the dangerous condition or substance.

With a damned-if-you-do-damned-if-you-don't" result, the U.S. Supreme court considered just such a policy in the *UAW v. Johnson Controls, Inc.* case, (Supreme Court of the United States, NO. 89-1215, (1991)). A battery manufacturer had barred all fertile women from working in jobs which involved using lead. The policy was based on the premise that lead could harm an unborn baby. The High Court, in a majority opinion by Justice Blockman, called this a form of sex discrimination. In a divided opinion, the Court said, "Decisions about the welfare of future children must be left to the parents who conceive, bear, support and raise them rather than the employers who hire those parents." The court continued, "It is no more appropriate for the courts than it is for individual employers to decide whether a woman's reproductive role is more important to herself and her family than her economic role. Congress has left this choice to the woman as hers to make."[215]

Query: If a pregnant woman chooses to work with the lead used in making batteries and her child is subsequently born with deformities which are traced to lead poisoning, would the battery company be able to use the UAW v Johnson Controls Supreme Court decision as an adequate defense? Don't count on it!

THREE TYPES OF MATERNITY POLICIES

Women in the Fire Service points out that there are three types of policies which should be considered related to maternity:

> **Maternity leave policies.** *These address the specific period of time when a woman is disabled as a result of childbirth. It could range from a few days to several months. The woman should be allowed sufficient time to have the baby and to physically recover afterwards. The policy should be flexible enough to accommodate the needs of the individual woman because recuperation periods vary greatly with different women. Leaves of absence for pregnancy are usually on a leave-without-pay basis, but may include continued accrual of seniority and fringe benefits. Some departments require the opinion of the woman's physician regarding how long she should continue working in her firefighting job. It may be useful to develop a standard release form which lists the requirements of the job for the information of the physician so s/he can make an informed decision with the woman's and her baby's medical interests in mind. The same form should be used for other temporary disabilities so that the pregnant woman is not treated in a disparate way.*

> **Pregnancy policies.** *Women firefighters need maternity leave just as pregnant workers in other occupations do, but because firefighting is a particularly hazardous occupation, their needs are more complex. The Pregnancy Discrimination Act guarantees women the right to work as long as they are able to perform all the functions of their jobs. In the case of a pregnant firefighter, however, there is real concern as to whether she SHOULD continue the usual job assignment while she is pregnant.*

Many chemical substances to which a firefighter is exposed could have an adverse affect on the health of the baby and the mother. Most physicians agree that women firefighters should stop fighting fires and engaging in other high risk work beginning at some point during their pregnancies. Because many environmental hazards are most dangerous to an unborn baby during the first three months of gestation, most fire departments will want the pregnant firefighter to leave the hazardous duty as soon as she knows she is pregnant. Most expectant mothers, concerned about the health of their babies, would probably want this option also. However, on the basis of the UAW v. Johnson Controls case, the fire department could not force the woman to take a less hazardous assignment if she does not wish to do so. The best policy is for the fire department to offer alternative, non-hazardous duty to the pregnant woman. Pregnant firefighters could work successfully and productively in training, public education, fire prevention, inspections, communications etc. Most pregnant women would probably readily accept such an alternate assignment. The heart of the PDA is that pregnant workers must be treated at least as well as other employees who are temporarily disabled.

Parental leave. Beyond the actual time required for physical recovery, a new mother or father may wish to spend extended time with a new-born or newly adopted child to promote bonding. To accommodate this desire, many fire departments offer parental leave. Parental leave should be available to both mothers and fathers. Four months of unpaid parental leave is usual, but some employers allow up to one year. Sometimes this type of leave is also available to employees who wish to take care of an elderly parent or an ill relative.

Women in the Fire Service offers the following suggested checklist for fire department policies regarding maternity and parental leave:

The specific terms of a fire department maternity policy will vary depending on the needs of the city or other employer. However, all comprehensive policies have some points in common. Keep the following guidelines in mind as you work to develop your department's policy. The policy should guarantee that the pregnant employee will not lose her job. Firing an employee because of pregnancy is a violation of the Pregnancy Discrimination Act of 1978.

- Alternate, non-hazardous duty should be available but not mandatory. The employee who transfers to this type of assignment should not lose any pay or benefits by doing so.
- It is the employer's responsibility to educate all employees, both male and female, about reproductive hazards in the workplace, to enable them to make informed decisions about conceiving and bearing children.

- *The employee's health care benefits should be maintained during any type of leave related to pregnancy.*
- *Pregnant employees should not be required to exhaust all sick leave, vacation time or other forms of personal leave, before being allowed to use maternity leave. Use of these other types of leave for the purpose of maternity should be at the employee's discretion.*
- *Pregnant employees should not lose seniority or eligibility for promotion due to any paid leaves or transfers to non-hazardous assignments.*
- *Unpaid leave should be an option, both during pregnancy and after the birth. Parental leave should be gender-neutral and equally available to both parents, regardless of their marital status, as well as to non-biological parents.*
- *If maternity/parental leave is part of contract language, and the contract does not cover recruit and probationary firefighters, the department's policy should address what options are available to the recruit or probationary firefighter who becomes pregnant or becomes a parent.*

FACILITIES FOR MEN AND WOMEN

Since many fire stations were constructed before there were women in the fire service, most do not have separate restroom, shower or bunkhouse facilities for men and women. Obviously, if new stations are to be constructed or substantially rehabilitated, provision for these separate facilities should be made.

In the absence of this opportunity, fire departments have coped with this problem in a variety of ways. A single restroom could be labeled "men/women" or "occupied/unoccupied." Airplane lavatories accommodate both men and women by the simple device of having a light activated by the locking mechanism to indicate "occupied" or "vacant". Fire departments could use the same technique for restrooms and shower rooms. Bunkhouse facilities could be partitioned with a curtain such as is found in hospital wards to provide some minimal measure of privacy or the male and female sections could be subdivided by a row of lockers.

Some men and women argue that separate facilities should not be provided, pointing out that a combined bunkroom promotes cohesion and esprit de corps in the crew and facilitates the information-sharing process when a fire call comes in. However, some men and women are not comfortable in sharing the same bunkroom. This dilemma presents an important policy decision for the fire department's management.

If it is practicable, the ideal solution would seem to be to provide privacy for both men and women with separate male/female changing areas, restrooms, showers and individual sleeping cells. Not only do individual sleeping cubicles ensure privacy, but they solve the problems of snoring and some members talking or reading with a light on when others wish to sleep. If the partitions do not extend all the way to the ceiling, this would allow alarms to be easily heard as well as allow for the exchange of information when a fire call comes in.

CHAPTER TWENTY-THREE

HAZARDOUS MATERIALS

S tates, Indian tribes and local governments may apply for planning and training grants from the U.S. Environmental Protection Agency to deal with hazardous materials emergencies. (49 CFR 110.)

EPA REIMBURSES FOR HAZMAT MEASURES

EPA has a program for reimbursement of local governments for temporary measures to prevent or mitigate injury to human health or the environment as authorized under the Comprehensive Environmental Response Compensation and Liability Act of 1980 (CERCLA), (Public Law 96-510, 42 USC 9601-75) as amended by the Superfund Amendments and Reauthorization Act of 1986 (SARA), (Public Law 99-499, 42 USC 9601). This program is intended to alleviate significant financial burdens on local governments for response to releases or threatened releases of hazardous substances, pollutants or contaminants. Reimbursement does not apply for routine firefighting.

PUBLIC MUST BE PROVIDED HAZMAT INFORMATION

The law requires that the public and the media be provided important information regarding hazardous chemicals in the community to enhance community awareness and to facilitate development of effective emergency response plans. Any facility which has a Material Safety Data Sheet (MSDS) for a hazardous chemical under OSHA must report on certain hazardous chemicals present at the facility over certain quantities. The owner or operator of such a facility must also provide a revised MSDS to various agencies, including the local fire department, within three months of acquiring significant new information concerning the chemical for which the MSDS had been submitted. The owner or operator must also submit an inventory of hazardous chemicals present at the facility. The fire department must also be allowed to conduct on-site inspections and must be given information regarding the specific location of the hazardous chemicals at the facility. The information regarding the MSDS must also be furnished to any person who requests it.

CERTAIN RECORDS MUST BE KEPT FOR EPA

Manufacturers, importers, processors of chemical substances are required to keep certain records and report on them to EPA under the Comprehensive Assessment Information Rule (CAIR) (40 CFR 700).

To obtain approval, a state program must be consistent with the federal program. The state can then enforce the program including regulation of transportation of hazardous wastes.

Those within the fire department who have responsibility for hazardous materials should familiarize themselves with state and federal laws regarding their handling and shipment.

If an employee violates a statute, such as those regulating hazardous materials, s/he could face serious legal problems. Not only might the person be held personally responsible, but in the criminal infraction, employer-paid defense attorneys, which might normally be available, would probably not be provided.

The hazmat accident prevention, preparedness and response system is a complex spider web of statutory and regulatory systems involving at least twenty different federal laws and agency regulations.

Hazmat safety regulations for accident prevention cover transportation, worker protection, environmental protection and protection of the public at large.

The following federal agencies bear major responsibility for hazmat matters:

Department of Transportation (DOT), including its Research and Special Programs Administration (RSPA) and the Coast Guard (USCG)

Environmental Protection Agency (EPA)

Occupational Safety and Health Administration (OSHA) of the Department of Labor

DEPARTMENT OF TRANSPORTATION

The hazmat regulations of the Federal Department of Transportation state that no person may transport hazardous materials unless they are "properly closed, described, packaged, marked, labeled and in condition for shipment" as required by law. They must be transported and handled according to DOT's regulations. They must also be loaded and unloaded according to the regulations.

DOT tries through its regulations to prevent accidents by requiring proper containerization and handling of hazardous materials and, when hazmat accidents do occur, to facilitate easy identification of the substances for the benefit of emergency responders. In addition to regulating the transportation of hazardous materials, DOT/RSPA also regulates pipeline safety. The main statutes giving DOT regulatory authority for transportation of hazardous materials include:

Hazardous Materials Transportation Act

Hazardous Materials Transportation Uniform Safety Act, Natural Gas Pipeline Safety Act

Hazardous Liquids Pipeline Safety Act

Oil Pollution Act

Comprehensive Environmental Response, Compensation and Liability Act.

U.S. COAST GUARD

The U.S. Coast Guard has regulatory authority over bulk transport by water. Its regulations control vessel design, operations, pollution prevention, personnel qualifications and a number of other categories. Safety and accident prevention are important aspects of the Coast Guard's work. The main statutes from which the USCG derives its authority are:

Port and Waterway Safety Act

Tanker Safety Act

Oil Pollution Act

Federal Water Pollution Control Act

Ports and Harbors Safety Act.

ENVIRONMENTAL PROTECTION AGENCY

EPA's regulatory authority is divided among a number of different statutes, some of which address specific media and long-term environmental problems and others which address accidents and the response to them.

EPA's regulatory authority comes from the following laws:

Comprehensive Environmental Response, Compensation and Liability Act (CERCLA)

Clean Air Act (CAA)

Resource Conservation and Recovery Act (RCRA)

Oil Pollution Act of 1990 (OPA)

Federal Water Pollution Control Act (FWPCA)

Federal Insecticide, Fungicide, and Rodenticide Act (FIFRA)

Toxic Substances Control Act (TSCA)

Superfund Amendments and Reauthorization Act (SARA)

Emergency Planning and Community Right-to-Know Act (EPCRA).

EPA's Chemical Safety Audit Program is designed to identify problems and effective safety practices to prevent accidents and to share this information with industry, local and state governments and the general public. EPA also shares the information it collects under its Accidental Release Information Program which targets facilities with significant hazmat releases and gathers and disseminates information regarding the causes of accidents and possible precautions to prevent reoccurrences.

OCCUPATIONAL SAFETY AND HEALTH ADMINISTRATION

OSHA's regulatory oversight is designed to protect workers. Consequently, OSHA is concerned with all aspects of hazmat handling: development, production, storage, transportation, use, recycling and disposal of hazardous materials because workers are involved in every phase of this spectrum.

OSHA's total system is called the Process Safety Management Standard which is designed to prevent unwanted releases of hazardous materials and is concerned with the technology, operations, maintenance, emergency preparedness, training and all other relevant activities related to this concern. This standard includes a list of highly hazardous chemicals, identifying the toxic, flammable, highly reactive and explosive substances. OSHA targets employers who use and store hazardous chemicals which have the potential for causing a catastrophic accident. It considers its Process Hazards Analysis the most important aspect of its Safety Management Program.

OSHA standards include: Hazard Communication Standard and the Hazardous Waste Operations and Emergency Response Standard. The latter is intended to protect workers who respond to, and clean up hazmat accidents. OSHA gets its regulatory authority from:

Occupational Safety and Health Act of 1970

Superfund Amendments and Reauthorization Act of 1986
(Section 126)

Clean Air Act Amendments of 1990.

PREPAREDNESS AND RESPONSE

State and local governments have the primary responsibility for preparing for and responding to hazmat incidents. Some federal statutes mandate state and local contingency plans as a condition for receiving financial assistance from the federal government. The federal statutes which address these requirements include: CERCLA, SARA, EPCRA, RCRA, FWPCA, OPA, Stafford Act, CAA, Federal Civil Defense Act and the Nuclear Regulatory Commission Appropriations Act of 1980.

The National Oil and Hazardous Substances Pollution Contingency Plan (also referred to as the National Contingency Plan) is designed to address hazardous substances and oil pollution incidents. The Federal Radiological Emergency Response Plan addresses radiological emergencies. The Federal Response Plan was developed to cover any type of catastrophic disaster.

EPA also provides support for local contingency planning efforts of local and state responders when incidents exceed the response capabilities of local and state governments.

OTHER AGENCIES WHICH INVESTIGATE HAZMAT INCIDENTS

The National Transportation Safety Board investigates hazmat and radiological accidents in transportation and pipelines as well as general transportation accidents. The Defense Nuclear Facility Safety Board investigates process and design safety at department of energy defense nuclear facilities and the Chemical Safety and Hazard Investigation Board investigates chemical accidents at fixed facilities.

The Coast Guard has jurisdiction for accidents in the coastal zone and has trained personnel and equipment for responding to these incidents.

DISASTER HELP IS AVAILABLE FOR LOCAL AND STATE GOVERNMENTS

The Disaster Relief Act of 1974 authorizes the President to help states and local governments to cope with disasters which exceed their response and recovery capabilities. In 1988 the Robert T. Stafford Disaster Relief and Emergency Assistance Act amended the Disaster Relief Act to address catastrophic disasters. Upon the request of a governor, the President may declare a major disaster and appoint a Federal Coordinating Officer. In the event of a catastrophic disaster — which could be a hazmat incident — the Federal Response Plan (FRP) would be activated. In the event of a catastrophic hazardous materials incident, the National Contingency Plan would be activated because, by definition, this type of emergency would require additional federal assistance, so a state's governor would probably request a disaster declaration and the Federal Response Plan would be activated.

The Hazard Mitigation Grant Program (HMGP) under the Stafford Act assists state and local governments in implementing long-term hazard mitigation measures following a major disaster declaration. The Hazard Mitigation and Relocation Assistance Act increased federal funding of HMGP projects from 50% to 75% of total eligible costs. The state and local match does not have to be in cash. In-kind services or materials may be used. Eligible jurisdictions must apply through the state since the states are responsible for administering the program.

FEDERALLY MANDATED STATE AND LOCAL CONTINGENCY PLANS

Several different statutes mandate state and local contingency planning for emergency response to hazmat incidents. The Emergency Planning and Community Right-to-Know Act (EPCRA) requires local hazardous materials contingency planning. Through this Act's provisions, local planners obtain information about hazardous materials in their communities.

Training help is available through a number of different federal agencies which provide a variety of training programs for state and local contingency planning and for emergency response to hazmat incidents.

GRANTS ARE AVAILABLE FOR TRANSPORTATION SAFETY

The Hazardous Materials Transportation Uniform Safety Act (HMTUSA) created a funding mechanism, a grant program for emergency planning and training. EPCRA and HMTUSA both support local and state contingency planning and emergency response responsibilities. The Oil Pollution Act also requires area contingency planning that must involve state and local participation on the planning committee.

The Civil Defense Act is used by the Federal Emergency Management Agency to authorize the development of local and state all-hazards emergency operations plans. Similarly, the Nuclear Regulatory Commission Appropriations Act is used by FEMA to authorize and maintain a state and local contingency planning system in communities surrounding commercial nuclear power plants. The Department of Defense provides funding to FEMA for grants to state and local governments which have chemical weapons storage and incineration sites to finance preparedness programs. This program is known as the Chemical Stockpile Emergency Preparedness Program.

Each regulatory agency charged with controlling hazardous materials has developed an accident reporting system and related databases to accommodate its specific accident notification requirements and data needs.

NATIONAL FIRE INCIDENT REPORTING SYSTEM

The U.S. Fire Administration (USFA) operates the National Fire Incident Reporting System which gathers data from the states about fires and hazardous materials incidents. Among the data collected regarding hazmat incidents are the type of site, (fixed or transportation), the chemical involved and the injuries and deaths, if any, which resulted from the incident.

HIGH COURT UPHOLDS CONVICTION OF U.S. EMPLOYEES IN HAZMAT CASE

Statutes can involve civil sanctions or criminal fines and imprisonment. The U.S. Supreme Court in 1991 affirmed a decision of the Fourth Circuit Court of Appeals upholding a conviction of three federal employees who allowed hazardous materials to be stored on government property, a criminal violation of the Resource Conservation and Recovery Act. They received prison terms which were later changed to community service.[216]

While several federal agencies are active in the hazardous materials arena, the primary responsibility for preparing for, and responding to, hazardous materials incidents lies with state and local government as well as with private users of those materials.

ICS USE REQUIRED

All emergency response agencies are required by federal law to use the Incident Command System in responding to hazardous materials incidents (29 CFR 1910.120). Although this provision specifically applies to OSHA states, all EPA states are also instructed to adhere to this requirement (40 CFR 311.1).

CHAPTER TWENTY-FOUR

TIPS FOR MINIMIZING LIABILITY

B e a good manager and learn as much as you can about the law as it affects you and your operations. Don't become paralyzed by fear of liability. Approach it as you approach all other management challenges.

Most important: If you're sued for something you did during the scope of your employment, does your government provide lawyers to defend you and, if you lose the case, does the government pay? If not, IMMEDIATELY initiate action to gain this protection. (Usually, even if your government does provide this protection, it will not be available to you in the case of gross negligence or intentional torts.)

Minimize sources of liability and maximize protection (immunity/legislation).

Find out what your immunity is in a given situation. Even if your state does grant immunity, the good manager will not rely on it, but will always try to act in a non negligent manner.

Learn the limitations on your authority and stay within these legal parameters.

Satisfy all laws.

Perform all duties imposed on you by law and make sure your performance measures up to applicable standards.

Provide due process in all of your activities.

Be fair and follow appropriate procedures regarding personnel, contracts and administrative functions.

Train, train and train.

Demand professional performance from all your personnel. Put people with the appropriate training, education and practical experience in decision-making positions. Have the best trained, most competent people making the hardest decisions.

Consult experts and make sure they have all the relevant information to study before they give you advice.

Have access to a competent attorney whom you consult regularly BEFORE you have a legal problem.

Be pro-active in drafting and seeking new laws and regulations. If you don't like what the law or regulations are, work to change them.

Follow up on violations.

Take prompt action on complaints.

Document your actions, including personnel matters. Build a record. Where appropriate, take photographs. Keep records especially of all important events and decisions. Always do this no matter how pressed you might be for time. Good records might become useful evidence during legal proceedings. (Remember, however, that under discovery, your legal opponent will be able to obtain access to all your files and records which are pertinent to the case.) Don't let paperwork get lost. Be attentive to the paper trail. Keep logs and make them as accurate and professional as possible. Keep records of personnel training, accommodations made for disabled applicants or employees, validation of entry level tests, equipment purchases and maintenance, incident reports, meeting minutes, after-action reports, complaints and actions thereon. Keep detailed records of everything important.

Know the aspects of, and ramifications of, constitutional torts. (Civil rights cases and infringement of rights.) Re-read the chapters on Sexual Harassment and Drug Testing. Remember, if you have state immunity, it will not protect you in the case of a constitutional tort.

Before making a decision, take advantage of all the time available to you. Collect all relevant facts. Assess risks and benefits. Even in emergency situations when time is tight, make sure you use whatever time is available to consider the alternatives open to you before you make a decision. When you do make a decision, make it clear so that there is no doubt on anyone's part as to what that decision is.

Know your job, do your job and stay within the scope of your job.

Treat all employees in the same, fair, even-handed way.

Follow all provisions of the personnel law, labor relations code, employee handbook and union contract regarding drug testing, promotions, stealing, discrimination, progressive discipline etc.

Follow rules regarding competitive bidding and contracting.

Know the statutes and court cases in your state which affect the fire service as well as the federal laws related to ADA, ADEA, sexual harassment, civil rights, equal pay, drug testing etc.

It is very important to know what the law is in your state as it relates to ALL of your activities.

Profit from the mistakes of other jurisdictions. Monitor legal and professional publications to stay abreast of liability problems, new legislation and court decisions. Stay in touch with your fire service friends throughout the country and borrow ideas from them.

Choices you have for handling legal risk:

Eliminate the risk— Stop doing the thing which entails the risk. This may affect the services you are expected to render, however, and may not be a practical solution.

Assume the risk — Self-insurance. Do what you should or must do and let the chips fall where they may.

Reduce the risk — By gaining additional knowledge to help you recognize potential legal problems, by devising controls, by analyzing your forms and procedures from a legal point of view, by training, (not only on technical subjects, but on liability issues as well), by pre-planning, etc. Good forms can help to insure that all the bases are covered, that everyone operates in a uniform manner. This will help to control the activities of personnel. Good forms also help you to document your actions for future reference.

Transfer the risk — Get insurance. Ask your insurance carrier to study optimum coverage for you and/or your department. Many fire service executives have bought extra insurance to protect them against personal liability. The costs are reasonable and might be a good investment in peace of mind. It could be a rider on your existing insurance policies.

LEGAL AUDIT

To arm yourself to cope with legal problems, it is important to do a legal audit. You may be amazed at what you find. Have the city/county attorney do it, a pro bono lawyer, a contract lawyer, a law student, etc., but DO IT! The legal audit might bring to your attention new legislation you want to promote.

Study your programs, policies and procedures and measure them against statutes, constitutional requirements, etc.

Analyze all of your forms and procedures from a legal perspective and use them as tools to minimize exposure to liability.

During the audit analyze the legal problems you have had in the past and implement a program to prevent them from happening in the future.

Existing statutory provisions might conflict with each other. They may be inconsistent. There may be some you are not following because you are not aware of them. You might not have the authority to do some of the things you are now doing.

There may be overlapping jurisdictions and there may be glaring gaps.

There is no greater frustration for a fire official than to discover in the midst of an emergency that no adequate authority exists to perform a necessary function, or to find out that there are conflicting authorities among various governmental units. Laws can impede or facilitate the management process during emergencies.

It might be useful to have abstracts compiled of the relevant provisions sprinkled throughout the various statutes so they can be updated and codified.

Compare your statutes with model statutes or statutes of other states and other local governments.

It is essential that laws provide appropriate legal and policy enablements and immunities under which fire personnel can carry out their responsibilities.

Because it is impossible for emergency legislation to contemplate the infinite range of situations which might occur, the statutes should, whenever possible, provide flexibility.

Some questions to ask during your legal audit:

Do we have all the authority we think we have? What authority would we like to have that we don't have?

What immunities do we have under state law? Who is covered and under what circumstances? The government? Employees? Volunteers? Are death and injury covered? Damage to property? Are planning and preparedness actions covered?

What is the status of our mutual aid agreements? Are they up-to-date? Responsive? Do we have second and third tier responders? Do they provide us the type of specialized help we might need? Are they fully authorized by law? Do the responders come as self-contained units or will we be required to provide equipment and services to them? Where do they fit into the chain of command, ours or their own? Whose worker's compensation applies if they are hurt? Are we responsible for paying them overtime if they are entitled to it under their labor contract?

Identify your risks — legal as well as substantive.

DEALING WITH YOUR LAWYER

Use your attorney:

- for advice, not only for court action. Fit him/her into your plans before the time for making decisions.
- to plan a system for handling foreseeable emergencies.
- to help develop policies and procedures.
- as a continuing resource person.
- to counsel you on constitutional and statutory requirements.
- to conduct your legal audit.
- to explain changes in laws.
- to help draft new legislation or regulations.
- to seek judicial support for your actions to forestall possible future opposition from the courts.
- to foresee the ramifications of proposed legislation on your operations.
- to put your ideas in the proper legal and procedural form to help survive future attack in the courts.

- to be on the scene during emergencies to give you advice regarding your actions.
- to help you resolve possible turf battles with other departments.

Don't let unreasonable fear of liability interfere with doing your job. Laws are designed to protect us from negligence and intentional wrongdoing, not to block you from doing your job. Use the authority you have, but don't exceed it. Use it creatively to perform your duties.

Managers face numerous challenges and they devise appropriate actions to cope with them. The law is simply one of many challenges for the manager and should be approached from that perspective.

Good luck!

GLOSSARY

A

Abatement A lessening or reduction, as in a lawsuit. In common law, it is the complete ending of a suit; in equity, it is merely a suspension, and the suit may be revived.

Abet To actively assist another individual in the actual or attempted commission of a crime.

Ab Initio Latin, "from the beginning."

Abrogate To annul, revoke, set aside, cancel, or repel.

Accomplice One who voluntarily helps to commit or attempts to commit a crime.

Accrue To accumulate, as in the amount of time allowed for the prosecution of a suit before the Statute of Limitations takes effect.

Acquit To exonerate; to formally clear of guilt.

Action A proceeding in a court of law.

Actionable Any conduct which is sufficient grounds for a law suit.

Adduce To offer as example or proof; to cite.

Ad Hoc Latin, "toward this;" for a particular purpose.

Adjudicate To setting through the use of a judge; to act as a judge.

Adjudication The determination of the solution to a controversy, based on evidence presented.

Administrative Law Refers to that body of rules and regulations adopted and promulgated by regulatory agencies on the federal, state and local levels. They are quasi-judicial bodies and their pronouncements have the force of law.

Affidavit A voluntary, written statement given under oath and sworn to before some person legally authorized to administer it, such as a notary public.

Agency A relation in which one person acts on behalf of another, with the latter's consent.

Agent One who, by mutual consent, acts on behalf of another.

Agreements to Agree Not a contract; agreements in which two people decide to form a contract at some future time.

Allegation An accusation or charge.

Allege To assert without proof. An alleged criminal is one who is, as yet, unconvicted.

Ambient Air Quality Standards The legal requirements for general air quality in the environment.

Amicus Curiae Latin, "friend of the court." One who gives information to the court.

Anarchy Lawless society; society where there is no one with the power to make rules binding all.

Annotated Footnotes further explaining something in a text; for example, a code would be annotated to cases interpreting words in the code.

Annotated Code A lengthy statute that refers to particular cases interpreting it.

Answer The document a defendant files with a court after being sued civilly, which "answers" plaintiff's complaint.

Apparent Authority An agent's authority resulting from something the principal does that leads third persons to believe the agent has authority.

Appeal A filing with a higher court, questioning a ruling by a lower court in hopes that the decision will be reversed.

Appellant the party who appeals a decision.

Appellee The party who argues against the reversal of the decision on appeal.

Arbitrary, Capricious, and Unreasonable Test An evidentiary test used to decide if certain agency actions are supported by enough facts.

Arbitration A way to settle disputes by staying out of court; the disputants hire a third party who makes a decision the parties agree to be bound to.

Arraign To accuse of a wrong; to call into custody a person against whom an indictment has been handed down.

Arraignment The initial step in the criminal process in which the accused is formally charged with an offense, given a copy of the accusation, and informed of his or her Constitutional rights.

Arrest Seizure and detention of an alleged or suspected offender to answer for a crime.

Arson The malicious burning of property.

Arsonist One who commits the crime of arson.

Assault An intentional attempt to commit a violent physical act, placing one in reasonable apprehension of physical harm.

Assumption of Risk A defense used by the defendant in a negligence suit, claiming that the plaintiff had prior knowledge of a dangerous condition and thus is responsible for voluntarily exposing him/herself to the danger. It requires that the plaintiff knowingly accept a certain risk of his or her own free will.

Attempt An endeavor to commit an offense, not mere preparation, but falling short of actual commission of the offense.

Attorney or **Attorney at Law** Professional trained in the law who is admitted to practice law in one or more U.S. jurisdictions usually after passage of a bar exam.

Attorney General Society's attorney.

Attorney in Fact A written agency.

Aver To assert as a fact in a pleading.

Award The word describing an arbitrator's decision.

B

Bail Security, usually money, posted to ensure the appearance of a defendant at a legal proceeding.

Balancing Competing Claims The idea that law is a weighing of computing claims with the heavier prevailing.

Basis of the Bargain Damages The basic type of contractual recovery. Generally, this is a value of the performance as promised minus the value of the performance as received.

Battery An unpermitted touching.

Beneficiary Person(s) holding equitable or beneficial title to property in a trust.

Beyond a Reasonable Doubt The test for the amount of evidence the prosecutor must prove in a criminal case to get a conviction.

Bilateral Contract A contract formed by each of two parties exchanging promises ("a promise for a promise").

Bill A proposed law that has not yet been passed by the legislature nor signed by the chief executive.

Bind To subject to legal duties or obligations.

Bind Over To order that a defendant be placed in custody until the decision of the proceeding (usually criminal) is made.

Binding As used in statute, commonly means obligatory.

Binding Arbitration Where a third party, not a court, is selected by parties to a dispute to settle it after hearing both parties' claims and whose decision the parties agree to follow.

Binding Precedent Highest court in the state controls decisions of lower courts.

Blue Sky Laws State laws regulating the sale of securities.

Body Politic The society subject to particular government; the people and businesses subject to a rule.

Boulwareism Term used in labor law where management studies employee demands and makes its best offer at the start of bargaining. It represents a rejection of bargaining given its initial refusal to modify its offer.

Breach To break or violate a law, duty, contract, or right, either by commission or omission.

Breach of Contract A failure to perform a duty under a contract.

Breach of Duty The first element of a negligence case—usually, the failure to act as a reasonable person.

Breach of Warranty of Authority An agent's breaking his/her duty to third persons by believing to have and claiming to have authority to represent a principal when such authority does not exist.

Breaking The forcible attempt to enter a building.

Bribery The giving of something of value to influence performance of an official duty.

Brief An attorney's written argument, citing legal points and authorities to convey to the court the essential facts of the case; also, a way to summarize and digest law cases.

Burden of Proof The duty of a party to prove disputed facts or allegations, either to avoid dismissal of the suit or to win the case.

Burglary The entering of a building with the intent to commit a felony or steal property of value.

Business Judgment Rule A rule that holds harmless corporate management from stockholder suits for good faith errors which would injure the corporation.

Business Trust A business organized in the form of a trust; similar in many ways to a corporation.

Bylaws Rules made by a corporation for its own government.

C

C.F.R. *Code of Federal Regulations.*

Case A case is any judicial or administrative proceeding in which facts of a controversy are presented in technical legal form for decision making. The object of such a proceeding is to enforce rights and remedy wrongs. Seeking money for a breach, money for damages and injuries, prosecuting defendant for crime, divorce, child custody, injunction. A case is also the opinion and decision of an appellate court deciding appeals. A case has two functions: 1) authoritatively decides the particular controversy; 2) it establishes a precedent (or a possible precedent) for the resolution of future controversies with similar facts and issues.

Case Style *see* "Style of a Case"

Cause That which brings about a result.

Cause of Action A claim in a law and fact sufficient to demand judicial attention entitling the plaintiff to recover if certain elements are proven.

Caveat Emptor Let the buyer beware.

Caveat Venditor Let the seller beware.

Cease and Desist Orders Orders from administrative agencies telling someone (such as a business) to stop doing something.

Certiorari A writ used to ask an appellate court to review a case.

Chancellor The "judge" in a chancery (or equity) court.

Chancery Court A type of civil court where a case can start; the chancellor (judge) in such a court applies equitable principles (ideas of fairness) to decide the cases.

Charter State authorization for someone to act in a certain capacity (for instance, as a corporation) .

Checkoff A term used in labor law to describe a system for collecting union dues which involves the employer's deducting the union dues from the employee/union member's check and forwarding this amount to the union; a union security device.

Circumscribe To constrict the range or activity of something.

Circumstantial Evidence Indirect evidence; secondary facts from which a principal fact may be rationally inferred.

Civil Disobedience The idea that people break the positive law when it violates much more important natural law.

Civil Law Law that determines rights and duties between private persons; or law of certain European countries relying mainly on codes of law instead of judicial opinions as the main source of law; that branch of the law dealing with noncriminal matters.

Claim The assertion of a right to property or money.

Claimant One who makes a claim.

Clayton Act A federal antitrust law passed in 1914 that addresses several antitrust matters including interlocking directorates, race discrimination, exclusive dealing, tying contracts, requirements contracts, and mergers. The Clayton Act creates only civil not criminal violations.

Close Corporation A corporation whose stock is held by very few persons. Often "family" corporations.

Code A compilation of laws (such as the U.S. code or a state code) or all on a particular subject (such as the Uniform Commercial Code).

Code of Federal Regulations (CFR) A set of federal books organized so that all regulations of a particular agency are together in one (or several) books.

Commerce Clause The clause of the U.S. Constitution giving Congress power to regulate commerce among the several states (interstate commerce). The commerce clause is the main constitutional basis for the extensive congressional regulation of American society that exists today. It also is an implicit check on state laws that unduly hinder or burden interstate commerce.

Commercial Paper Promissory notes (two party instruments), draft and checks (three party instruments), and certificates of deposit (acknowledgment by a bank of receipt of money with a promise to repay it).

Commercial Speech Speech intended to advance the economic interests of the speaker and his or her audience. Advertising is the best example. This sort of speech now receives First Amendment protection, but not as much protection as other forms of First Amendment speech.

Common Law Law made by judges when deciding a case not governed by other kinds of law.

Common Law System Distinctive system of law adapted in the United States which was developed in England hundreds of years ago in which the king's judges made the law by deciding cases which came before them. Earlier cases become precedents for deciding later cases with similar issues and facts. These accumulated cases became a vast body of case law. Now statutory law has become as important as case law. Cases apply and interpret statutes.

Common Pleas Court A low-level court with power to try both small civil and criminal cases.

Comparative Fault Often used as a synonym for comparative negligence. But, the term primarily refers to a doctrine whereby the plaintiff's recovery is reduced in proportion to his percentage share of the overall fault (not just negligence) causing his injury.

Comparative Negligence The proportional sharing of compensation for injuries between plaintiff and defendant. Damages suffered by the plaintiff are reduced in proportion to the plaintiff's fault. Not all states recognize the doctrine.

Compensation Remuneration for work done or compensation for injury.

Compensatory Something designed to restore one person injured by another.

Compensatory Damages Money a losing defendant must pay to a legally injured plaintiff to put him/her in the position s/he was in before the injury occurred.

Complaint An accusation or charge against a person alleged to have committed an injury or offense.

Complaint Case A type of lawsuit brought under the Wagner Act in which a claim is made by either an employer or the union that an unfair labor practice has been committed.

Concealment A type of fraud where defendant misrepresents by covering up or not disclosing a fact s/he has a duty to disclose.

Conciliation Discussions Attempts to settle informally a Title VII Civil Rights Act of 1964 case.

Concurrent Powers Those powers which both Congress and the states can exercise.

Concurrent Resolution Rules which legislatures pass to govern their internal workings. Technically they bind no one outside the legislature and thus are not laws.

Condemnation Process by which government takes private property. The government must have a public purpose and pay just compensation.

Confession An admission of guilt or other incriminating statement made by an accused.

Conflict of Laws A body of law which has been developed to resolve disputes between parties from different states when the law in one state differs from the law in the other state.

Consent To agree, in allowing something to be done.

Consent Decree A judgment, ruling, or order made by a court or an administrative agency with the consent of the parties. Generally, its terms are the result of a prior agreement between the parties settling the matter before the court or agency.

Consent Order A kind of consent decree in which an alleged violator of a law agrees to stop a certain business practice, but does not admit any violation of the law.

Consequential Damages Refers to those damages caused by the breach of a contract which are more remote from the breach than basis of the bargain damages. They include personal injury, property damage, and indirect economic loss (e.g., lost profits) caused by the breach.

Consideration Something given in return for something else to bind a contract which creates a benefit and a detriment to the parties.

Consignment(s) Legal arrangement where seller of goods keeps title (ownership) but transfers possession to distributor who agrees to resell them; a type of bailment (separation of title and possession of goods).

Constitution A constitution is a fundamental political and legal charter for the people of a particular jurisdiction. It defines the character of the government by specifying the nature and extend of sovereign power; by allocating this power (separation of powers) and by setting forth basic principles and limitations for the exercise of this power by the branches of government and listing the rights of the people.

Constitutional Government Government limited by law and answerable to the people ruled.

Constitutional Law That branch of law dealing with interpretation of the Constitution.

Contempt An act tending to interfere with the orderly administration of justice or to impair the dignity of the court. Those found in contempt may be punished by fine or imprisonment or both.

Contingent Fee A lawyer's fee that is based on a percentage of the money the lawyer wins for a client in a civil lawsuit.

Contract An agreement by which one person agrees to do or refrain from doing something in exchange for some action or other thing of value being given or done by the other party. There must be a "meeting of the minds" between the parties and there must be a "consideration." In other words, something of value must be done or given by each party to make the bond. Mutual aid agreements are contracts. Compacts between states are contracts. Your personnel have an employment contract with your government.

Contract Bar Rule A rule of labor law that prevents a representation election's being held while a union-employer collective bargaining agreement is in effect.

Contract Clause A provision in Article I, section 10 of the U.S. Constitution preventing the states from impairing the obligations of existing contracts, both private and governmental.

Contravene To oppose or act contrary to.

Contributory Negligence Conduct by the plaintiff in failing to exercise due care which contributed to the plaintiff's harm. It involves the plaintiff's failure to act as a reasonably self-protective person. It is offered by a defendant as a defense in a negligence case.

Conversion An intentional tort involving wrongful interference with the dominion and control by one person of another's personality. The type of lawsuit one person brings against another if the other steals goods (movables) from him or her.

Convey To transfer property from one individual to another.

Coroner A public official who investigates the causes and circumstances of deaths occurring within a specific jurisdiction and reports the findings of the investigation in a Coroner's Inquest.

Corporate Law That branch of the law dealing with corporations.

Corporate Enabling Act A statute that lets people form a corporation if they follow a certain procedure.

Corporate Promoter One who organizes corporations and interests others in subscribing to its stock.

Corporation An artificial legal person endowed by law with legal rights and the capacity for perpetual succession and which is entirely distinct from the individuals who compose it. A popular business organizational form.

Corporation by Estoppel A court-created doctrine which says that if persons act like corporations even though they are unincorporated and third persons deal with these corporate pretenders, then the third persons may not later challenge the lack of corporateness.

Corpus Deliciti Latin, "the body of the crime." The objective proof that a crime has been committed.

Corroborate To support or confirm with evidence or authority.

Counterclaim A legal claim a civil defendant makes against the person suing.

Counsel A lawyer.

Counter-offer An offer made by the offeree back to the original offeror and in response to the original offer; it ends the original offer and is an offer back to the original offeror.

County Court Low level court which hears disputes involving small amounts of money and other small matters.

Court A tribunal established for the public administration of justice.

Court of Appeals A court having jurisdiction to review prior decisions of lower courts and either affirm or reverse those decisions.

Court of Equity A court having jurisdiction in cases in which adequate or complete remedy cannot be had in a case of law. Injunction is a form of specific remedy often used in equity.

Court of Original Jurisdiction Court where a case starts.

Court of Record Any court in which proceedings are permanently recorded and which may fine or imprison those in contempt of its authority.

Covert Secret.

Crime A wrong against society.

Criminal One who has been convicted of a violation of criminal law.

Criminal Conspiracy Agreement to commit a lawful act in an unlawful way or an unlawful act; requires some significant step to carry out this agreement.

Criminal Law That branch of law concerned with crimes and their prosecution.

Criminal Procedure Prescribed legal process used in criminal law including the arresting, prosecuting, and convicting of alleged offenders.

Cross-examination The process of by which an attorney subjects witnesses for the opposing party to vigorous questioning in an effort to undermine their credibility.

Cross-examine To test a witness by asking a series of questions designed to check or discredit the answers to previous questions.

Culpable At fault, deserving of moral blame.

Curtesy A man's interest in his dead wife's property.

Curtilage The enclosed space of ground and buildings immediately surrounding a dwelling house.

D

Damages Monetary compensation awarded to one who has been injured by another's action.

Damnum Absque Injuria Latin for "a loss has occurred but there is no legal way to obtain recovery"; loss without a remedy.

Decertification Election In labor law refers to election at unionized employer's facility where unionized employees vote to decide whether to continue having the union represent them.

Decision A judgment given by a qualified tribunal.

Declaratory Judgment A judgment in which the court simply declares the rights of the parties or expresses the court's opinion on a question of law. It must deal with a real dispute between parties. In that sense it is not an advisory opinion which some administrative agencies render. It does not seek execution or performance from either of the parties.

Deed A writing under seal used to transfer property, usually real estate.

"Deep Pocket Rule" Refers to the unwritten tradition that when there is a choice of defendants, the one who has the most financial resources is the one against whom the decision is made because of a better ability to pay.

Defamation Injuring a person's character, fame, or reputation, either by writing or orally.

Default Judgment What plaintiff wins in civil lawsuit because the defendant fails to answer the complaint within the required time.

Defendant The party against whom a legal action is brought.

Defer To put off or delay.

Deferral States States which have met certain federal requirements entitling them to enforce the Federal Age Discrimination in Employment Act.

Defined Pension Plans Pension plans that promise retirees a certain amount when they retire.

Defraud To deprive a person of property by fraud, deceit, or artifice.

Defunctness A rule of labor law that says that if a union is a collective bargaining representative but it does not function actively as such, a representation election can be held even though an unexpired collective bargaining agreement exists.

Del Credere Agent An agent who sells for a principal and who guarantees to pay the principal if the buyer does not.

Democracy Government in which the ruled make the rules.

Demur To object to the pleading of the opposite party as insufficient to support an action or defense and request that it be dismissed.

Demurrer A formal allegation that the facts of a pleading, even if they are true, are not legally sufficient for a case to proceed any further. A defendant's responsive pleading to plaintiff's complaint that, in effect, says plaintiff does not state a cause of action.

De Novo Latin, "anew; from the beginning." Any legal proceeding, such as a trial or hearing, that is held a second time but is treated as though it had never been held before. The consideration of the dispute starts fresh.

De Novo Trial New trials, starting from scratch; typically this happens when a case has been tried at an administrative hearing and then is appealed to a trial court.

Deposition Testimony in question-and-answer form made under oath by witnesses in the presence of a judicial officer, such as an attorney, which is taken down in a transcript prior to the actual trial.

Derivative Suit A civil lawsuit where the plaintiff sues on behalf of someone else (who is often unable to sue on its behalf, such as a corporation against its board of directors).

Detention Being held in custody.

Deterrence Discouraging persons who did not commit crimes from community crime by punishing criminals; one of the four basic objectives of criminal law.

Dictum (Dicta) A statement or observation in a judicial opinion not central to the decision of the case. It is not legally binding on subsequent court decisions. (See Obiter Dicta). Remarks a judge makes in a legal opinion that are not necessary to decide a case. (Dictum is singular; dicta, plural).

Direct Evidence Evidence that applies immediately to the fact being proved as, for example, if A saw B set A's house on fire.

Directed Verdict Where a trial judge tells the jury what its verdict will be.

Discharge In bankruptcy, the objective of going bankrupt; a discharge is a defense (usually successful) against a creditor's attempting to collect a discharged debt.

Disclaimer A term in a contract (often a contract for the sale of goods) by which one party (usually a seller) attempts to prevent the other party from recovering under a particular theory of recovery. If a disclaimer is enforced, its effect is to block *all* recovery under that theory of recovery.

Discovery A pretrial procedure through which one party forces the other to disclose vital information concerning the case.

Discretion The reasonable exercise of a power or right to act in an official capacity and make choices from among several possible courses of action.

Dismiss To remove a case from a court, or terminate a case.

Dismissal Word used in labor law to describe a way of settling an unfair labor practice case by, in effect, "throwing the case out."

Dissolution Selling one's assets and going out of business. In antitrust, breaking up a business that restrains trade into several smaller businesses.

District Attorney Society's attorney

Diversity Cases *(See* Diversity of Citizenship).

Diversity of Citizenship A type of federal court jurisdiction that exists because the lawsuit is between people from different states or people from a state and citizens of another country.

Divest *see* Divestiture.

Divestiture In antitrust cases forcing a firm found restraining trade to sell off part of its operations.

Docket A list of cases to be tried in a court or other tribunal.

Domestic Corporation A corporation organized under the laws of the state one is speaking about.

Domestic Relations Court Divorce court (sometimes handle juvenile delinquency matters).

Double Jeopardy Being tried twice for the same offense, which is prohibited by the U.S. Constitution.

Dower A woman's interest in her dead husband's property.

Dual Agency Where an agent represents both parties to a contract.

Due Care A concept used in tort law to signify the standard of caution or legal duty one person owes to another. It is the degree of care that a reasonable and prudent person would exercise under the same circumstances.

Due Process A constitutional guarantee, contained in the Fifth and Fourteenth Amendments and thus applicable to both the federal government and the states, that basically requires government to provide fair procedures before it deprives a person of life, liberty, or property. Occasionally, however, due process has been given a substantive meaning and used as the basis for attacking laws regulating social and economic matters. It means that the government's action must be fundamentally fair and consistent.

Duress In contract such illegal force, or the threat of force, deprives a person of his or her free will to decide whether or not to enter a contract. A contract obtained through duress is not enforceable.

Duress Rule *see* Duress.

Duties When used in connection with the tort of negligence it refers to obligations people automatically have regarding others (such as the duty to be careful).

Duty to Indemnify Duty of a principal to pay the agent any losses the agent necessarily incurs while working on the principal's behalf.

Duty to Reimburse Duty of a principal to pay an agent for any money the agent has to spend to perform his/her duties for the principal .

E

EPA U.S. Environmental Protection Agency; federal agency whose main job is protecting the environment.

Economic Strikes Employee strikes motivated by employee economic demands.

Election Bar Rule A rule in labor law that prevents employees from voting on whether to have a union represent them within a certain period after a previous representation election.

Elicit To derive by logical processor. To draw forth or bring out.

Embezzlement The fraudulent appropriation for one's own use or benefit property entrusted to one by someone else.

Eminent Domain The governmental power to condemn property. This must be for public purposes and requires just compensation.

Employee One who works for another, usually for wages or salary, but often will also include a volunteer.

Employer One who uses or engages the services of others.

Empower To give official or legal authority.

Enabling Act A status that gives an agency authority to propose and promulgate regulations.

Enabling Statutes *see* Enabling Act.

Enact To authorize as law, as by a legislative body.

En Banc Term describing rare cases when all judges in a particular court of appeals hear a case; usually courts of appeals have several panels each of which is made up of three or some other number of judges who decide cases; when a case is very important, all judges from all panels of the particular court of appeals sit as judges in the case (this can be as many as seventeen judges).

Enjoin To command or instruct or to forbid.

Enumerated Powers Those congressional powers listed by Article 1, § 8 or some other law-making provision of the Constitution.

Equal Protection The requirement that~ government not deny the "equal protection of the laws." This applies both to the states and the federal government.

Equity That body of law which was developed in England and later in the United States in reaction to the inflexibility of the common law and its inability to cover certain situations. Equity qualifies or corrects the law In extreme cases, often with an injunction or other means of relief which was not available at common law. In most U.S. jurisdictions the equity and law litigations are now handled in the same court. In previous times, the two bodies of jurisprudence were litigated in separate courts. Fairness; a type of court which gives equitable remedies.

Equitable Remedies What certain courts (called equity or chancery courts) give to an injured person (such as specific performance) when legal remedies (usually money) are inadequate.

Equity Courts Courts which give equitable remedies (such as injunctions and specific performance) instead of legal remedies; also called chancery courts.

Escrow A writing, temporarily deposited with a neutral third party by the agreement of two parties.

Estoppel A legal doctrine which prevents a person from doing something or stating or denying a fact which contradicts that person's previous action, statement or denial.

Evidence Refers to all the ways by which any alleged fact is proved or disproved. There is direct evidence, circumstantial evidence, and hearsay evidence.

Excited Utterance A statement sounding as if it were an offer made when one is emotionally aroused by some event and which is not legally regarded as an offer because of the speaker's excited state of mind.

Exclusionary Rule The principle that provides that otherwise admissible evidence may not be used in a criminal trail if obtained through illegal conduct by the investigators.

Exempt Free from some requirement or liability to which others are subject.

Ex Parte Latin, "of or from one side or party." A court order handed down upon the application of one party to an action without notice to the other party of the action. Thus, an ex parte injunction is one granted without the adverse party having been previously notified. It also refers to a communication by one party without the other parties knowledge.

Ex Post Facto Latin, "after the fact." A term describing laws passed that punish or increase the punishment of crimes committed before the laws were passed. Such laws are expressly forbidden by the U.S. Constitution.

Exclusive Benefit Rule A rule under the Employee Retirement Income Security Act, which forbids pension assets from being used other than to fund pensions.

Exclusive Dealing A business arrangement by which one business agrees to handle only one particular brand of a product as a condition to getting the seller's desirable product.

Exculpatory Clauses Clauses put in contracts, which excuse someone from liability.

Executed Contract Contract that has been performed by both parties.

Execution Legal way creditors seize debtor's property to pay debts.

Executory Contract Formed but some elements are unperformed.

Exemptions Laws that prevent creditors from seizing certain property of a debtor.

Exempt Property *see* Exemptions.

Express Authority Power a principal spells out for an agent in oral or written words.

Express Contract A contract which the parties form by writing or saying their respective promises in words.

Express Warranty A contractual promise arising from a seller's actual words (or, in some cases, from conduct from which a promise can be readily inferred). As a general proposition, express warranties are made openly and voluntarily.

F

FOIA Exemptions The several types of federal agency information that a person may not get by filing an FOIA request.

FOIA Request A request by anyone pursuant to the federal Freedom of Information Act for information a federal agency has in its possession.

False Arrest Unlawful arrest. It is both a criminal offense and the basis for a civil suit.

False Imprisonment An intentional tort where defendant unjustifiably confines or restrains plaintiff.

False Light A type of invasion of privacy; an intentional tort occurring when defendant makes plaintiff appear other than how plaintiff actually is.

Fault Legal idea that someone is not liable unless s/he breached a legal duty resulting in harm to someone else.

Federal Of or pertaining to the United States government, as opposed to state or local government.

Federal Preemption A doctrine which makes certain state laws unconstitutional under the Supremacy Clause even though there is no direct verbal clash between state and federal law. The general idea is that congressional action preempted the states from acting within an area staked out by Congress for exclusive federal control.

Federal Question A type of federal court jurisdiction based on the lawsuits involving an interpretation of some federal law.

Federal Register A "newspaper-like" federal publication giving public notice of proposed and promulgated federal regulations, notice of agency meetings, presidential proclamations, and other agency proclamations.

Federal Register System The *U.S. Government Manual, Federal Register,* and *Code of Federal Regulations.*

Federal Reporter Set of books containing cases decided by all U.S. Courts of Appeals.

Federal Supplement Set of books containing cases decided by all U.S. Federal District Courts.

Federal Supremacy The doctrine that the Constitution, laws, and treaties of the United States defeat all inconsistent state laws in case of a clash between them.

Federalism Government organized so there are dual (two) sovereigns, state and federal, each with its appropriate sphere of operation.

Fee Simple Absolute A type of estate in realty; the estate in realty which confers the greatest number of rights on the holder of the estate.

Fellow Servant Rule A legal rule which says that, if a worker is injured by a coworker's negligence or misconduct, the injured employee may not recover from their employer.

Felon One convicted of committing a felony

Felony A serious crime, such as homicide, arson, rape, burglary, and larceny for which the punishment is usually imprisonment for more than one year.

Fiduciary A person having a duty to act primarily for the benefit of another in certain matters in trust or confidence.

Fiduciary Duty Duty of great faithfulness and confidence to avoid conflicts of interest or any other form of "double dealing."

Fiduciary Responsibility See Fiduciary Duty.

File To place material among official records as prescribed by

Finding A decision of a court determining a fact.

Fine A sum of money imposed as punishment for an offense.

Firm Offer A written offer made by a merchant in which the merchant assures by the offer's terms that the offer will be held open for a definite time up to 3 months or for a reasonable time up to 3 months if no definite time is stated, which offer is signed by the merchant; such offers may not be revoked for the term of the offer; such offers are legalized by Article 2-205 of the UCC.

Fiscal Of or pertaining to finance and financial transactions.

Five-to-Fifteen Year Graduated Vesting A formula for calculating vesting of employer contributions to employee retirement plans.

Fixed Fee A lawyer's fee that is definite and certain in total dollar amount and that does not depend on whether the lawyer wins or loses the lawsuit.

Fixture Property classification denoting personalty attached to realty with the intent of making the personalty permanently part of the realty; fixtures can regain their identity as personalty by the process known as severance.

Foreign Corporation A corporation organized under the laws of a state other than that one is speaking about.

Fraud Misrepresentation of a material fact with the intent that a person rely on it, that person does rely on the misrepresentation and is damaged thereby.

Fraudulent Characterized by or based upon fraud.

Fraudulent Conveyance Transfer of property that in some way misrepresents, such as a debtor's transfer of property to relatives pretending it is sold to them when in fact the relative intends to retransfer the property to the debtor after the creditors leave.

Freedom of Contract The idea that people can determine the course of their own lives by making their own contracts unprotected or unassisted by any government agency; this idea is most defensible if people have economic and political equality.

Freedom of Speech A constitutional principle, contained in the First Amendment and made applicable to the states through the Fourteenth Amendment, that gives a very high degree of protection to "speech" in its many forms.

Fundamental Rights Certain rights which, when they are denied by government, are protected by requiring that the strict scrutiny test be met in order for the denial to be constitutional.

G

Garnishment A legal way for a creditor to get a debtor's property held by a third person (such as a bank account).

General Agency A person who acts in many activities for the benefit of and under the control of another.

General Duty Standard An OSHA standard requiring an employer to maintain a general safe place to work.

General Jurisdiction The power of a court to hear all types of lawsuits (for example, murder cases, breach of contract cases, and others).

General Partner A partner with unlimited personal liability in either a conventional partnership or limited partnership.

Gift A way to transfer title to property by delivery of the item, donative intent of the giver, and acceptance by the recipient.

Golden Parachutes High benefit severance contracts top management gives itself, particularly when the company becomes the target for a takeover attempt.

Government Manual A book containing names of all federal agencies plus much more information about them.

Grand Jury Jury used in criminal law to determine if there is enough evidence to show probable cause that a crime has been committed and that the person arrested committed it. Grand juries issue "true bills" or "indictments" if they believe the prosecutor has established probable cause or "no true bills" if the prosecutor fails to establish probable cause.

Grant To give, consent, allow, or transfer something to another.

Grantee One to whom a grant is made.

Grantor One who makes a grant.

Gratuitous Agent A person (agent) who does something for another expecting to receive no pay.

Gross Negligence Failure to use even slight care.

Guilty Having been found by a judge or jury to have committed a crime or civil wrong.

H

HIDC Rule Rule that an HIDC (holder in due course) of a negotiable instrument takes it free of personal defenses (such as fraud or duress).

Habeas Corpus Latin, "you have the body." It obtains a judicial determination of the legality of the custody of an individual. In criminal cases it requires that an individual be brought before the court and charged promptly.

Hearing A proceeding in which evidence is taken in order to decide issues of law or fact and reach a decision based on those facts. It takes place before a judge or magistrate sitting without a jury.

Hearsay Evidence Evidence not based on a witness's personal knowledge but on matters told to the witness by someone else. Hearsay is generally inadmissible in court as evidence.

Holder in Due Course (HIDC) Good faith purchaser of a negotiable instrument for value without notice of any defense that would prevent the HIDC from collecting on it.

Holding A ruling of the court or in property law, being in possession of property.

Homestead Exemption A debtor's right to keep creditors from taking all or part of the debtor's equity (ownership) in his or her principal residence.

Homicide Any killing of one human being by another.

Hung Jury A jury which cannot agree on a verdict, thus ending the trial without resolving the dispute.

I

ICC The Interstate Commerce Commission; a federal administrative agency.

Ibid Latin, "in the same place, time, or manner." A term used in source citations to mean "on the same page" or "in the same book," thus avoiding unnecessary repetition of source data.

Illegal Not authorized by law, or contrary to the law.

Illegal Contracts Contracts whose formation or performance violates some positive law.

Immunity The right to exemption from a duty or penalty. Immunity from prosecution might be granted to a witness who answers questions that otherwise might incriminate him/her; the insulation against civil suits given to governments or individuals.

Implied Created by operation of law, often by inference from the relevant facts and circumstances.

Implied Authority Power an agent gets as a result of things necessary to do to carry out the agent's express authority.

Implied Contract A contract formed when the parties' actions but not words indicate that they intend to form a contract; contract formed by conduct, not words.

Implied Powers Those powers "necessary and proper" for effectuating Congress's inherent or enumerated powers. An example would be the power to regulate intrastate matters affecting interstate commerce.

Implied Warranty A warranty arising by operation of law. Implied warranties do not depend on the consent of the party giving the warranty. They contravene "freedom of contract" principles and are usually imposed to protect the buyer.

Imprison To put in prison.

In Pari Delicto Where two persons are equally culpable or guilty, in which case the law leaves them where it finds them helping neither.

In Personam Jurisdiction Power of a court (or other legal authority) over a particular person.

In Rem Jurisdiction Power over property; ("rem" is Latin for thing").

Incapacity When used in a legal sense it refers to the law's decision that certain people (such as minors, insane people, and drunks) cannot protect themselves so it allows them to escape (avoid) their contracts.

Incorporate The act of creating a corporation.

Incriminate To charge with or implicate in a crime or fault.

Indemnified Repaid for a loss.

Independent Contractor A non-employee who agrees to do a piece of work for another, and who retains control of the means and method of doing the work.

Independent Constitutional Checks Limits on congressional power which exist outside the provision granting Congress power to act in the first place. The Bill of Rights and the Fourteenth Amendment are examples. These limits are the principal checks on congressional law-making power today.

Indicia Indications, signs, or circumstances tending to show that something as probable.

Indictment A formal written accusation by a grand jury charging someone with a crime; the "complaint" in a criminal lawsuit.

Information In criminal procedure the information serves the same purpose as the indictment but for a minor offense.

Injunction An order from a court compelling someone to do something or to refrain from doing something. For example, a court could order that fire code violations be corrected or that exits not be blocked or locked. Failure to comply with an injunction can be punished by contempt of court, which is a criminal offense. Whereas the court might be reluctant to jail someone for ignoring an order of the fire department, it may be less reluctant to jail someone who is in contempt for violating an order of the court.

Injury Any wrong or damage done to a person's body, rights, representation, or property.

Innocent Free from guilt or blame.

Innocent Misrepresentation If someone unknowingly states something is a fact when it is not which induces someone else to enter a contract, this is called innocent misrepresentation. This is a basic for the party to whom the representation is made to escape the contract in question.

Inquest A judicial inquiry or examination, most commonly used to refer to the inquiry made by a coroner's jury.

In Rem Latin, "of the thing." Proceedings against a thing rather than a person.

Intent The state of mind in which a person knows and wishes to bring about the result of an action.

Intentional Torts Civil wrongs where defendant intends to do the harm.

Intentional Infliction of Mental Distress An intentional tort in which defendant knowingly causes plaintiff extreme mental suffering; usually accompanies physical injury although in extreme cases mental distress alone can be recoverable.

Interest in the Partnership The right to receive profits and surplus from a partnership.

Interpretative Regulations Rules of administrative agencies that usually take effect as soon as they are made and are not binding as law.

Interrogatories Questionnaires which one party sends to potential witnesses in the case in question-and-answer form to gather information with respect to the evidence in the case.

Interstate Commerce Clause The portion of Article I, §8 giving Congress the power to regulate commerce "among the several states." This clause is now probably the single most important source of congressional lawmaking authority.

Interstitial Lawmaking Judicial lawmaking by filling gaps in statutes.

Intrusion Upon Solitude or Seclusion A type of invasion of privacy; an intentional tort.

Invasion of Privacy An intentional tort where defendant does one of four things to plaintiff 1) invades plaintiff's peace and solitude; 2) holds plaintiff up in a false light; 3) unauthorizedly uses plaintiff's name or likeness; or 4) defendant publicly discloses some private fact about plaintiff.

Invitations to Negotiate Statements or actions which one party makes trying to induce the other party to make an offer; not an offer.

Involuntary Bankruptcy A debtor's being forced into bankruptcy by his/her/its creditors against the debtor's will.

Involuntary Manslaughter Homicide in which the death of the victim usually results from gross negligence or recklessness, but contains no malice. Example: an automobile accident.

Ipso Facto Latin, "by the fact itself; in and of itself."

J

Jeopardize To expose to death, loss, injury, hazard, or danger.

Joint Resolution A law passed by the legislature; similar in effect to a statute.

Joint Tenants Two or more persons who own equal shares in realty at the same time and having the "right of survivorship" feature.

Judge A public official authorized to preside over legal trials and hearings and to decide questions of law.

Judgment The final determination of a court.

Judgment Creditor A person who, when a lawsuit ends, is owed money as a result of the lawsuit.

Judgment Debtor A person who, when a lawsuit ends, owes money as a result of the lawsuit.

Judgment Notwithstanding the Verdict Where a trial judge, after a jury has given its verdict, gives a judgment different from what the verdict is on the grounds that no reasonable jury could have come to the conclusion which this jury came to.

Judgment Proof When a person is so poor that he/she is not worth suing because a judgment against him/her could not be paid; a person with few if any assets.

Judicare A plan for providing legal assistance to members of the public much as medicare provides medical help to the public.

Judicial Of or relating to a judgment, the judiciary, or the administration of justice.

Judicial Review The power of courts to declare the acts of other government bodies unconstitutional.

Jurisdiction The power, right, or authority to interpret and apply the law or the place within which that authority may be exercised.

Jurisprudence The study of different legal philosophies (ideas about what law is or is supposed to be).

Jury A group of people summoned and sworn to decide the facts at issue in a trial.

Justice The maintenance or administration of what is right or just, especially by the impartial adjudication of conflicting claims; a judge.

Justice of the Peace Courts Low level courts dealing with minor civil or criminal matters.

Justifiable Homicide The killing of one human being by another without blame.

L

Laboratory Conditions Word used in labor law to describe the neutral atmosphere that is supposed to exist when employees vote to decide whether to have a union represent them.

Laissez Faire Economy where there are few, if any, government restrictions on business.

Larceny the unlawful taking of another's property with the intention of depriving the owner of its use.

Last Clear Chance A doctrine holding that defendant may still be liable for the injuries caused to a plaintiff, even though the plaintiff is guilty of contributory negligence. If the defendant observed the plaintiff in a position of danger and failed to avoid injuring him/her.

Law A rule to guide the actions of individuals in a society, the breach of which is subject to punishment or loss of property or a right.

Law Courts Courts which give legal remedies (such as money damages) instead of equitable remedies (such as specific performance) .

Lawyers Edition Set of privately published books containing U.S. Supreme Court decisions in chronological order.

Lead Agency Words used in certain environmental law matters where several agencies are involved in a matter, but one fills out an EIS for all agencies involved.

Lease A contract by which one conveys an interest in real estate, equipment, or facilities for a specified time and a specified remuneration.

Leasehold A temporary possessory interest in real estate; a lease.

Legal Of or concerning the law, that which is allowed or authorized by law.

Legal Duty Something that the law specifies must be done by a particular person or group.

Legal Fiction Something the law says exists or is true even though our senses tell us it does not exist; for example, the idea that corporations are people. Legal fictions are created to achieve some desirable social objective such as justice.

Legal Guardian Someone a court appoints to manage someone else's affairs. That someone else is called a "ward."

Legal Remedies The usual recovery an injured person receives in a civil lawsuit; damages (money) is the most common legal remedy.

Legislation A law or laws enacted by a legislative body which have the force of authority.

Legislative Concerning that branch of government charged with enacting laws, levying and collecting taxes, and making financial appropriations; relating to the process of making laws.

Legislative Regulations Agency regulations that are legally binding; also called substantive regulations.

Legislative Veto When a legislature nullifies an agency regulation.

Legislature An organized body of person with the power to enact laws for a particular geographic jurisdiction.

Lessee One who leases property from a landlord. The landlord is the lessor.

Levy To raise, collect, or assess, such as to levy a tax.

Liable To be legally responsible for.

Liability A legal obligation to do something or refrain from doing it.

Libel A malicious defamation of a living person, expressed in print, recording, writing, pictures, or effigy, or other permanent form which tends to hold a person up to hatred, contempt or ridicule.

License A right granted to a person giving him/her permission to do something not allowed without the permission.

Licensee One who has been granted a license.

Lien A hold or claim upon the property of another as security for a debt.

Life Estate A type of estate in realty which lasts as long as the holder of the estate is alive.

Limited Liability The idea that a person's financial responsibility for something has limits; said of corporate shareholders with respect to corporate debts.

Limited Partner Partner in a limited partnership where liability is limited.

Limited Partnership A partnership formed pursuant to a state statute that requires only one general partner and lets at least one other partner limit his/her liability for limited partnership debts.

Litigants Plaintiff(s) and defendant(s) in a lawsuit.

Litigation The process through which legal matters are resolved in a court, a contest in court to determine legal rights and duties.

Loans Where creditors give a borrower money (instead of goods) to use for a period of time and charge a fee for this service.

Lockout In labor law, the employer's temporary refusal to allow employees to work by locking up the plant.

Long-Arm Statute Allows a court to obtain jurisdiction over a defendant even though the process is served beyond its borders if the defendant: 1) committed a tort in the state; 2) owns property in the state; 3) entered into a contract in the state or transacted the business which is the subject matter of the lawsuit within the state.

M

Mailbox Rule Rule of contract law which says that acceptance is effective when sent ("put in mailbox") by offeree (not when received by offeror).

Magistrate A public official entrusted with some part of the administration of the law.

Mala in Se Latin, "evil in and of themselves." Offenses that are bad in themselves, based on principles of natural and moral law. Reckless driving is, for example, a crime *Malum in se. Malum* is singular; *mala* is plural.

Mala Prohibita latin, "wrong because they are prohibited." Offenses that are unlawful simply because statutes forbid them. Exceeding the speed limit, for example, is an offense *malum prohibitum.* They would not be wrong except for the statute. *Malum* is singular; *mala* is plural.

Malfeasance The doing of an act that is wrong and unlawful.

Malice Evil intent; the state of mind accompanying the intentional doing of a wrongful act.

Malingering False claims of injury or illness to escape work or increase damage recovery from someone.

Malpractice Failure to perform a professional duty or to exercise appropriate professional skill by one rendering a professional service (such as a physician or attorney) which results in loss, injury, or damage.

Malum in Se Singular of *mala in se.*

Malum Prohibitum singular of *mala prohibita.*

Mandamus Latin, "we order; we compel." A writ issued by a legal court to compel an official to perform a ministerial act recognized by the law as a legal duty.

Mandamus Orders Court orders telling someone to do something (such as telling an administrative agency to do a nondiscretionary duty).

Mandate A judicial command; to issue a judicial command.

Mandatory Obligatory, containing a command.

Manifest System Term used in environmental law to refer to a way to trace hazardous wastes by keeping a paper record (called a "manifest system") of them.

Manslaughter The unlawful killing of another without malice, as in an automobile accident or in the sudden heat of passion, upon provocation.

Massachusetts Trust Business organization form in the form of a trust; similar in many ways to a corporation.

Mayhem Willful and permanent crippling, disfiguration, or mutilation.

Mens Rea Latin "a guilty mind." The mental state accompanying a prohibited act.

Merger When two companies combine in such a way that one company continues to exist after the merger.

Minors Persons below a certain age set by each state designed to protect young people from making unwise contracts and taking part in other activities society deems is unwise for them.

Mirror Image Rule of contract law which says that acceptances must conform exactly to the terms of the offer or else the acceptance is ineffective to form a contract.

Mischief Offenses of the law that annoy or irritate, but are usually too minor to be designated crimes.

Misdemeanor A class of criminal offenses not as serious as felonies and punished by less severe crimes.

Misfeasance Doing a lawful act in a wrongful manner. The proper performance of the act would have been lawful.

Mistrial A trial terminated and declared void prior to the verdict due to some extraordinary circumstance (such as death or illness), or because of a fundamental error prejudicial to one of the parties or because the jury cannot reach a verdict. The latter is also called a "hung jury."

Mitigate To make less severe or harsh. Mitigating circumstances do not free a person from guilt or crime, but they may lessen it.

Model Statutes Suggested laws that a legislature might pass; often suggested by some organization of scholars or practitioners.

Modus Operandi Latin, "the method of operation." The way an act is accomplished. For example, the *modus operandi* of the arsonist was to use an electrical timing device to ignite containers of gasoline.

Monarchy Government in which only one person makes the rules.

Monopolies Economy where there is only one seller.

Mortgage A conveyance of lands or other subjects of property as security for a debt on the condition that if the debt is paid, the conveyance shall be void.

Motion A request to the court asking that the court rule in favor of the party making the request.

Motion for Summary Judgment A motion a party to a lawsuit makes that, in effect, says "There are no fact issues, just legal issues, so you decide the case without a jury trial, judge."

Motion to Dismiss A legal argument a civil defendant makes to cause the judge to throw plaintiff's case out of court.

Municipal Of or relating to the affairs of a municipality.

Municipal Courts Low-level courts dealing with minor civil or criminal matters.

Murder The malicious killing of one human being by another; it requires premeditated intent to kill.

N

National Conference of Commissioners on Uniform State Laws A group of legal scholars who draft suggested laws for states; suggested laws—called model acts—are designed to simplify, make uniform, and modernize the law.

National Environmental Policy Act (NEPA) A federal law that tells federal administrative agencies to take environmental factors into account when they decide to do major things that affect the quality of the human environment.

Negligence A type of tort (civil wrong) that holds a person liable for damages proximately caused by carelessness. The failure to exercise that degree of care which a person of ordinary prudence would exercise under similar circumstances. Negligence is perhaps the most important area of the law from the fire department point of view, because most liability cases will arise from negligence.

Negligence Per Se In negligence law, a breach of duty based on the violation of a statute.

Negotiable Note A type of negotiable instrument; a signed writing having an unconditional promise to pay a sum certain (exact amount) of money (any country's currency) at an exact date or on demand (when asked) that has certain words (called words of negotiability) implying that the person originally issuing the note intended that person(s) other than the person originally receiving the note could legally come into possession of it.

Nolo Contendere A plea of "no contest" in a criminal case. Latin for "I do not wish to contend." A plea that a defendant in a criminal prosecution enters neither admitting nor denying guilt, but not contesting the prosecution's case. The judge can mete out punishment but a *nolo contendere* plea does not show on the record as a conviction.

Nominal Damages Symbolic damages that are a small amount of money a law-breaker pays to a victim to show a wrong was done but the victim was not harmed.

Nondiscretionary Duty Used in connection with administrative agencies to describe agency duties that must be done and which a court will order an agency to do if it does not do them voluntarily.

Nonexistent Principal A principal that does not legally exist, such as a corporation not yet organized.

Nonfeasance The failure to perform a required duty.

N.O.V. *Non obstante veredicto*, Latin for "notwithstanding the verdict." A judgment n.o.v. is one that reverses the decision of the jury when it is obvious that the judgment had no support on the basis of the evidence or was not lawful.

Nonlitigation Legal work not involving lawsuits.

Nuisance A type of tort (civil wrong) where someone or something injures or annoys another or his/her property; anything that renders the enjoyment of life and property uncomfortable.

Null and Void Invalid; having no legal force or effect.

O

OSHA Walkaround Privilege Right of an employer and employee subject to OSH Act to accompany the OSHA inspector during a plant inspection.

Oath A declaration of truth which, if made by one who knows it to be false, renders that person liable for perjury.

Obiter Dicta (Dictum; dicta) Discussion in an opinion which is not necessary to the holding. It is extraneous, not germane, a gratuitous comment by the judge.

Objective Intent The intent attributed to a person who appears to a reasonable third person to be making an offer to contract.

Objective Standard A test of liability based on the defendant's behavior or some other physical fact, and not on what the actor (defendant) thinks.

Objective Test Regarding contracts, test to determine if a contract exists based on what a reasonable third person would think the people allegedly making the contract have done.

Objective Theory of Contracts Way to determine if a contract exists; theory says that if reasonable person would say that if two parties appear to have made a contract, a contract exists in the eyes of the law even though the parties did not subjectively intend to form a contract.

Offense A violation of the law.

Offer A definite, communicated expression one party makes to another objectively intending to form a contract stating what the offeror will do and what the offeror expects in return from the offeree.

Offeree Person to whom an offer is made.

Offeror Person who makes an offer.

Oligarchy Government in which a small group makes the rules.

Open Meeting Law A law forcing certain agency meetings to be open to the public. Also called "Sunshine Acts."

Operation of Law The idea that the law does something automatically when certain events occur (such as ending an agency when the agent dies).

Opinion The reasoning behind a court's decision of a case.

Option An agreement to hold an offer open for a specified time in return for a consideration.

Options Contracts whose subject matter is an offer.

Ordinance A term usually used to describe a law enacted by a governmental authority at the local level; a law passed by a municipality (city and town).

Ordinary Negligence Failure to use the normal care that an ordinary person would use in similar circumstances.

Organic Acts Statutes giving an administrative agency the legal right to exist.

Overrule To overturn or make void the holding of a prior case; to deny as by the court a motion or point raised during a case in court by one of the parties.

P

Paper Remedies Remedies that theoretically exist ("exist on the pages of the books") but which, as a practical matter, do not exist.

Partially Disclosed Principal Agency where the identity of the agent and fact of agency are known but the principal's identity is unknown.

Partnership Two or more persons carrying on as co-owners a business for profit. A form of business.

Partnership by Estoppel Where a person holds out another as a partner, and a third person reasonably relies on this holding out and is damaged as a result. The person holding out or permitting another to be held out as a partner is stopped from saying the person held out is not a partner.

Party Either litigant (plaintiff or defendant) in a lawsuit; a person directly interested in the subject matter of a case; a person who enters into a contract, lease, deed, etc.

Penalty The punishment specified for a particular crime or offense.

Pending Not yet completed.

Per Curiam Latin "by the court." An opinion written by the court whose author is not named. A phrase used to distinguish an opinion of the whole court from an opinion written by only one judge. Sometimes it denotes an opinion written by the chief justice or presiding judge.

Per Se Offenses Certain antitrust offenses that are so obviously harmful that the government does not have to show any particular economic harm to prove its case (harm is presumed).

Percentage of Disability A way of paying injured employees covered by worker's compensation when a loss occurs not expressly covered in the worker's compensation schedule of recoveries.

Peremptory Challenges Challenges to a prospective juror where the challenging lawyer does not have to tell why the prospective juror is excused from jury duty.

Perfected Security Interest A creditor's claim on certain property of a debtor where the claim is superior to certain other creditors' claims to the same property.

Perjury The criminal offense of making false statements under oath.

Permanent Injunction A court order telling someone to do something or never to do something; punishment for breaking such an injunction can be being jailed.

Perpetrate To cause to happen.

Perpetrator One who causes something to happen.

Personalty Moveable property such as clothing, vehicles, furniture.

Personnel A body of persons employed in an organization.

Persuasive Precedent Decision in other state, although with the same facts, is not binding on the home state court.

Petition A formal written request.

Plaintiff The one who initiates action in civil law; the complianant.

Plea Bargain A deal between the prosecutor and a criminal defendant often where a guilty defendant pleads guilty to a lesser offense. The government is relieved of the cost of trial.

Pleadings Statements in logical and legal form made by the parties in a court case that constitute the reasons for supporting or defeating the suit.

Pocket Parts Pamphlets inserted in the back of law books (statute books and case books) to update them with recent changes in the laws or court cases.

Police Power The power to regulate for the public health, safety, morals and welfare. All the states possess this power. For all practical purposes, Congress possesses this power also, especially under the interstate commerce clause.

Power of Attorney Written agency under which one person may do future acts for another person.

Precedent A previously decided case recognized as an authority for the decision of future cases.

Preempted Where a federal statue so completely dominates an area of law that the state has no legal power to pass laws in the subject matter area.

Premeditated Having considered a matter before taking an action.

Preponderance of the Evidence A test in civil law which a jury uses to decide if enough evidence exists to find for the plaintiff.

Prima Facie Case ("at first view") The plaintiff must introduce sufficient evidence at trial to establish his or her cause of action. Only then will the judge allow the case to go to a jury for fact finding; a case strong enough to enable the plaintiff to win the suit unless contradicted by persuasive evidence from the defendant.

Privilege Defense to defamation; basically says that even if plaintiff's allegations about defendant are true, certain policy reasons (such as absolute free expression on the floor of the legislature) exempt defendant's remarks from a successful defamation suit; privilege can be qualified or absolute. Also means an advantage not enjoyed by all.

Privity of Contract Contractual relationship between two parties. If A sells goods to B, who in turn sells the goods to C, both A and B, as well as B and C, are "in" privity of contract, but A and C are not.

Pro Se *See* Pro Se Litigant

Pro Se Litigant A party to a lawsuit who handles his/her own case without a lawyer.

Probable Cause Cause to believe a crime was committed and the person arrested committed it; also the test for a grand jury's issuing an indictment; the grounds to justify a search warrant.

Probate Court Court that deals with wills and estates.

Probation A procedure in which the sentence of a convicted offender is suspended and the offender is allowed freedom as long as he maintains good behavior under the supervision of a court-appointed person.

Procedure Legal method; the body of rules and practice by which legal justice is conducted. Each court in each jurisdiction has rules of procedure which must be followed. Failure to follow proper procedures can lose a case.

Proceeding The succession of events by which a case is begun, carried through, and terminated; a general term for any of these events.

Procedural Due Process A basic requirement of procedural fairness imposed on the federal government and the states by the Fifth and Fourteenth amendments; notice and fair hearing.

Procedural Law Rules governing steps in lawsuits or legal matters.

Procedural Rules Rules establishing how the legal system, courts, etc. are to operate.

Procedural Unconscionability A form of unconscionability involving either the way contract terms are packaged (e.g., fine print), or the factors that made a party enter into a contract (e.g., "sales hype").

Promissory Notes A written promise to pay a sum certain in money (any country's currency) at a future certain date or on demand. A two party document (that is, only two persons—a person promising to pay called the note's "maker" and a second called the "payee" are mentioned on the front ("face") of the note.

Proof The cogency of evidence that compels acceptance of that evidence as fact or truth.

Property Something real or personal that can be owned or possessed.

Proposed Rulemaking When a federal agency publishes a regulation in the *Federal Register* which the agency wants to make binding as law at a future time unless the public can convince the agency the proposed regulation is unwise.

Proprietary Characterized by having legal right to or ownership of something.

Prosecute To institute legal proceedings in criminal law by the government.

Prosecution The process of conducting a criminal trial; the party initiating a criminal trial i.e., the state.

Prosecutor A public official (representing the people, or the state, or city, or federal government) whose duty it is to prepare and conduct the case against persons accused of committing crimes. S/He is society's attorney for criminal matters.

Proximate Close to near; imminent, about to happen.

Proximate Cause That which in natural and unbroken sequence directly produces an event; a term used to describe the legal issue presented by the degree of proximity or closeness between breach of duty and injury required in a negligence case. In most jurisdictions, it is based on the reasonable foreseeability of the plaintiff's injury having resulted from the defendant's breach.

Public Law Law that deals with matters affecting all or a large part of society.

Public Nuisance Conduct or the use of property in a way that annoys or interferes with public rights that offends the public at large or a segment of the public, and interferes with the general health, safety, peace or welfare.

Puffing Seller's talk that is legally only opinions and not facts and therefore these remarks may not be the basis for a lawsuit based on fraud.

Punitive Used to describe laws that punish. Criminal laws are punitive.

Punitive (or Exemplary) Damages These are damages imposed on a party as a sort of penalty for particularly egregious behavior. Usually, some sort of recklessness or malicious behavior is required for their imposition. Punitive damages are most common in tort cases.

Q

Quantum Meruit Literally, "as much as he deserves." An obligation imposed by law which gives a party the value s/he deserves notwithstanding the lack of an agreement between the parties.

Quash To annul, overthrow, or vacate by judicial decision.

Quasi Latin, "as it were; approximately."

Quasi Contract An obligation imposed by law to do justice or to prevent the unjust enrichment of one party at the expense of another. The obligation is inferred from the circumstances.

Quasi in Rem *see* Quasi in Rem Jurisdiction.

Quasi in Rem Jurisdiction Power of a court or other legal authority over a person because the court has physical control over some item of the person's property.

R

Ratification Retroactive authority a principal gives someone who, without authority, purports to act for another and whose acts the purported principal fully approves when s/he learns of them.

Rational Basis Test The basic equal protection test and the test applied to economic regulation. It requires only that the challenged law have a reasonable relationship with a valid governmental purpose. It is not a great obstacle to government regulation.

Real Estate Investment Trust (REIT) A trust device used to own real estate. A business organization form.

Realty Property classification denoting realty and buildings; immovables.

Reasonable Doubt A term for the lack of certainty required in the mind of a juror for that juror to determine a criminal defendant innocent.

Reasonable Person Standard A phrase used to denote a hypothetical person who has "those qualities of attention, knowledge, intelligence, and judgment which a society requires of its members for the protection of their own interest and the interests of others"; the basic test used to determine whether a breach of duty has occurred in a negligence case. The defendant's behavior is compared to the way a hypothetical reasonable person would have behaved under the circumstances.

Recidivism In criminal law it refers to repeating criminal conduct.

Record To preserve in writing, printing, on tape, or film; a precise history of a suit from beginning to end, including transcriptions of all proceedings.

Recovery The fact of obtaining a final legal judgment; the amount of a judgment rendered in one's favor.

Redress To remove the cause of a grievance or to compensate for an injury or loss.

Registered Agent A person named by corporate organizers on whom process can be served if someone wishes to sue the particular corporation.

Registration Statement A document that must be filed with the Securities and Exchange Commission before one can offer certain securities for public sale.

Regulations Law made by an administrative agency; they have the status of law; legal rules authorized by statute and issued by an agency or commission for the governance of matters within its jurisdiction.

Regulatory Occupational License A type of permit one needs to do some types of work, where the main purpose of the occupational license is to protect the public from incompetent, unethical persons.

Rejection Unqualified refusal by the offeree communicated to the offeror that the offer is not satisfactory; ends an offer.

Remand To send back for further deliberation.

Remedy Limitation A term in a contract (often a contract for the sale of goods) by which one party (usually a seller) attempts to block the other party's (usually the buyer's) recovery of certain kinds of damages. Remedy limitations are often confused with disclaimers, but differ because they attempt to restrict a party's *remedies*, not the party's *liability* under a particular theory of recovery.

Remuneration A sum of money paid as compensation for loss, injury, or service.

Render To hand down, as in a judgment or a verdict.

Reports Books containing judges' opinions deciding lawsuits.

Res Property in a trust.

Rescind To take back, annul, or cancel, as a contract.

Rescue Doctrine A rule of torts that holds a defendant liable for the victim's rescuer should the rescuer injure himself or herself during a reasonable rescue attempt. (The Fireman's Rule is an exception to the Rescue Doctrine.)

Res Ipsa Loquitur Latin for "the thing speaks for itself." A way to prove a breach of duty in negligence cases where the accident would not ordinarily have occurred without someone's negligence, and the thing causing the harm is exclusively within the control of the defendant. The doctrine assists the plaintiff because it enables proof of the breach where this would not otherwise be possible.

Res Judicata Latin for "the thing has been decided." Once a particular claim is finally decided by a court with jurisdiction over the claim and the parties, that court's judgment and the factual and legal issues underlying it are considered conclusively decided and may not be re-litigated by the parties. It bars endless relitigation of claims.

Rescission Most often, a contract remedy whereby the court returns the parties to their original precontractual position.

Respondeat Superior Latin for, "let the superior reply." The doctrine governing the relationships between masters and servants, employers and employees, principal and agents. Under this doctrine the employer or principal is liable for the acts or ommisions of the employee or agent.

Restitution The act of making good, or giving the equivalent for, any loss, injury, or damage; returning something to someone else; a type of contract or tort remedy.

Restraint Imprisonment of a criminal.

Retribution Punishment of a criminal.

Revocation When an offeror calls back his/ her offer, a way to end an offer.

Right-to-See Letter A procedural requirement in Title VII Civil Rights Act of 1964 cases.

Rights Things to which one has a just claim, such as power, privilege, or property.

Rights and Duties "Rights" are those things to which one party has a just claim, such as power, privileges or property. "Duty" is the flip side of nights. We owe a duty to all other people in society to respect their rights. A breach of that legal duty can give rise to a suit for money damages or to mandate some specific action or lack of action.

Rules Law made by administrative agencies.

S

Scope-of-Employment Limit of one's job.

Search and Seizure A practice in which a person or place is searched and evidence is seized. The Constitution specifies that search and seizure must be reasonable. It also covers arrests.

Search Warrant An order issued by a judicial official authorizing a search of certain premises for certain things or persons suspected to be there.

Sentence The punishment inflicted on a convicted criminal.

Slander Oral defamation of another; spoken words which tend to damage the reputation of another.

Sole Proprietor One person who owns all of the assets and is solely liable for all of its liabilities. A legal form of business.

Solicitation An offense that consists of enticing, inciting, or begging another to commit a felony.

Sovereign The highest political authority in a society; the state (or government); supreme lawmaker in a society.

Sovereign Immunity A doctrine that precludes bringing a suit against the government without the government's consent.

Special Agent Someone who acts for and under the control of another in a very limited matter (such as an agent hired to do one thing).

Specific Intent Crimes Crimes in which the intent required is different that the intent to do the act that is the crime (such as "assault with intent to murder").

Specific Performance Equity court order forcing a person to perform a contract exactly as agreed. An equitable remedy.

Standard Tells you what to do and how to do it, or how not to do it.

Standing The legal right of a person to challenge the conduct of another.

Stare Decisis A latin expression which literally means "to stand by what was decided." It is the principle which courts use to help decide the case which is presently before the court by studying what other courts in that jurisdiction have decided in previous cases where the circumstances and facts were similar; the decision of a appellate court in one case provides a precedent for deciding future cases in the same jurisdiction with similar facts and issues. (It is not absolute. Courts change their minds.)

State Action The requirement that, in order for certain independent constitutional checks to apply, there must be action by a governmental body, state or federal. This doctrine generally frees private bodies from having to obey these constitutional requirements.

State Blue Book A book published by each state that explains its government regulations and includes a list of state administrative agencies.

State Exemptions Property that cannot be taken by creditors to pay a person's debts because state law exempts it.

State Right-to-Work Laws State laws that allow employees at unionized employer to refuse to join the union or pay dues.

State Usury Laws State laws setting limits on the rate of interest creditors can charge borrowers.

Statute A law enacted by the legislative branch of government. "Statute" usually refers to laws enacted at the state or federal level, but actually all enactments by any legislative body are statutes.

Statute of Descent A "statutory will" the legislature has made for people who die without a will. They are said to die intestate.

Statute of Frauds Sets forth those types of contracts which must be in writing, or evidenced by a written memorandum signed by the person being sued.

Statute of Limitations Specifies the time within which a legal action must be initiated or it is barred forever. There are limitations in both civil and criminal law. The lengths of time vary depending on the offense or legal rights. The period may begin to run at different times depending on the cause of action in question.

Strict Liability Tort liability without fault. That is, liability imposed irrespective of whether the defendant intended to cause harm, was reckless, or was negligent.

Strict Liability Criminal Statue A criminal statute that requires only an act but no intent to commit a crime.

Strike In labor laws, employees' temporary refusal to work.

Style (of a Case) Names of plaintiff(s) and defendant(s) in a lawsuit.

Subject Matter Jurisdiction The power of a court or other legal authority to deal with the type of case in question. For example, a probate court may decide the validity of wills but may not try murder cases.

Subpoena A writ issued on court authority to compel a person to appear in court at a judicial proceeding or to produce material, the disobedience of which is punishable as contempt of court.

Substantial Evidence Test An evidentiary test used to decide if certain agency actions are supported by enough facts.

Substantive Due Process A doctrine according to which certain state or federal substantive laws might be unconstitutional because they interfere with certain protected liberties and cannot be justified by important governmental interests.

Substantive Law Laws governing our relationships with other people and things in society; different than procedural law.

Substantive Regulations Agency rules that have the effect of law; they must be made in accordance with the authority granted to the agency.

Substantive Rules Legal rules setting out rights, duties, and standards of behavior applying to the public at large.

Sue to seek justice from another by legal process; to initiate a legal action.

Suit A proceeding in a civil court in which a person seeks remedy for an injury or loss.

Summary Judgment A decision rendered by a judge in response to a motion claiming that there is no factual dispute in a case, and that hence there is no need to send the case to a jury because the dispute involves only a question of law, not questions of fact.

Summons An order requiring the defendant in a case to appear in court.

Sunset Laws Laws that cause an administrative agency to end at a certain time.

Sunshine Exeptions Certain agency meetings that do not have to be open to the public under open meeting or "sunshine" laws.

Superior Agent Rule A rule holding corporate principals liable only for crimes of high-level employees.

Superior Courts State courts of general jurisdiction where major civil or criminal lawsuits usually start. These are courts of record with possible jury trials unlike municipal, justice of the peace, domestic relations, common pleas, juvenile, and probate courts.

Supra Latin, "above." A term that, in a written work, directs the reader to a preceding portion of the work or previous citation.

Supreme Court Reporter A set of privately published books containing U.S. Supreme Court decisions in chronological order.

Survival Statutes Statutes that keep a cause of action from dying when the person who has the cause of action dies; for example survival statutes give the dependents of a person killed in an auto accident a claim against the person who killed him or her.

Suspect Classifications Governmental discrimination based on race or national origin, sex, illegitimacy and alienage. These classifications all get some form of equal protection strict scrutiny.

Sustain To support or approve as true, legal, or just.

T

Taxing Power Congressional power to lay and collect taxes. Basically used to raise revenue, this congressional power is also an important regulatory tool.

Tenant One who holds land by any kind of title, license, right, or permission.

Tenant in Partnership The way partners own partnership property.

Tenet A principle, belief, or doctrine generally held to be true.

Testify To give evidence; to make a statement in a legal proceeding.

Testimony A statement made by a witness under oath in a legal proceeding.

Title the evidence of one's right to property and defining the extent of one's interest in that property; property concept used to identify owner of property.

Title Documents Pieces of paper which tell who owns certain types of property such as vehicles and real estate.

Tort A legal wrong against an individual to whom a duty was owed. The object is to compensate. It is a private or civil wrong as opposed to a public wrong resulting from the breach of a legal duty owed to an individual or group if individuals, not society as a whole. It is distinguished from a breach of contract.

Tort-feasor One who commits a tort.

Treble Damages Triple damages.

Trespass Wrongful entry onto the property of another; a type of tort (civil wrong) in which one person unlawfully interferes with one's property or person.

Trial Court Consists of a judge with or without a jury which performs four basic functions in the administration of legal justice: 1) fact-finding — determines the facts that are in conflict in particular controversies that are brought before it in technical legal form. Decides which witnesses are believable; 2) determines which rules of law from which areas of law apply; 3) applies the rules to the facts; 4) does all of this following prescribed procedures. Some courts have general powers; other courts have special powers.

Tribunal A court or other body having authority to adjudicate matters.

Trust Legal way to hold property in which there are two titles, one legal the other equitable.

Trustee Person(s) holding legal title to property in a trust.

U

Ultra Vires Latin for , "beyond the powers." An action taken which is beyond the power authorized by law for an entity. It is often applied to corporations that exceed the scope of their authorized powers or an administrative agency which exceeds the authority granted to it by statute.

Unannotated Without footnotes explaining the text (said of codes or statutes where no footnotes exist referring to cases interpreting or refining the code).

Unannotated Code A lengthy statute that does not refer to particular cases interpreting it.

Unconscionability An open-ended contract doctrine under which courts refuse to enforce especially one-sided contracts or contract terms. The conscience of the court is shocked.

Undisclosed Principal A type of agency relationship where the third person knows neither the fact that an agency exists nor that the person dealt with is an agent nor the principal's identity.

Undue Influence Where a trust relationship (although not necessarily a formal trust) exists between two people who make a contract and the one in whom the other places trust "double-crosses" the trusting person making a contract to the disadvantage of the trusting party which are grounds for a court to declare the contract voidable (not enforceable).

Uniform Guidelines on Employee Selection Procedures (UGESP) Equal Employment Opportunity Guidelines that help employers comply with federal civil rights laws.

Uniform Limited Partnership Act A model statute governing the formations of limited partnerships.

Unilateral Contract A contract formed when one party promises something in exchange for the other's act ("a promise for an act").

United States Reports A set of books containing cases in chronological order decided by the U.S. Supreme Court; publisher is U.S. government.

U.S. Claims Court A federal court where people can sue the U.S. government.

U.S. Court of Appeals A federal, multi-judge appellate court.

U.S. Court of International Trade A federal court with the power to hear cases involving tariff matters.

Use Immunity A rule of criminal procedural law that says if the prosecutor gives a criminal suspect immunity from prosecution in exchange for the suspect's information about crimes, the prosecutor will not in any way use that information to convict the person who gave the information.

V

"V" Versus, Latin for "against." A word used in the title of a case between the parties who oppose each other in the case, as in the State v. Smith.

Valid Having legal efficacy or force; true.

Venue A term for the jurisdiction in which a case may be tried.

Verdict The decision of the jury, or the judge acting as the tier of fact, on a question of fact.

Vested Fixed.

Vested Benefits Benefits that are someone's property and which cannot be taken away unless the person is given due process of law.

Vesting Giving someone property rights or the equivalent.

Vicarious Liability Responsibility of one person for the actions of another. It is present in the concept of respondent superior, in which an employer may be held liable for the unlawful actions of an employee.

Void Having no legal force or effect; said of illegal contracts.

Voidable In contract matters, a contract that because of some legal flaw can be set aside by at least one of the parties to it.

Voidable Title A flawed ownership in property which allows the holder of the property to pass "good," i.e., flawless, title to good faith purchasers for value without notice of the flaw; the flaw in the holder of the voidable title would let someone to claim superior title to the property and take the property away from the person with the voidable title.

Violation An infringement of the law.

Voir Dire Questioning prospective jurors to see if they are appropriate to sit on the jury in a particular case.

Voluntary Manslaughter A homicide in which there is an intentional killing committed under circumstances that, although they do not justify the killing, they lessen the seriousness of it.

W

Waive Voluntarily give up a legal right, such as "waive jury trial."

Ward Someone not qualified to manage his or her own affairs (a legal incompetent); also a political subdivision; someone who has a guardian appointed to hand his/her legal matters.

Warrant A written order, issued by a lawful authority directing the arrest of a certain person or persons, or writ authorizing the doing of a certain act, such as a search.

Warranty A contractual promise as to the nature of a thing sold.

Withdrawal Words used in labor law to decide a way of settling an unfair labor practice case by, in effect, the complaining party's dropping its complaint.

Witness One who testifies in a case under oath or who was present at an event so as to be able to say what has taken place.

Worker's Compensation A system of paying workers for damages resulting from job-related injuries, illness, or death; the system abandons fault as a basis for assigning liability for such work-related losses; the employer bears the cost of such losses.

Writ A mandatory order issued by an authorized court or tribunal which compels a person to do something or refrain from doing something.

Writ of Execution A legal device that lets a public official take certain property of a debtor, sell it, and use the proceeds to pay off a debt owed a creditor.

Wrong An injurious, unfair, or illegal act.

Wrongful Interference with Contractual Relations An intentional tort in which a third party knows that two other persons have a contract and knowingly interferes with that contract.

THE CONSTITUTION OF THE UNITED STATES OF AMERICA

We the People of the United States, in Order to form a more perfect Union, establish Justice, insure domestic Tranquility, provide for the common defence, promote the general Welfare, and secure the Blessings of Liberty to ourselves and our Posterity, do ordain and establish this Constitution for the United States of America.

Article I
Section 1

All legislative Powers herein granted shall be vested in a Congress of the United States, which shall consist of a Senate and House of Representatives.

Section 2

The House of Representatives shall be composed of Members chosen every second Year by the People of the several States, and the Electors in each State shall have the Qualifications requisite for Electors of the most numerous Branch of the State Legislature.

No Person shall be a Representative who shall not have attained to the Age of twenty five Years, and been seven Years a Citizen of the United States, and who shall not, when elected, be an Inhabitant of that State in which he shall be chosen.

Representatives and direct Taxes shall be apportioned among the several States which may be included within this Union, according to their respective Numbers, which shall be determined by adding to the whole Number of free Persons, including those bound to Service for a Term of Years, and excluding Indians not taxed, three fifths of all other Persons. The actual Enumeration shall be made within three Years after the first Meeting of the Congress of the United States, and within every subsequent Term of ten Years, in such Manner as they shall by Law direct. The Number of Representatives shall not exceed one for every thirty Thousand, but each State shall have at Least one Representative; and until such enumeration shall be made, the State of New Hampshire shall be entitled to chuse three, Massachusetts eight, Rhode Island and Providence Plantations one, Connecticut five, New York six, New Jersey four, Pennsylvania eight, Delaware one, Maryland six, Virginia ten, North Carolina five, South Carolina five, and Georgia three.

When vacancies happen in the Representation from any State, the Executive Authority thereof shall issue Writs of Election to fill such vacancies.

The House of Representatives shall ch use their Speaker and other Officers; and shall have the sole Power of Impeachment.

Section 3

The Senate of the United States shall be composed of two Senators from each State, chosen by the Legislature thereof, for six Years; and each Senator shall have one Vote.

Immediately after they shall be assembled in Consequence of the first Election, they shall be divided as equally as may be into three Classes. The Seats of the Senators of the first Class shall be vacated at the Expiration of the second Year, of the second Class at the Expiration of the fourth Year, and of the third Class at the Expiration of the sixth Year, so that one third may be chosen every second Year; and if Vacancies happen by Resignation, or otherwise, during the Recess of the Legislature of any State, the Executive thereof may make temporary Appointments until the next Meeting of the Legislature, which shall then fill such Vacancies.

No Person shall be a Senator who shall not have attained to the Age of thirty Years, and been nine Years a Citizen of the United States, and who shall not, when elected, be an Inhabitant of that State for which he shall be chosen.

The Vice President of the United States shall be President of the Senate but shall have no Vote, unless they be equally divided.

The Senate shall chuse their other Officers, and also a President pro tempore, in the Absence of the Vice President, or when he shall exercise the Office of President of the United States.

The Senate shall have the sole Power to try all Impeachments. When sitting for that Purpose, they shall be on Oath or Affirmation. When the President of the United States is tried the Chief Justice shall preside: And no Person shall be convicted without the Concurrence of two thirds of the Members present.

Judgment in Cases of Impeachment shall not extend further than to removal from Office, and disqualification to hold and enjoy any Office of honor, Trust or Profit under the United States; but the Party convicted shall nevertheless be liable and subject to Indictment, Trial, Judgment and Punishment, according to Law.

Section 4

The Times, Places and Manner of holding Elections for Senators and Representatives, shall be prescribed in each State by the Legislature thereof; but the Congress may at any time by Law make or alter such Regulations, except as to the Places of chusing Senators.

The Congress shall assemble at least once every Year, and such Meeting shall be on the first Monday in December, unless they shall by Law appoint a different Day.

Section 5

Each House shall be the Judge of the Elections, Returns and Qualifications of its own Members, and a Majority of each shall constitute a Quorum to do Business; but a smaller Number may adjourn from day to day, and may be authorized to compel the Attendance of absent Members, in such Manner and under such Penalties as each House may provide.

Each House may determine the Rules of its Proceedings, punish its Members for disorderly Behaviour, and, with the Concurrence of two thirds, expel a Member.

Each House shall keep a Journal of its Proceedings, and from time to time publish the same, excepting such Parts as may in their Judgement require Secrecy; and the Yeas and Nays of the Members of either House on any question shall, at the Desire of one fifth of those Present, be entered on the Journal.

Neither House, during the Session of Congress, shall, without the Consent of the other, adjourn for more than three days, nor to any other Place than that in which the two Houses shall be sitting.

Section 6

The Senators and Representatives shall receive a Compensation for their Services, to be ascertained by Law, and paid out of the Treasury of the United States. They shall in all Cases, except Treason, Felony and Breach of the Peace, be privileged from Arrest during their Attendance at the Session of their respective Houses and in going to and returning from the same; and for any Speech or Debate in either House, they shall not be questioned in any other Place.

No Senator or Representative shall, during the Time for which he was elected, be appointed to any civil Office under the Authority of the United States, which shall have been created, or the Emoluments whereof shall have been increased during such time, and no Person holding any Office under the United States, shall be a Member of either House during his Continuance in Office.

Section 7

All Bills for raising Revenue shall originate in the House of Representatives; but the Senate may propose or concur with amend-ments as on other Bills.

Every Bill which shall have passed the House of Representatives and the Senate, shall, before it become a Law, be presented to the President of the United States, If he approve he shall sign it, but if not he shall return it, with his Objections to that House in which it shall have originated, who shall enter the Objections at large on their Journal, and proceed to reconsider it. If after such Reconsideration two thirds of that House shall agree to pass the Bill, it shall be sent, together with the Objections, to the other House, by which it shall likewise be reconsidered, and if approved by two thirds of that House, it shall become a Law. But in all such Cases the Votes of both Houses shall be determined by Yeas and Nays, and the Names of the Persons voting for and against the Bill shall be entered on the Journal of each House respectively. If any Bill shall not be returned by the

President within ten Days (Sundays excepted) after it shall have been presented to him, the Same shall be a Law, in like Manner as if he had signed it, unless the Congress by their Adjournment prevent its Return, in which Case it shall not be a Law.

Every Order, Resolution, or Vote to which the Concurrence of the Senate and House of Representatives may be necessary (except on a question of Adjournment) shall be presented to the President of the United States; and before the Same shall take Effect, shall be approved by him, or being disapproved by him, shall be repassed by two thirds of the Senate and House of Representatives, according to the Rules and Limitations prescribed in the Case of a Bill.

Section 8

The Congress shall have Power to lay and collect Taxes, Duties, Imposts and Excises, to pay the Debts and provide for the common Defence and general Welfare of the United States; but all Duties, Imposts and Excises shall be uniform throughout the United States;

To borrow Money on the credit of the United States;

To regulate Commerce with foreign Nations, and among the several States, and with the Indian Tribes;

To establish an uniform Rule of Naturalization, and uniform Laws on the subject of Bankruptcies throughout the United States;

To coin Money, regulate the Value thereof, and of foreign Coin, and fix the Standard of Weights and Measures;

To provide for the Punishment of counterfeiting the Securities and current Coin of the United States;

To establish Post Offices and post Roads;

To promote the Progress of Science and useful Arts, by securing for limited Times to Authors and Inventors the exclusive Right to their respective Writings and Discoveries;

To constitute Tribunals inferior to the supreme Court;

To define and punish Piracies and Felonies committed on the high Seas, and Offences against the Law of Nations;

To declare War, grant Letters of Marque and Reprisal, and make Rules concerning Captures on Land and Water;

To raise and support Armies, but no Appropriation of Money to that Use shall be for a longer Term than two Years;

To provide and maintain a Navy;

To make Rules for the Government and Regulation of the land and naval Forces;

To provide for calling forth the Militia to execute the Laws of the Union, suppress Insurrections and repel Invasions;

To provide for organizing, arming, and disciplining, the Militia, and for governing such Part of them as may be employed in the Service of the United States, reserving to the States respectively, the Appointment of the Officers, and the Authority of training the Militia according to the discipline prescribed by Congress;

To exercise exclusive Legislation in all Cases whatsoever, over such District (not exceeding ten Miles square) as may, by Cession of particular States, and the Acceptance of Congress, become the Seat of the Government of the United States, and to exercise like Authority over all Places purchased by the Consent of the Legislature of the State in which the Same shall be, for the Erection of Forts, Magazines, Arsenals, dock-Yards, and other needful Buildings;—And

To make all Laws which shall be necessary and proper for carrying into Execution the foregoing Powers, and all other Powers vested by this Constitution in the Government of the United States, or in any Department or Officer thereof.

Section 9

The Migration or Importation of such Persons as any of the States now existing shall think proper to admit, shall not be prohibited by the Congress prior to the Year one thousand eight hundred and eight, but a Tax or duty may be imposed on such Importation, not exceeding ten dollars for each Person.

The Privilege of the Writ of Habeas Corpus shall not be suspended, unless when in Cases of Rebellion or Invasion the public Safety may require it.

No Bill of Attainder or ex post facto Law shall be passed.

No Capitation, or other direct, Tax shall be laid, unless in Proportion to the Census or Enumeration herein before directed to be taken.

No Tax or Duty shall be laid on Articles exported from any State.

No Preference shall be given by any Regulation of Commerce or Revenue to the Ports of one State over those of another; nor shall Vessels bound to, or from, one State, be obliged to enter, clear or pay Duties in another.

No Money shall be drawn from the Treasury, but in Consequence of Appropriations made by Law; and a regular Statement and Account of the Receipts and Expenditures of all public Money shall be published from time to time.

No Title of Nobility shall be granted by the United States: And no Person holding any Office of Profit or Trust under them, shall, without the Consent of the Congress, accept of any present, Emolument, Office, or Title, of any kind whatever, from any King, Prince or foreign State.

Section 10

No State shall enter into any Treaty, Alliance, or Confederation; grant Letters of Marque and Reprisal; coin Money; emit Bills of Credit; make any Thing but gold and silver Coin a Tender in Payment of Debts; pass any Bill of Attainder, ex post facto Law, or Law impairing the Obligation of Contracts, or grant any Title of Nobility.

No State shall, without the Consent of the Congress, lay any Imposts or Duties on Imports or Exports, except what may be absolutely necessary for executing its inspection Laws: and the net Produce of all Duties and Imposts, laid by any State on Imports or Exports, shall be for the Use of the Treasury of the United States, and all such Laws shall be subject to the revision and Controul of the Congress.

No State shall, without the Consent of Congress, lay any Duty of Tonnage, keep Troops, or Ships of War in time of Peace, enter into any Agreement or Compact with another State, or with a foreign Power, or engage in War, unless actually invaded, or in such imminent Danger as will not admit of delay.

Article II
Section 1

The executive Power shall be vested in a President of the United States of America. He shall hold his Office during the Term for four Years, and, together with the Vice President, chosen for the same Term. be elected, as follows:

Each State shall appoint, in such Manner as the Legislature thereof may direct, a Number of Electors, equal to the whole Number of Senators and Representatives to which the State may be entitled in the Congress: but no Senator or Representative, or Person holding an Office of Trust or Profit under the United States, shall be appointed an Elector.

The Electors shall meet in their respective States, and vote by Ballot for two Persons, of whom one at least shall not be an Inhabitant of the same State with themselves. And they shall make a List of all the Persons voted for and of the Number of Votes for each; which List they shall sign and certify, and transmit sealed to the Seat of the Government of the United States, directed to the President of the Senate. The President of the Senate shall, in the Presence of the Senate and House of Representatives, open all the Certificates, and the Votes shall then be counted. The Person having the greatest Number of Votes shall be the President, if such Number be a Majority of the whole Number of Electors appointed; and if there be more than one who have such Majority, and have an equal Number of Votes then the House of Representatives shall immediately chuse by Ballot one of them for President, and if no Person have a Majority, then from the five highest on the List the said House shall in like Manner chuse the President. But in chusing the President, the Votes shall be taken by States, the Representation from each State having one Vote; a quorum for this Purpose shall consist of a Member or Members from two thirds of the States, and a Majority of all the States shall be necessary to a Choice. In every Case, after the Choice of the President, the Person having the greater Number of Votes of the Electors shall be the Vice President. But if there should remain two or more who have equal Votes, the Senate shall chuse from them by Ballot the Vice President.

The Congress may determine the Time of chusing the Electors, and the Day on which they shall give their Votes; which Day shall be the same throughout the United States.

No Person except a natural born Citizen, or a Citizen of the United States, at the time of the Adoption of this Constitution, shall be eligible to the Office of President; neither shall any Person be eligible to that Office who shall not have attained to the Age of thirty five Years, and been fourteen years a Resident within the United States.

In Case of the Removal of the President from Office, or of his Death, Resignation, or Inability to discharge the Powers and Duties of the said Office, the Same shall devolve on the Vice President, and the Congress may by Law provide for the Case of Removal, Death, Resignation or Inability, both of the President and Vice President, declaring what Officer shall then act as President, and such Officer shall act accordingly, until the Disability be removed, or a President shall be elected.

The President shall, at stated Times, receive for his Services, a Compensation, which shall neither be increased nor diminished during the Period for which he shall have been elected, and he shall not receive within that Period any other Emolument from the United States, or any of them.

Before he enter on the Execution of his Office, he shall take the following Oath or Affirmation:—"I do solemnly swear (or affirm) that I will faithfully execute the Office of President of the United States, and will to the best of my Ability, preserve, protect and defend the Constitution of the United States."

Section 2

The President shall be Commander in Chief of the Army and Navy of the United States, and of the Militia of the several States, when called into the actual Service of the United States; he may require the Opinion, in writing, of the principal Officer in each of the executive Departments, upon any Subject relating to the Duties of their respective Offices, and he shall have Power to grant Reprieves and Pardons for Offences against the United States, except in Cases of Impeachment.

He shall have Power, by and with the Advice and Consent of the Senate, to make Treaties, provided two thirds of the Senators present concur; and he shall nominate, and by and with the Advice and Consent of the Senate, shall appoint Ambassadors, other public Ministers and Consuls, Judges of the supreme Court, and all other Officers of the United States, whose Appointments are not herein otherwise provided for, and which shall be established by Law: but the Congress may by Law vest the Appointment of such inferior Officers, as they think proper, in the President alone, in the Courts of Law, or in the Heads of Departments.

The President shall have Power to fill up all Vacancies that may happen during the Recess of the Senate, by granting Commissions which shall expire at the End of their next Session.

Section 3

He shall from time to time give to the Congress Information of the State of the Union, and recommend to their Consideration such Measures as he shall judge necessary and expedient; he may, on extraordinary Occasions, convene both Houses, or either of them, and in Case of Disagreement between them, with Respect to the Time of Adjournment, he may adjourn them to such Time as he shall think proper; he shall receive Ambassadors and other public Ministers; he shall take Care that the Laws be faithfully executed, and shall Commission all the Officers of the United States.

Section 4

The President, Vice President and all Civil Officers of the United States shall be removed from Office on Impeachment for, and Conviction of, Treason, Bribery, or other high Crimes and Misdemeanors.

Article III

Section 1

The judicial Power of the United States, shall be vested in one supreme Court, and in such inferior Courts as the Congress may from time to time ordain and establish. The Judges, both of the supreme and inferior Courts, shall hold their Offices during good Behaviour, and shall, at stated Times, receive for their Services, a Compensation, which shall not be diminished during their Continuance in Office.

Section 2

The judicial Power shall extend to all Cases, in Law and Equity, arising under this Constitution, the Laws of the United States, and Treaties made, or which shall be made under their Authority;—to all Cases affecting Ambassadors, other public Ministers and Consuls;—to all Cases of admiralty and maritime Jurisdiction;—to Controversies to which the United States shall be a Party;—to Controversies between two or more States;—between a State and Citizens of another State;—between Citizens of different States;—between Citizens of the same State claiming Lands under Grants of different States, and between a State, or the Citizens thereof, and foreign States, Citizens or Subjects.

In all Cases affecting Ambassadors, other public Ministers and Consuls, and those in which a State shall be a Party, the Supreme Court shall have original Jurisdiction. In all the other Cases before mentioned, the supreme Court shall have appellate Jurisdiction, both as to Law and Fact, with such Exceptions, and under such Regulations as the Congress shall make.

The Trial of all Crimes, except in Cases of Impeachment, shall be by Jury; and such Trial shall be held in the State where the said Crimes shall have been committed; but when not committed within any State, the Trial shall be at such Place or Places as the Congress may by Law have directed.

Section 3

Treason against the United States, shall consist only in levying War against them, or in adhering to their Enemies, giving them Aid and Comfort. No Person shall be convicted of Treason unless on the Testimony of two Witnesses to the same overt Act, or on Confession in open Court.

The Congress shall have Power to declare the Punishment of Treason, but no Attainder of Treason shall work Corruption of Blood or Forfeiture except during the Life of the Person attainted.

Article IV

Section 1

Full Faith and Credit shall be given in each State to the public Acts, Records, and judicial Proceedings of every other State. And the Congress may by general Laws prescribe the Manner in which such Acts, Records and Proceedings shall be proved and the Effect thereof.

Section 2

The Citizens of each State shall be entitled to all Privileges and Immunities of Citizens in the several States.

A Person charged in any State with Treason, Felony, or other Crime, who shall flee from Justice, and be found in another State, shall on Demand of the executive Authority of the State from which he fled, be delivered up, to be removed to the State having Jurisdiction of the Crime.

No Person held to Service or Labour in one State, under the Laws thereof, escaping into another, shall, in Consequence of any Law or

Regulation therein, be discharged from such Service or Labour, but shall be delivered up on Claim of the Party to whom such Service or Labour may be due.

Section 3

New States may be admitted by the Congress into this Union; but no new State shall be formed or erected within the Jurisdiction of any other State; nor any State be formed by the Junction of two or more States, or Parts of States, without the Consent of the Legislatures of the States concerned as well as of the Congress .

The Congress shall have Power to dispose of and make all needful Rules and Regulations respecting the Territory or other Property belonging to the United States; and nothing in this Constitution shall be so construed as to Prejudice any Claims of the United States, or of any particular State.

Section 4

The United States shall guarantee to every State in this Union a Republican Form of Government, and shall protect each of them against Invasion; and on Application of the Legislature, or of the Executive (when the Legislature cannot be convened) against domestic Violence.

Article V

The Congress, whenever two thirds of both Houses shall deem it necessary, shall propose Amendments to this Constitution, or, on the Application of the Legislatures of two thirds of the several States, shall call a Convention for proposing Amendments, which, in either Case, shall be valid to all Intents and Purposes, as Part of this Constitution, when ratified by the Legislatures of three fourths of the several States, or by Conventions in three fourths thereof, as the one or the other Mode of Ratification may be proposed by the Congress; Provided that no Amendment which may be made prior to the Year One thousand eight hundred and eight shall in any Manner affect the first and fourth Clauses in the Ninth Section of the first Article; and that no State, without its Consent, shall be deprived of its equal Suffrage in the Senate.

Article VI

All Debts contracted and Engagements entered into, before the Adoption of this Constitution, shall be as valid against the United States under this Constitution, as under the Confederation.

This Constitution, and the Laws of the United States which shall be made in Pursuance thereof; and all Treaties made, or which shall be made, under the Authority of the United States, shall be the supreme Law of the Land; and the judges in every State shall be bound thereby, any Thing in the Constitution or Laws of any State to the Contrary notwithstanding.

The Senators and Representatives before mentioned, and the Members of the several State Legislatures, and all executive and judicial Officers, both of the United States and of the several States, shall be bound by Oath or Affirmation, to support this Constitution; but no religious Test shall ever be required as a Qualification to any Office or public Trust under the United States.

Article VII

The Ratification of the Conventions of nine States, shall be sufficient for the Establishment of this Constitution between the States so ratifying the Same.

Amendment I [1791]

Congress shall make no law respecting an establishment of religion, or prohibiting the free exercise thereof; or abridging the freedom of speech, or of the press; or the right of the people peaceably to assemble and to petition the Government for a redress of grievances.

Amendment II [1791]

A well regulated Militia, being necessary to the security of a free State, the right of the people to keep and bear Arms, shall not be infringed.

Amendment III [1791]

No Soldier shall, in time of peace be quartered in any house, without the consent of the Owner, nor in time of war, but in a manner to be prescribed by law.

Amendment IV [1791]

The right of the people to be secure in their persons, houses, papers, and effects, against unreasonable searches and seizures, shall not be violated, and no Warrants shall issue, but upon probable cause, supported by Oath or affirmation, and particularly describing the place to be searched, and the persons or things to be seized.

Amendment V [1791]

No person shall be held to answer for a capital, or otherwise infamous crime, unless on a presentment or indictment of a Grand Jury, except in cases arising in the land or naval forces, or in the Militia, when in actual service in time of War or public danger; nor shall any person be subject for the same offence to be twice put in jeopardy of life or limb; nor shall be compelled in any criminal case to be a witness against himself, nor be deprived of life, liberty, or property, without due process of law; nor shall private property be taken for public use, without just compensation.

Amendment VI [1791]

In all criminal prosecutions, the accused shall enjoy the right to a speedy and public trial, by an impartial jury of the State and district wherein the crime shall have been committed, which district shall have been previously ascertained by law, and to be informed of the nature and cause of the accusation; to be confronted with the witnesses against him; to have compulsory process for obtaining Witnesses in his favor, and to have Assistance of Counsel for his defence.

Amendment VII [1791]

In Suits at common law, where the value in controversy shall exceed twenty dollars, the right of trial by jury shall be preserved, and no fact tried by a jury, shall be otherwise reexamined in any Court of the United States, than according to the rules of the common law.

Amendment VIII [1791]

Excessive bail shall not be required nor excessive fines imposed, nor cruel and unusual punishments inflicted.

Amendment IX [1791]

The enumeration in the Constitution, of certain rights, shall not be construed to deny or disparage others retained by the people.

Amendment X [1791]

The powers not delegated to the United States by the Constitution, nor prohibited by it to the States, are reserved to the States respectively, or to the people.

Amendment XI [1798]

The Judicial power of the United States shall not be construed to extend to any suit in law or equity, commenced or prosecuted against one of the United States by Citizens of another State, or by Citizens or Subjects of any Foreign State.

Amendment XII [1804]

The Electors shall meet in their respective states and vote by ballot for President and Vice President, one of whom, at least, shall not be an inhabitant of the same state with themselves; they shall name in their ballots the person voted for as President, and in distinct ballots the person voted for as Vice-President, and they shall make distinct lists of all persons voted for as President, and of all persons voted for as Vice-President, and of the number of votes for each, which lists they shall sign and certify, and transmit sealed to the seat of the government of the United States, directed to the President of the Senate,—The President of the Senate shall, in the presence of the Senate and House of Representatives, open all the certificates and the votes shall then be counted;—The person having the greatest number of votes for President, shall be the President, if such number be a majority of the whole number of Electors appointed; and if no person have such majority, then from the persons having the highest numbers not exceeding three on the list of those voted for as President, the House of Representatives shall choose immediately, by ballot, the President. But in choosing the President, the votes shall be taken by states, the representation from each state having one vote; a quorum for this purpose shall consist of a member or members from two-thirds of the states, and a majority of all the states shall be necessary to a choice. And if the House of Representatives shall not choose a President whenever the right of choice shall devolve upon them, before the fourth day of March next following, then the Vice-President shall act as President, as in the case of the death or other constitutional disability of the President—The person having the greatest number of votes as Vice-President, shall be the Vice-President, if such number be a majority of the whole number of electors appointed, and if no person have a majority, then from the two highest numbers on the list, the Senate shall choose the Vice-President; a quorum for the purpose shall consist of two-thirds of the whole number of Senators, and a majority of the whole number shall be necessary to a choice. But no person constitutionally ineligible to the office of President shall be eligible to that of Vice-President of the United States.

Amendment XIII [1865]
Section 1

Neither slavery nor involuntary servitude, except as punishment for crime whereof the party shall have been duly convicted, shall exist within the United States, or any place subject to their jurisdiction.

Section 2

Congress shall have power to enforce this article by appropriate legislation.

Amendment XIV [1868]
Section 1

All persons born or naturalized in the United States and subject to the jurisdiction thereof, are citizens of the United States and of the State wherein they reside. No State shall make or enforce any law which shall abridge the privileges or immunities of citizens of the United States; nor shall any State deprive any person of life, liberty, or property, without due process of law; nor deny to any person within its jurisdiction the equal protection of the laws.

Section 2

Representatives shall be apportioned among the several States according to their respective numbers, counting the whole number of persons in each State, excluding Indians not taxed. But when the right to vote at any election for the choice of electors for President and Vice

President of the United States, Representatives in Congress, the Executive and Judicial officers of a State, or the members of the Legislature thereof, is denied to any of the male inhabitants of such State, being twenty-one years of age, and citizens of the United States, or in any way abridged, except for participation in rebellion, or other crime, the basis of representation therein shall be reduced in the proportion which the number of such male citizens shall bear to the whole number of male citizens twenty-one years of age in such State.

Section 3

No person shall be a Senator or Representative in Congress, or elector of President and Vice President, or hold any office, civil or military, under the United States, or under any State, who, having previously taken an oath, as a member of Congress, or as an officer of the United States, or as a member of any State legislature, or as an executive or judicial officer of any State, to support the Constitution of the United States, shall have engaged in insurrection or rebellion against the same, or given aid or comfort to the enemies thereof. But Congress may by a vote of two-thirds of each House, remove such disability.

Section 4

The validity of the public debt of the United States, authorized by law, including debts incurred for payment of pensions and bounties for services in suppressing insurrection or rebellion, shall not be questioned. But neither the United States nor any State shall assume or pay any debt or obligation incurred in aid of insurrection or rebellion against the United State, or any claim for the loss or emancipation of any slave; but all such debts, obligations and claims shall be held illegal and void.

Section 5

The Congress shall have power to enforce, by appropriate legislation, the provisions of this article.

Amendment XV [1870]

Section 1

The right of citizens of the United States to vote shall not be denied or abridged by the United States or by any State on account of race, color, or previous condition of servitude.

Section 2

The Congress shall have power to enforce this article by appropriate legislation.

Amendment XVI [1913]

The Congress shall have power to lay and collect taxes on incomes, from whatever source derived, without apportionment among the several States, and without regard to any census or enumeration.

Amendment XVII [1913]

The Senate of the United States shall be composed of two Senators from each State, elected by the people thereof, for six years; and each Senator shall have one vote. The electors in each State shall have the qualifications requisite for electors of the most numerous branch of the State legislatures.

When vacancies happen in the representation of any State in the Senate, the executive authority of such State shall issue writs of election to fill such vacancies: *Provided,* That the legislature of any State may empower the executive thereof to make temporary appointments until the people fill the vacancies by election as the legislature may direct.

This amendment shall not be so construed as to affect the election or term of any Senator chosen before it becomes valid as part of the Constitution.

Amendment XVIII [1919]
Section 1

After one year from the ratification of this article the manufacture, sale, or transportation of intoxicating liquors within, the importation thereof into, or the exportation thereof from the United States and all territory subject to the jurisdiction thereof for beverage purposes is hereby prohibited.
Section 2

The Congress and the several States shall have concurrent power to enforce this article by appropriate legislation.

Section 3

This article shall be inoperative unless it shall have been ratified as an amendment to the Constitution by the legislatures of the several States, as provided in the Constitution, within seven years from the date of the submission hereof to the States by the Congress.

Amendment XIX [1920]

The right of citizens of the United States to vote shall not be denied or abridged by the United States or by any State on account of sex.

Congress shall have power to enforce this article by appropriate legislation.

Amendment XX [1933]

Section 1

The terms of the President and Vice President shall end at noon on the 20th day of January, and the terms of Senators and Representatives at noon on the 3d day of January, of the years in which such terms would have ended if this article had not been ratified; and the terms of their successors shall then begin.

Section 2

The Congress shall assemble at least once in every year, and such meeting shall begin at noon on the 3d day of January, unless they shall by law appoint a different day.

Section 3

If, at the time fixed for the beginning of the term of the President, the President elect shall have died, the Vice President elect shall become President. If the President shall not have been chosen before the time fixed for the beginning of his term, or if the President elect shall have failed to qualify, then the Vice President elect shall act as President until a President shall have qualified; and the Congress may by law provide for the case wherein neither a President elect nor a Vice President elect shall have qualified, declaring who shall then act as President, or the manner in which one who is to act shall be selected, and such person shall act accordingly until a President or Vice President shall have qualified.

Section 4

The Congress may by law provide for the case of the death of any of the persons from whom the House of Representatives may choose a President whenever the right of choice shall have devolved upon them, and for the case of death of any of the persons from whom the Senate may choose a Vice President whenever the right of choice shall have devolved upon them.

Section 5

Sections 1 and 2 shall take effect on the 15th day of October following the ratification of this article .

Section 6

This article shall be inoperative unless it shall have been ratified as an amendment to the Constitution by the legislatures of three-fourths of the several States within seven years from the date of its submission.

Amendment XXI [1933]

Section 1

The eighteenth article of amendment to the Constitution of the United States is hereby repealed.

Section 2

The transportation or importation into any State, Territory, or possession of the United States for delivery or use therein of intoxicating liquors, in violation of the laws thereof, is hereby prohibited.

Section 3

This article shall be inoperative unless it shall have been ratified as an amendment to the Constitution by conventions in the several states, as provided in the Constitution, within seven years from the date of the submission hereof to the States by the Congress.

Amendment XXII [1951]

Section 1

No person shall be elected to the office of the President more than twice, and no person who has held the office of President, or acted as President, for more than two years of a term to which some other person was elected President shall be elected to the office of the President more than once. But this Article shall not apply to any person holding the office of President when this Article was proposed by the Congress, and shall not prevent any person who may be holding the office of President, or acting as President, during the term within which the Article becomes operative from holding the office of President or acting as President during the remainder of such term.

Section 2

This article shall be inoperative unless it shall have been ratified as an amendment to the Constitution by the legislatures of three-fourths of the several States within seven years from the date of its submission to the States by the Congress.

Amendment XXIII [1961]

Section 1

The District constituting the seat of Government of the United States shall appoint in such manner as the Congress may direct:

A number of electors of President and Vice President equal to the whole number of Senators and Representatives in Congress to which the District would be entitled if it were a State, but in no event more than the least populous State; they shall be in addition to those appointed by the States, but they shall be considered, for the purposes of the election of President and Vice President, to be electors appointed by a State; and they shall meet in the District and perform such duties as provided by the twelfth article of amendment.

Section 2

The Congress shall have power to enforce this article by appropriate legislation.

Amendment XXIV [1964]

Section 1

The right of citizens of the United States to vote in any primary or other election for President or Vice President, for electors for President or Vice President, or for Senator or Representative in Congress, shall not be denied or abridged by the United States or any State by reason of failure to pay any poll tax or other tax.

Section 2

The Congress shall have power to enforce this article by appropriate legislation.

Amendment XXV [1967]

Section 1

In case of the removal of the President from office or of his death or resignation, the Vice President shall become President.

Section 2

Whenever there is a vacancy in the office of the Vice President, the President shall nominate a Vice President who shall take office upon confirmation by a majority vote of both Houses of Congress.

Section 3

Whenever the President transmits to the President pro tempore of the Senate and the Speaker of the House of Representatives his written declaration that he is unable to discharge the powers and duties of his office, and until he transmits to them a written declaration to the contrary, such powers and duties shall be discharged by the Vice President as Acting President.

Section 4

Whenever the Vice President and a majority of either the principal officers of the executive departments or of such other body as Congress may by law provide, transmit to the President pro tempore of the Senate and the Speaker of the House of Representatives their written declaration that the President is unable to discharge the powers and duties of his office, the Vice President shall immediately assume the powers and duties of the office as Acting President .

Thereafter, when the President transmits to the President pro tempore of the Senate and the Speaker of the House of Representatives his written declaration that no inability exists, he shall resume the powers and duties of his office unless the Vice President and a majority of either the principal officers of the executive department or of such other body as Congress may by law provide, transmit within four days to the President pro tempore of the Senate and the Speaker of the House of Representatives their written declaration that the President is unable to discharge the powers and duties of his office. Thereupon Congress shall decide the issue, assembling within forty-eight hours for that purpose if not in session. If the Congress, within twenty-one days after receipt of the latter written declaration, or, if Congress is not in session, within twenty-one days after Congress is required to assemble, determines by two-thirds vote of both Houses that the President is unable to discharge the powers and duties of his office, the Vice President shall continue to discharge the powers and duties of his office, the Vice President shall continue to discharge the same as Acting President; otherwise, the President shall resume the powers and duties of his office.

Amendment XXVI [1971]
Section 1

The right of citizens of the United States, who are eighteen years of age or older to vote shall not be denied or abridged by the United States or by any State on account of age.

Section 2

The Congress shall have power to enforce this article by appropriate legislation.

ENVIRONMENTAL LAWS & REGULATIONS ON ACCIDENT REPORTING AND OTHER INFORMATION COLLECTION PROVISIONS

Specification of the hazardous materials that may be released, the amount released, who must report the release, when it must be reported and to whom the reports must be made are among the primary provisions for accident reporting included in most environmental laws. Other provisions giving agencies authority to collect additional information are also included in environment statutes. Following is a discussion of the specific accident reporting components of each environmental law.

COMPREHENSIVE ENVIRONMENTAL RESPONSE, COMPENSATION, AND LIABILITY ACT (CERCLA)

Reporting and Information Collection Authorities: Section 102(a) and (b), 103(a),(b) and (f), 104(b)(1) and (e)

What Must Be Reported: CERCLA is one of the primary environmental laws with provisions for accident reporting as well as other federal information gathering activities. Section 101(14) of CERCLA defines hazardous substances as substances designated under section 102 and by reference to other environmental statutes including substances already designated under CAA section 112,CWA section 311 and section 307(a), RCRA section 3001 and TSCA section 7. Petroleum products or derivatives are specifically excluded from the CERCLA definition unless the substance is specifically listed.

CERCLA section 102 authorizes EPA to establish reportable quantities (RQs) for release of all CERCLA hazardous substances; section 102(a) also authorizes EPA to designate additional substances and section 102(b) imposes statutorily reportable quantities of one pound until EPA takes regulatory action adjusting the reportable quantities.

CERCLA contains the following exemptions:

- Any federally permitted release as defined by CERCLA section 101(10);
- Any release which is continuous as defined by CERCLA section 103(f), 40 CFR 302.3;
- Certain releases that result in exposure solely within a workplace, CERCLA section 101(22);
- Certain releases from FIFRA registered pesticides in accordance with its purpose, CERCLA section 103(e);
- Emissions from engine exhaust of a motor vehicle, rolling stock, aircraft, or pipeline pumping station, CERCLA section 101(22);
- Normal application of fertilizer, CERCLA section 101(22); and
- Releases of source, byproduct, or special nuclear material from a nuclear incident at a facility subject to the Price Anderson Act requirements, CERCLA section 101(22).

Who Must Report, When, and To Whom: Section 103(a) of CERCLA requires any person in charge of any facility who has knowledge of a release equal to or in excess of the reportable quantity to immediately notify to the National Response Center of the release. Exceptions to the §103(a) reporting requirements are noted above. Section 103(f) provides reporting relief from requirements of notification if the release is a continuous one and meets certain additional requirements.

Follow-up Reports: The law does not specify any additional reporting requirements after immediate notification to the NRC. For continuous release, follow-up notification is required by 40 CFR 302.8.

Other Reporting Authorities: Section 104(b)(1) provides broad authority to undertake investigations, monitor, survey, test, or gather other information deemed necessary whenever there is reason to believe that a release has occurred or is about to occur, or that illness, disease or complaints may be attributable to exposure and a release may have occurred or be occurring. Section 104(e), inter alia, grants persons designated by the President with the authority to obtain information from any person who has or may have information relevant to a release or threat of release to provide such information or documents relating to such matters. The information required may include the identification, nature and quantity of materials which have been or are generated, treated, stored transported or disposed, the nature and extent of the release or threatened release, and information relating to the ability of a person to pay for or to perform a cleanup.

EMERGENCY PLANNING AND COMMUNITY RIGHT-TO-KNOW (EPCRA OR SARA TITLE III)

Reporting Authorities: Section 304

What Must Be Reported: EPCRA section 302 required EPA to publish a specified list of extremely hazardous substances (EHSs), chemicals that could cause serious and irreversible damage to health and the environment. Some chemicals listed as extremely hazardous substances under EPCRA are also included in the CERCLA lists of hazardous substances.

EPCRA section 304 contains provisions for what must be reported in the event of a release. There are two lists of chemicals subject to the section 304 emergency release notification requirement. These are: l) Extremely Hazardous Substances identified under EPCRA section 302; and 2) CERCLA Hazardous Substances reportable under section 103(a) of CERCLA. The release must occur in a manner which would require notification under CERCLA section 103. Thus, the threshold question for section 304 is whether the release would be reportable — if this substance were covered under CERCLA.

The notification must include: the chemical name or identity of any substance involved in the release; an indication of whether the substance is extremely hazardous; an estimation of the quantity released; the time and duration of the release; the media or medium into which the release occurred; any known or anticipated acute or chronic health risks associated with the emergency; precautions needed as a result of the release; and the names and telephone numbers of persons to be contacted for further information.

Section 304 contains the following additional exemption: any release resulting in exposure to persons solely within the site or sites on which the facility is located.

Failure to report under section 304 is subject to civil or criminal penalties.

Who Must Report, When, and To Whom: Section 304 requires the owner or operator of a facility at which a hazardous chemical (as defined by OSHA's Hazard Communication Standard and EPCRA section 311) is produced, used, or stored to immediately report any releases of an EHS or a CERCLA hazardous substance in excess of its reportable quantity. Reports must be made to the State Emergency Response Commissions and to the Emergency Coordinator of the Local Emergency Planning Committees established under section 301 of EPCRA.

The owner or operator of a facility for which there is a transportation related release over the substance's reportable quantity may meet section 304 reporting requirements by providing the required information to the 911 emergency service or to a telephone operator if the 911 service is not used in the area. For transportation related releases in which the substance is also a CERCLA hazardous substance reports must also be made to the National Response Center. For purposes of section 304, section 329 includes motor vehicles, rolling stock, and aircraft in its definition of facility.

Follow-up Reports: As soon as practicable after a release, the owner or operator of a facility is required to submit written follow-up reports to the appropriate LEPC and SERC. These reports must include all the information required in the initial notice and updated or new information for: 1) actions taken to respond to and contain the release; 2) any known or anticipated acute or chronic health risks associated with the release; and, 3) advise, where possible, regarding appropriate medical procedures. EPA strongly recommends that the cause of the release also be reported.

CHEMICAL SAFETY AND HAZARD INVESTIGATION BOARD

The Chemical Safety and Hazard Investigation Board, authorized in the Clean Air Act §112(r)(6)(A)-(S), may investigate (or cause to be investigated) any accidental release resulting in a fatality, serious injury, or serious property damage. The Board may also establish accident reporting rules. After an investigation, the Board must determine and report to the public:

- The facts, conditions, and circumstances; and
- The cause or probable cause of any accidental release.

TOXIC SUBSTANCES CONTROL ACT (TSCA): SECTION 8(E)

What Must Be Reported: Must meet the criteria expressed by the statute.

Under TSCA 8(a), EPA may promulgate rules requiring manufacturers to keep records and make reports to EPA on chemical information such as categories of use, amounts produced, by-products, disposal methods, and the chemical's environment and health effects. The final regulation implementing section 8(a) at 47 Federal Register 26992 requires reporting of production, release, and exposure data for 245 specific chemicals on the inventory list. EPA also issued a final inventory update 51 Federal Register 21438 under 8(a) to require manufacturers of chemicals on the inventory list to report current data on the production, volume, plant size, site-limited status of the substances. TSCA 7 addresses chemicals presenting an "imminent hazard" and allows EPA to suspend production of a chemical through the courts, an authority which has not been exercised.

Section 8(e) of TSCA requires the submission to EPA of information that reasonably supports the conclusion that a chemical substance or mixture presents a substantial risk of injury to health or the environment. A substantial risk of injury to health or the environment is a risk of considerable concern because of: (a) the seriousness of the effect, and (b) the fact or probability of its occurrence. This provision does not require EPA to act (either regulate or otherwise request) to get information.

Who Must Report, When and To Whom: Substantial risk information must be submitted immediately, (within fifteen days after a person receives information, except in the case of emergency incidents of environmental condemnation, which must be reported by telephone as soon as a person subject to section 8(e) has knowledge of such information.) Section 8(e) applies only to:

- Persons or businesses engaged in manufacturing, processing, or distribution in commerce of chemical substances; and
- Information that is not already known to EPA, published in scientific literature, or submitted to EPA pursuant to mandatory reporting requirements under TSCA or any other EPA administered statutory authority, including notifications required under section 311 of the CWA. Information that corroborates well-established adverse effects is also not reportable.

The EPA interpretation of 8(e) was published in the Federal Register March 16,1978. Revisions are being considered.

RESOURCE CONSERVATION AND RECOVERY ACT (RCRA) 3007(A), 3001 ET SEQ.

Who Must Report, When and To Whom: Since all listed hazardous wastes under RCRA are CERCLA hazardous substances, they are subject to the CERCLA and EPCRA section 304 notification requirements. Treatment, storage and disposal facilities generally must follow the CERCLA and EPCRA 304 notification procedures for any release of a hazardous waste that could threaten health or the environment, even if the amount released is below the reportable quantity. If the release is below the RQ, however, notification may be made to the regional on-scene coordinator instead of the NRC.

Under 40 CFR section 264.196, tank treatment, storage and disposal facilities must notify the EPA regional of office within 24 hours of any hazardous waste release unless the release is under one pound and immediately recovered or it has already been reported under CERCLA (40 CFR 302). Any leak or drop in the level of a surface impoundment must be reported by the owner or operator in the same manner as a tank leak 40 CFR 264.226. Small generators (those that generate between 100 and 1000 kilograms of hazardous waste per month) storing waste onsite are required to notify the NRC in the event of any release that threatens health outside the facility or reaches surface water, 40 CFR 262.34. In this case also, the amount of the release requiring notification may be less than the reportable quantity. Transporters of hazardous waste must immediately report releases to the National Response Center, 40 CFR 263.30.

Follow-up Reports: For transporter of hazardous wastes, under 40 CFR section 263.30, written follow-up reports must be submitted within 15 days to the EPA regional office. A written notice to EPA is also required within 15 days of any incident that necessitated implementation of a RCRA contingency plan developed under 40 CFR Part 264 and 265, Subpart D.

A written follow-up report to the EPA regional office regarding tank treatment, storage and disposal facility releases must be made within 30 days. Generators of 100 to 1000 kg of hazardous wastes per month are not required to provide a written follow up report.

Other Reporting Authorities: RCRA section 3007 provides EPA the authority to collect information from persons who handle or have handled hazardous waste, and to inspect any establishment or other place where hazardous wastes are generated, stored, treated, disposed of, or transported from and obtain records and other information and to obtain samples. This section has been used by EPA as partial authority to develop the ARIP program.

TRANSPORTATION STATUTES AND REGULATIONS FOR ACCIDENT REPORTING, AND OTHER RELEVANT INFORMATION COLLECTION AUTHORITIES

A number of different transportation and pipeline safety laws contain provisions for accident reporting, including reporting of hazardous materials transportation accidents. The following section identifies and describes the accident reporting requirements of transportation laws.

Hazardous Materials Transportation Act (HMTA)

Each of the DOT modal administrations keeps separate modal accident data and several agencies keep data that include minimal information on releases of hazardous materials. However, RSPA is the official DOT repository of hazardous materials release information. A transportation-related incident or release is defined in DOT regulations as any unintentional release of a hazardous material during transportation, or during loading/unloading or temporary storage related to transportation. Every release, except for those from bulk water transporters and solely intrastate motor carriers that do not transport hazardous substances and wastes and cryogenic liquids, must be reported to RSPA in writing as prescribed in 49 CFR 171.16,174.45(rail), 175.45(air), and 176.48(marine vessels). The only other exceptions are consumer commodities that present a limited hazard during transportation, such as electric storage batteries and certain paints and materials. Notification of the NRC is required for those transportation releases that meet DOT's telephone reporting requirements, 49 CFR 171.15.

Who Must Report, When, and To Whom: Carriers are required to make an immediate telephone report to the National Response Center (NRC) when a spill has resulted in one or more of the following circumstances as a direct result of a hazardous material: a fatality; a serious injury requiring hospitalization; estimated carrier damage or other property damage exceeding $50,000; an evacuation of the public lasting one or more hours; the operational flight plan or routine of an aircraft is altered; fire, breakage, or suspected condemnation involving shipment of radioactive materials or etiologic agents; or a situation of such a nature that the carrier judges that a report should be made even though it does not meet the reporting criteria. Notice involving etiological agents may be given to the Director, Center for Disease Control.

Follow-up Reports: A written response must be prepared by the carrier on DOT Form F5800.1, for all incidents for which a telephone notice was made. A written report is also required whenever there is any unintentional release of a hazardous material during transportation. This report must be submitted to RSPA within 30 days of discovery of the release.

MODAL TRANSPORTATION AUTHORITIES WITH ACCIDENT REPORTING PROVISIONS AND REGULATIONS

Independent of the RSPA release reporting system are several accident reporting systems maintained by various modal administrations. In this instance, the term accident refers to a vehicular accident. Other than human error, the most frequently cited cause of hazardous materials transportation incidents, most hazardous materials transport releases are not caused by vehicular accidents themselves, but instead are due to other causes such as faulty valves or closures. The accident reporting systems designed by the modal administrations typically cover all transportation accidents under the jurisdiction of the particular modal administration, not just those involving hazardous materials. In many cases, however special identifiers have been placed in the reporting format that permit designation of an accident involving hazardous materials. Accident reports required by the various modal administrations are usually based on reporting procedures independent of RSPA due to differences in statute, regulation and information needs. Following are brief synopses of the statutory language and regulatory approach taken by each of the modal administrations for accident reporting.

OFFICE OF MOTOR CARRIERS, FORMERLY THE BUREAU OF MOTOR CARRIER SAFETY (BMCS)

The statutory and regulatory authority governing the accident reporting system for the Office of Motor Carriers is 49U.S.C.App 2505;49U.S.C.504 and 3102; and 49CFR 1.48. Regulations describing accident reporting are included in 49 CFR 394. A reportable accident means an occurrence involving a commercial motor vehicle engaged in the interstate, foreign, or intrastate operations of a motor carrier, who is subject to the DOT Act, resulting in a fatality or injury requiring medical treatment away from the scene of the accident; or total damage to all property aggregating $4,400. The rules do not apply to farm-to-market agricultural transportation of an occurrence in the course of the operation of a passenger car by a motor carrier not transporting passengers for hire or hazardous materials that require placarding.

What Must Be Reported: Any motor carrier accident in which a fatality or injury occurred or for which at least $2,000 in property damage was incurred must be reported. Reports are filed on Form 50-T, which requests carrier identification and address, location of the incident, characteristics of the event, cause, information on the cargo, and consequences of the accident. The carrier identification, cargo description and certain accident characteristics are also recorded.

Who Must Report, When, and To Whom: Immediate notification (within 24 hours) by telephone or in person to the Director, Regional Office of Motor Carriers of the Federal Highway Administration Region in which the carrier's principal place of business is located is required if a fatality is involved. Within 30 days after the accident, a written report must be filed.

FEDERAL RAILROAD ADMINISTRATION (FRA)

The statutory authority for railroad accident reporting was originally provided by the Accident Reports of 1910. This Act, was last amended by the Railroad Safety Act of 1970 and the FRA promulgated new accident reporting regulations in 1974. The statutory authority for railroad accident reporting is found in 45 U.S.C 38, 42, 43, and 43a, as amended; 45 U.S.C.431,437,and 438, as amended; Pub.L. 100-342;and 49 CFR 1.49(c) and (m). The regulations are found in 49 CFR Part 225. The purpose of the regulations is to provide FRA with information concerning hazardous conditions on the nation's railroads. Issuance of these regulations preempts states from prescribing accident or incident reporting requirements, although states may require that railroads submit copies of the accident or incident reports filed with FRA. Telephonic reports are required whenever an accident or incident arising from the operation of the railroad results in the death of a rail passenger or employee or the death or injury of five or more persons.

Written reports must be submitted to FRA monthly. The criteria for submitting a report is:

- Any impact between railroad on track equipment and an automobile, bus, truck, motorcycle, bicycle, farm vehicle, or pedestrian at a rail-highway grade crossing;

- Any collision, derailment, fire, explosion, act of God, or other event involving operation of railroad on track equipment (standing or moving) that results in more than S6.300in damages to railroad on-track equipment, signals, track, track structures, and roadbed;
- Any event arising from the operation of a railroad which results in death, injury to a non-railroad employee requiring medical treatment; injury to an employee requiring medical treatment or results in restriction of work or motion, one or more lost work days, transfer to another job, termination of employment, loss of consciousness, or occupational illness of a railroad employee as diagnosed by a physician.

Information is requested on the reporting form on derailed or damaged hazardous materials cars including whether the accident resulted in evacuation, explosion, fire, escape of hazardous materials and number evacuated.

NATIONAL HIGHWAY TRANSPORTATION SAFETY ADMINISTRATION (NHTSA)

NHTSA's Fatal Accident Reporting System (FARS) contains data on a census of fatal traffic crashes within the 50 States, the District of Columbia, and Puerto Rico. To be included in FARS, a crash must involve a motor vehicle travelling on a trafficway customarily open to the public and result in the death of a person (occupant of a vehicle or a non-motorist) within 30 days of the crash. There is one question on the FARS reporting form involving hazardous materials — "Hazardous Cargo - 0-no; 1-yes; 2-unknown.

FEDERAL AVIATION ADMINISTRATION (FAA)

FAA regulations (49 CFR 175.45) require each operator who transports hazardous materials to report hazardous materials incidents telephonically and in writing using the same criteria established by RSPA, as does the Coast Guard (CG).

OCCUPATIONAL SAFETY AND HEALTH ACT (OSHA) REPORTING: 29 CFR Part 1910

OSHA has standards dealing with a substantial number of substances as "toxic and hazardous substances~ in 29 CFR 1910.1000, that set forth limits for employee exposure to these substances.

Inspections and information gathering are covered by section 8 of the OSH Act, 29 USC 657. Employers must maintain accurate records and make them available to OSHA on request. OSHA is also authorized to prescribe regulations describing what records must be kept, including records about work-related deaths, injuries and illnesses, and exposure to potentially toxic materials.

Employers are also required to make available to employees their own records indicating exposure to toxic materials and, under certain OSHA standards applicable to toxic substances, to notify employees whenever they are exposed to toxic materials in excess of prescribed standards, and state what corrective actions are being taken. Employees will have access to their own records on exposure to toxic material or harmful physical agents.

OSHA has issued a number of individual occupational safety and health standards for various classes of hazardous substances. Typically, each standard includes reporting requirements, including the requirement that releases be reported to OSHA — e.g., 29 CFR 1910.1003(f), 29 CFR 1910.1004(f), 29 CFR 1910.1006(f).

Employers must report any accidental release of a hazardous substance to OSHA when there has been a death or when five or more workers have been hospitalized. States with approved OSHA plans may have more stringent reporting requirements.

DOE NOTIFICATION PROCEDURES

The notification and reporting component of the Emergency Management System (EMS) is primarily implemented by two Orders: DOE Order 5000.3A and DOE Order 5500.2B. The notification and reporting process is supplemented by the Occurrence Reporting and Processing System (ORPS), an operational data base used to transmit, update, and approve occurrence reports required under DOE Order 5000.3A.

DOE Order 5oo0.3A establishes the comprehensive system for reporting information related to operations occurring at DOE-owned or operated facilities. As part of this system, DOE Order 5500.2B requires all DOE facilities to: (1) promptly and accurately categorize all occurrences; (2) determine the appropriate category and class of events categorized as emergencies; and (3) make appropriate notifications and reports.

The Order specifies the criteria for categorizing emergencies. All operational occurrences must be categorized within 2 hours of identification. Emergencies are continually monitored for the purpose of reclassification as they evolve towards increased or decreased severity. It should be noted that the emergency classes outlined in the Order 5000.2B have been expanded to include a reference standard for non-radiological releases as well as radiological releases. EPA has promulgated Protective Action Guides to which the exposure level resulting from releases of radiological material are compared to determine the appropriate emergency class. Because EPA has not promulgated exposure levels for releases of non-radiological material, this Order uses the Emergency Response Planning Guidelines (ERPGs) developed and approved by the American Industrial Hygiene Association in determining the appropriate emergency class for releases of non-radiological material. Generally comparable exposure levels (i.e., no irreversible health effects) is considered for both radiological and non-radiological hazards. Further guidance on use of ERPGs is being developed to assist in the implementation of this Order.

Internal DOE reporting requirements vary according to each category of operational occurrence, as follows:

- Off-Normal Occurrence. These are abnormal or unplanned events or conditions that could adversely affect the safety, security, environment, or health protection performance or operation of a facility. Facility managers must provide written notification to DOE within 24 hours of the occurrence.
- Unusual Occurrence. These are non-emergency occurrences that have a potential for significant impact on a facility's safety, environment, health, security, or operations (e.g., a release of radioactive or hazardous materials above established limits).

Facility managers must verbally notify DOE as soon as possible, but no later than 2 hours after categorization, and must provide written notification within 24 hours.

- Emergency Occurrence . These include any actual or potential release of radioactive material or other material to the environment which could result in significant off-site consequences (e.g., need to relocate people, major wildlife kills, aquifer contamination, etc.). For emergency occurrences, facility managers must promptly notify state, tribal, local, and other federal agencies and must provide verbal notification to DOE within 15 minutes of emergency categorization. Written notification to DOE should be made as soon as practicable, but within 24 hours of emergency categorization.

Notifications are intended to promptly inform state, tribal, local, DOE, and other federal agencies of events categorized as emergencies. Additional notification requirements established in DOE Order 5500.2B include follow-up notification and communications between Field Elements and HQ Program Offices to meet the reporting requirements of other Orders (particularly DOE 5000.3A). Facility managers must make verbal follow-up notifications to DOE if any category of occurrence results in continued degradation in the level of safety at the facility or other worsening conditions, any change from one emergency action level to another, or if an emergency terminates. In addition, the DOE reporting requirements do not relieve DOE facilities from the notification and reporting requirements legally mandated by, or negotiated with, other federal, state, tribal, or local agencies. Specifically, the reporting requirements pursuant to CERCLA, RCRA, and EPCRA must be followed.

All notifications also must be made to DOE's Emergency Operations Center (EOC), which serves as the focal point for all emergency notifications and reports. The EOC receives, coordinates, and disseminates emergency information to HQ program elements and program office emergency points of contact, Congressional offices, the White House Situation Room, and other federal agencies.

The notification and response elements of the EMS are supplemented by the ORPS, which is an operational data base used by DOE contractor and Department Elements to transmit, update, and approve occurrence reports required under DOE Order 5000.3A.

REGULATIONS REGARDING BLOOD-BORNE PATHOGENS

STATE BLOOD SHIELD STATUTES

Ala. Code § 7-2-314(4)

Alaska Stat. § 45.02.316(e)

Ariz. Rev. Stat. § 36-1151

Ariz. Kev. Stat. Ann. § 32-1481

Ark. Code Ann. § 82-1608

Cal. Health & Safety Code §1606

Colo. Rev. Stat. §13-22-104

Conn. Gen. Stat. Ann. § 19a-280

Del. Code Ann. tit. 6, 2-316(5)

D.C., Fisher v. Sibley Memorial Hospital, 403 A.2d 1130 (D.C. App. 1979)

Fla. Stat. Ann. § 672.316(5), (6)

Ga. Code Ann. § 105-1105

Haw. Rev. Stat. § 327-51

Idaho Code § 39-3702

Ill. Ann. Stat. ch. 111 1/2, para. 5102

Ind. Code §16-8-7-2

Iowa Code Ann. § 142A.8

Kan. Stat. Ann. § 65-3701

Ky. Rev. Stat. Ann. §139.125

La. Rev. Stat. Ann. § 9:2797

La. Civ. Code Ann. art. 2322.1

Me. Rev. Stat. Ann. tit. 11, § 2-108

Md. Health-Gen. Code Ann. §18-402

Mass. Gen. Laws Ann. ch. 106, § 2-316(5)

Mich. Comp. Laws Ann. § 333.9121

Minn. Stat. Ann. §§ 525.928 and 525.921

Miss. Code Ann. § 41-41-1

Mo. Ann Stat. § 431.069

Mont. Code Ann §§ 50-33-101 through 50-33-104

Neb. Rev. Stat. § 71-4001

Nev. Rev. Stat. § 460.010

N.H. Rev. Stat. Ann. § 507-8b

N.J., Brody v. Overlook Hospital, 66 N.J. 448, 332 A.2d 596 (1975)

N.M. Stat. Ann. § 24-10-5

N.Y. Pub. Health Law § 580(4)

N.C. Gen. Stat. § 90-220.10 (1981)

N.D. Cent. Code §§ 43-17-40 and 41-02-33(3)(d)

Ohio Rev. Code Ann. § 2108.11

Okla. Stat. tit. 63, § 2151

Or. Rev. Stat. § 97.300

Pa. Stat. Ann. tit. 42, § 8333

R.I. Gen. Laws § 23-17-30

S.C. Code Ann. § 44-43-10

S.D. Codified Laws Ann. § 57A-2-315.1

Tenn. Code Ann. § 47-2-316(5)

Tex. Civ. Prac. & Rem. Code §§ 77.001-.004;

Tex. Rev. Civ. Stat. Ann. art.2.316 (Tex. U.C.C.)

Utah Code Ann. § 26-31-1

Va. Code Ann. § 32.1-297

Wash. Rev. Code Ann. § 70.54.120

W. Va. Code §16-23-1

Wis. Stat. §146-31

Wyo. Stat. § 34-21-233(c) (iv) (1977)

RECOMMENDATIONS FOR PREVENTING TRANSMISSION OF HUMAN IMMUNODEFICIENCY VIRUS AND HEPATITIS B VIRUS TO PATIENTS DURING EXPOSURE-PRONE INVASIVE PROCEDURES, 40 MMWR RR-8 (1991).

This document has been developed by the Centers for Disease Control (CDC) to update recommendations for prevention of transmission of human immunodeficiency virus (HIV) and hepatitis B virus (HBV) in the health-care setting. Current data suggest that the risk for such transmission from a health-care worker (HCW) to a patient during an invasive procedure is small; a precise assessment of the risk is not yet available. This document contains recommendations to provide guidance for prevention of HIV and HBV transmission during those invasive procedures that are considered exposure-prone.

Introduction

Recommendations have been made by the Centers for Disease Control (CDC) for the prevention of transmission of the human immunodeficiency virus (HIV) and the hepatitis B virus (HBV) in health-care settings. These recommendations emphasize adherence to universal precautions that require that blood and other specified body fluids of all patients be handled as if they contain blood-borne pathogens.

Previous guidelines contained precautions to be used during invasive procedures and recommendations for the management of HIV- and HBV-infected health-care workers (HCWs). These guidelines did not include specific recommendations on testing HCWs for HIV or HBV infection, and they did not provide guidance on which invasive procedures may represent increased risk to the patient.

The recommendations outlined in this document are based on the following considerations:

- Infected HCWs who adhere to universal precautions and who do not perform invasive procedures pose no risk for transmitting HIV or HBV to patients.

- Infected HCWs who adhere to universal precautions and who perform certain exposure-prone procedures pose a small risk for transmitting HBV to patients.
- HIV is transmitted much less readily than HBV.

In the interim, until further data are available, additional precautions are prudent to prevent HIV and HBV transmission during procedures that have been linked to HCW-to-patient HBV transmission or that are considered exposure-prone .

BACKGROUND

INFECTION-CONTROL PRACTICES
Previous recommendations have specified that infection-control programs should incorporate principles of universal precautions (i.e., appropriate use of hand washing, protective barriers, and care in the use and disposal of needles and other sharp instruments) and should maintain these precautions rigorously in all health-care settings. Proper application of these principles will assist in minimizing the risk of transmission of HIV and HBV from patient to HCW, HCW to patient, or patient to patient.

As part of standard infection-control practice, instruments and other reusable equipment used in performing invasive procedures should be appropriately disinfected and sterilized as follows:

- Equipment and devices that enter the patient's vascular system or other normally sterile areas of the body should be sterilized before being used for each patient.
- Equipment and devices that touch intact mucous membranes but do not penetrate the patient's body surfaces should be sterilized when possible or undergo high-level disinfection if they cannot be sterilized before being used for each patient.
- Equipment and devices that do not touch the patient or that only touch intact skin of the patient need only be cleaned with a detergent or as indicated by the manufacturer.

Compliance with universal precautions and recommendations for disinfection and sterilization of medical devices should be scrupulously monitored in all heath-care settings. Training of HCWs in proper infection control technique should begin in professional and vocational schools and continue as an ongoing process. Institutions should provide all HCWs with appropriate inservice education regarding infection control and safety and should establish procedures for monitoring compliance with infection-control policies.

All HCWs who might be exposed to blood in an occupational setting should receive hepatitis B vaccine, preferably during their period of professional training and before any occupational exposures could occur.

TRANSMISSION OF HBV DURING INVASIVE PROCEDURES

Since the introduction of serologic testing for HBV infection in the early 1970s, there have been published reports of 20 clusters in which a total of over 300 patients were infected with HBV in association with treatment by an HBV-infected HCW. In 12 of these clusters, the implicated HCW did not routinely wear gloves; several HCWs also had skin lesions that may have facilitated HBV transmission. These 12 clusters included nine linked to dentists or oral surgeons and one cluster each linked to a general practitioner, an inhalation therapist, and a cardiopulmonary-bypass-pump technician. The clusters associated with the inhalation therapist and the cardiopulmonary-bypass-pump technician—and some of the other 10 clusters—could possibly have been prevented if current recommendations on universal precautions, including glove use, had been in effect. In the remaining eight clusters, transmission occurred despite glove use by the HCWs; five clusters were linked to obstetricians or gynecologists, and three were linked to cardiovascular surgeons. In addition, recent unpublished reports strongly suggest HBV transmission from three surgeons to patients in 1989 and 1990 during colorectal (CDC, unpublished data), abdominal, and cardiothoracic surgery.

Seven of the HCWs who were linked to published clusters in the United States were allowed to perform invasive procedures following modification of invasive techniques (e.g., double gloving and restriction of certain high risk procedures). For five HCWs, no further transmission to patients was observed. In two instances involving an obstetrician/gynecologist and an oral surgeon, HBV was transmitted to patients after techniques were modified.

Review of the 20 published studies indicates that a combination of risk factors accounted for transmission of HBV from HCWs to patients. Of the HCWs whose hepatitis B e antigen (HBeAg) status was determined (17 of 20), all were HBeAg positive. The presence of HBeAg in serum is associated with higher levels of circulating virus and therefore with greater infectivity of hepatitis-B-surface-antigen (HBsAg)-positive individuals; the risk of HBV transmission to an HCW after a percutaneous exposure to HBeAg-positive blood is approximately 30%. In addition, each report indicated that the potential existed for contamination of surgical wounds or trauma-tized tissue, either from a major break in standard infection-control practices (e.g., not wearing gloves during invasive procedures) or from unintentional injury to the infected HCW during invasive procedures (e.g., needlesticks incurred while manipulating needles without being able to see them during suturing).

Most reported clusters in the United States occurred before awareness increased of the risks of transmission of blood-borne pathogens in healthcare settings and before emphasis was placed on the use of universal precautions and hepatitis B vaccine among HCWs. The limited number of reports of HBV transmission from HCWs to patients in recent years may reflect the adoption of univer-sal precautions and increased use of HBV vaccine. However, the limited number of recent reports does not preclude the occurrence of undetected or unreported small clusters or individual instances of transmission; routine use of gloves does not prevent most injuries caused by sharp instruments and does not eliminate the potential for exposure of a patient to an HCW's blood and transmission of HBV.

TRANSMISSION OF HIV DURING INVASIVE PROCEDURES

The risk of HIV transmission to an HCW after percutaneous exposure to HIV-infected blood is considerably lower than the risk of HBV transmission after percutaneous exposure to HBeAg-positive blood (0.3% versus approximately 30%). Thus, the risk of transmis-sion of HIV from an infected HCW to a patient during an invasive procedure is likely to be proportionately lower than the risk of HBV transmission from an HBeAg-positive HCW to a patient during the same procedure. As with HBV, the relative infectivity of HIV probably varies among individuals and over time for a single individual. Unlike HBV infection, however, there is currently no readily available laboratory test for increased HIV infectivity.

Investigation of a cluster of HIV infections among patients in the practice of one dentist with acquired immunodeficiency syndrome (AIDS) strongly suggested that HIV was transmitted to five of the approximately 850 patients evaluated through June 1991. The investigation indicates that HIV transmission occurred during dental care, although the precise mechanisms of transmission have not been determined. In two other studies, when patients cared for by a general surgeon and a surgical resident who had AIDS were tested, all patients tested, 75 and 62, respectively, were negative for HIV infection. In a fourth study, 143 patients who had been treated by a dental student with HIV infection and were later tested were all negative for HIV infection. In another investigation, HIV antibody testing was offered to all patients whose surgical procedures had been performed by a general surgeon within 7 years before the surgeon's diagnosis of AIDS; the date at which the surgeon became infected with HIV is unknown. Of 1,340 surgical patients contacted, 616 (46%) were tested for HIV. One patient, a known intravenous drug user, was HIV positive when tested but may already have been infected at the time of surgery. HIV test results for the 615 other surgical patients were negative (95% confidence interval for risk of transmission per operation = 0.0%-0.5%).

The limited number of participants and the differences in procedures associated with these five investigations limit the ability to generalize from them and to define precisely the risk of HIV transmission from HIV-infected HCWs to patients. A precise estimate of the risk of HIV transmission from infected HCWs to patients can be determined only after careful evaluation of a substantially larger number of patients whose exposure-prone procedures have been performed by HIV-infected HCWs.

EXPOSURE-PRONE PROCEDURES

Despite adherence to the principles of universal precautions, certain invasive surgical and dental procedures have been implicated in the transmission of HBV from infected HCWs to patients, and should be considered exposure-prone. Reported examples include certain oral, cardiothoracic, colorectal (CDC, unpublished data), and obstetric/gynecologic procedures.

Certain other invasive procedures should also be considered exposure-prone. In a prospective study CDC conducted in four hospitals, one or more percutaneous injuries occurred among surgical personnel during 96 (6.9%) of 1,382 operative procedures on the general surgery, gynecology, orthopedic, cardiac, and trauma services. Percutaneous exposure of the patient to the HCW's blood may have occurred when the sharp object causing the injury recontacted the patient's open wound in 28 (32%) of the 88 observed injuries to surgeons (range among surgical specialties = 8%-57%; range among hospitals = 24%-42%).

Characteristics of exposure-prone procedures include digital palpation of a needle tip in a body cavity or the simultaneous presence of the HCW's fingers and a needle or other sharp instrument or object in a poorly visualized or highly confined anatomic site. Performance of exposure-prone procedures presents a recognized risk of percutaneous injury to the HCW, and—if such an injury occurs—the HCW's blood is likely to contact the patient's body cavity, subcutaneous tissues, and/or mucous membranes.

Experience with HBV indicates that invasive procedures that do not have the above characteristics would be expected to pose substantially lower risk, if any, of transmission of HIV and other blood-borne pathogens from an infected HCW to patients.

RECOMMENDATIONS

Investigations of HIV and HBV transmission from HCWs to patients indicate that, when HCWs adhere to recommended infection-control procedures, the risk of transmitting HBV from an infected HCW to a patient is small, and the risk of transmitting HIV is likely to be even smaller. However, the likelihood of exposure of the patient to an HCW's blood is greater for certain procedures designated as exposure-prone. To minimize the risk of HIV or HBV transmission, the following measures are recommended:

- All HCWs should adhere to universal precautions, including the appropriate use of hand washing, protective barriers, and care in the use and disposal of needles and other sharp instruments. HCWs who have exudative lesions or weeping dermatitis should refrain from all direct patient care and from handling patient-care equipment and devices used in performing invasive procedures until the condition resolves. HCWs should also comply with current guidelines for disinfection and sterilization of reusable devices used in invasive procedures.

- Currently available data provide no basis for recommendations to restrict the practice of HCWs infected with HIV or HBV who perform invasive procedures not identified as exposure-prone, provided the infected HCWs practice recommended surgical or dental technique and comply with universal precautions and current recommendations for sterilization/disinfection.

- Exposure-prone procedures should be identified by medical/ surgical/dental organizations and institutions at which the procedures are performed .

- HCWs who perform exposure-prone procedures should know their HIV antibody status. HCWs who perform exposure-prone procedures and who do not have serologic evidence of immunity to HBV from vaccination or from previous infection should know their HBsAg status and, if that is positive, should also know their HBeAg status.

- HCWs who are infected with HIV or HBV (and are HBeAg positive) should not perform exposure-prone procedures unless they have sought counsel from an expert review panel and been advised under what circumstances, if any, they may continue to perform these procedures.* Such circumstances would include notifying prospective patients of the HCW's seropositivity before they undergo exposureprone invasive procedures.

- Mandatory testing of HCWs for HIV antibody, HBsAg, or HBeAg is not recommended. The current assessment of the risk that infected HCWs will transmit HIV or HBV to patients during exposure-prone procedures does not support the diversion of resources that would be required to implement mandatory testing programs. Compliance by HCWs with recommendations can be increased through education, training, and appropriate confidentiality safeguards.

The review panel should include experts who represent a balanced perspective. Such experts might include all of the following: a) the HCW's personal physician(s), b) an infectious disease specialist with expertise in the epidemiology of HIV and HBV transmission, c) a health professional with expertise in the procedures performed by the HCW, and d) state or local public health official(s). If the HCW's practice is institutionally based, the expert review panel might also include a member of the infection-control committee, preferably a hospital epidemiologist. HCWs who perform exposureprone procedures outside the hospital/institutional setting should seek advice from appropriate state and local public health officials regarding the review process. Panels must recognize the importance of confidentiality and the privacy rights of infected HCWs.

HCWS WHOSE PRACTICES ARE MODIFIED BECAUSE OF HIV OR HBV STATUS

HCWs whose practices are modified because of their HIV or HBV infection status should, whenever possible, be provided opportunities to continue appropriate patient-care activities. Career counseling and job retraining should be encouraged to promote the continued use of the HCW's talents, knowledge, and skills. HCWs whose practices are modified because of HBV infection should be reevaluated periodically to determine whether their HBeAg status changes due to resolution of infection or as a result of treatment .

NOTIFICATION OF PATIENTS AND FOLLOW-UP STUDIES

The public health benefit of notification of patients who have had exposure-prone procedures performed by HCWs infected with HIV or positive for HBeAg should be considered on a case-by-case basis, taking into consideration an assessment of specific risks, confidentiality issues, and available resources. Carefully designed and implemented follow-up studies are necessary to determine more precisely the risk of transmission during such procedures. Decisions regarding notification and follow-up studies should be made in consultation with state and local public health officials.

ADDITIONAL NEEDS

- Clearer definition of the nature, frequency, and circumstances of blood contact between patients and HCWs during invasive procedures.
- Development and evaluation of new devices, protective barriers, and techniques that may prevent such blood contact without adversely affecting the quality of patient care.
- More information on the potential for HIV and HBV transmission through contaminated instruments.
- Improvements in sterilization and disinfection techniques for certain reusable equipment and devices.
- Identification of factors that may influence the likelihood of HIV or HBV transmission after exposure to HIV- or HBV-infected blood.

DEFINITION OF INVASIVE PROCEDURE

An invasive procedure is defined as "surgical entry into tissues, cavities, or organs or repair of major traumatic injuries" associated with any of the following: "1) an operating or delivery room, emergency department, or outpatient setting, including both physicians' and dentists' offices; 2) cardiac catheterization and angiographic procedures; 3) a vaginal or cesarean delivery or other invasive obstetric procedure during which bleeding may occur; or 4) the manipulation, cutting, or removal of any oral or perioral tissues, including tooth structure, during which bleeding occurs or the potential for bleeding exists."

Reprinted from: Centers for Disease Control. Recommendation for prevention of HIV transmission in health-care settings. MMWR 1987; 36 (suppl. no. 2S):6S-7S.

CENTERS FOR DISEASE CONTROL UNIVERSAL PRECAUTIONS: RECOMMENDATIONS FOR PREVENTION OF HIV TRANSMISSION IN HEALTH-CARE SETTINGS, 36 MMWR 2S (1987); UPDATE: UNIVERSAL PRECAUTIONS FOR PREVENTION OF TRANSMISSION OF HUMAN IMMUNODEFICIENCY VIRUS, HEPATITIS B VIRUS, AND OTHER BLOODBORNE PATHOGENS IN HEALTH-CARE SETTINGS, 37 MMWR 377 (1988).

INTRODUCTION

Human immunodeficiency virus (HIV), the virus that causes acquired immunodeficiency syndrome (AIDS), is transmitted through sexual contact and exposure to infected blood or blood components and perinatally from mother to neonate. HIV has been isolated from blood, semen, vaginal secretions, saliva, tears, breast milk, cerebrospinal fluid, amniotic fluid, and urine and is likely to be isolated from other body fluids, secretions, and excretions. However, epidemiologic evidence has implicated only blood, semen, vaginal secretions, and possibly breast milk in transmission.

The increasing prevalence of HIV increases the risk that health-care workers will be exposed to blood from patients infected with HIV, especially when blood and body-fluid precautions are not followed for all patients. Thus, this document emphasizes the need for health-care workers to consider all patients as potentially infected with HIV and/or other bloodborne pathogens and to adhere rigorously to infection-control precautions for minimizing the risk of exposure to blood and body fluids of all patients.

The recommendations contained in this document consolidate and update CDC recommendations published earlier for preventing HIV transmission in health-care settings: precautions for clinical and laboratory staffs and precautions for health-care workers and allied professionals; recommendations for preventing HIV transmission in the workplace and during invasive procedures; recommendations for preventing possible transmission of HIV from tears; and recommendations for providing dialysis treatment for HIV-infected patients. These recommendations also update portions of the "Guideline for Isolation Precautions in Hospitals" and reemphasize some of the recommendations contained in "Infection Control Practices for Dentistry." The recommendations contained in this document have been developed for use in health-care settings and emphasize the need to treat blood and other body fluids from all patients as potentially infective. These same prudent precautions also should be taken in other settings in which persons may be exposed to blood and other body fluids.

DEFINITION OF HEALTH-CARE WORKERS

Health-care workers are defined as persons, including students and trainees, whose activities involve contact with patients with blood or other body fluids from patients in a health-care setting.

HEALTH-CARE WORKERS WITH AIDS

As of July 10, 1987, a total of 1,875 (5.8%) of 32,395 adults with AIDS, who had been reported to the CDC national surveillance system and for whom occupational information was available, reported being employed in a health-care or clinical laboratory setting. In comparison, 6.8 million persons—representing 5.6% of the U.S. labor force—were employed in health services. Of the health-care workers with AIDS, 95% have been reported to exhibit high-risk behavior; for the remaining 5%, the means of HIV acquisition was undetermined. Health-care workers with AIDS were significantly more likely than other workers to have an undetermined risk (5% versus 3%, respectively). For both health-care workers and non-health-care workers with AIDS, the proportion with an undetermined risk has not increased since 1982.

AIDS patients initially reported as not belonging to recognized risk groups are investigated by state and local health departments to determine whether possible risk factors exist. Of all health-care workers with AIDS reported to CDC who were initially characterized as not having an identified risk and for whom follow-up information was available, 66% have been reclassified because risk factors were identified or because the patient was found not to meet the surveillance case definition for AIDS. Of the 87 health-care workers currently categorized as having no identifiable risk, information is incomplete on 16 (18%) because of death or refusal to be interviewed; 38 (44%) are still being investigated. The remaining 33 (38%) health-care workers were interviewed or had other follow-up information available. The occupations of these 33 were as follows: five physicians (15%), three of whom were surgeons; one dentist (3%); three nurses (9%); nine nursing assistants (27%); seven housekeeping or maintenance workers (21%); three clinical laboratory technicians (9%); one therapist (3%); and four others who did not have contact with patients (12%). Although 15 of these 33 health-care workers reported parenteral and/or other non-needlestick exposure to blood or body fluids from patients in the 10 years preceding their diagnosis of AIDS, none of these exposures involved a patient with AIDS or known HIV infection.

RISK TO HEALTH-CARE WORKERS OF ACQUIRING HIV IN HEALTH-CARE SETTINGS

Health-care workers with documented percutaneous or mucous-membrane exposures to blood or body fluids of HIV-infected patients have been prospectively evaluated to determine the risk of infection after such exposures. As of June 30, 1987, 883 health-care workers have been tested for antibody to HIV in an ongoing surveillance project conducted by CDC.9 Of these, 708 (80%) had percutaneous exposures to blood, and 175 (20%) had a mucous membrane or an open wound contaminated by blood or body fluid. Of 396 health-care workers, each of whom had only a convalescent-phase serum sample obtained and tested >90 days post-exposure, one—for whom heterosexual transmission could not be ruled out—was seropositive for HIV antibody. For 425 additional health-care workers, both acute- and convalescent-phase serum samples were obtained and tested; none of 74 health-care workers with nonpercutaneous exposures seroconverted, and three (0.9%) of 351 with percutaneous exposures seroconverted. None of these three health-care workers had other documented risk factors for infection.

Two other prospective studies to assess the risk of nosocomial acquisition of HIV infection for health-care workers with a total of 453 needlestick or mucous-membrane exposures to the blood or other body fluids of HIV infected patients were tested for HIV antibody at the National Institutes of Health. These exposed workers included 103 with needlestick injuries and 229 with mucous-membrane exposures; none had seroconverted. A similar study at the University of California of 129 health-care workers with documented needlestick injuries or mucous-membrane exposures to blood or other body fluids from patients with HIV infection has not identified any seroconversions. Results of a prospective study in the United Kingdom identified no evidence of transmission among 150 health-care workers with parenteral or mucous-membrane exposures to blood or other body fluids, secretions, or excretions from patients with HIV infection.

In addition to health-care workers enrolled in prospective studies, eight persons who provided care to infected patients and denied other risk factors have been reported to have acquired HIV infection. Three of these healthcare workers had needlestick exposures to blood from infected patients. Two were persons who provided nursing care to infected persons; although neither sustained a needlestick, both had extensive contact with blood or other body fluids, and neither observed recommended barrier precautions. The other three were health-care workers with non-needlestick exposures to blood from infected patients. Although the exact route of transmission for these last three infections is not known, all three persons had direct contact of their skin with blood from infected patients, all had skin lesions that may have been contaminated by blood, and one also had a mucous-membrane exposure.

A total of 1,231 dentists and hygienists, many of whom practiced in areas with many AIDS cases, participated in a study to determine the prevalence of antibody to HIV; one dentist (0.1%) had HIV antibody. Although no exposure to a known HIV-infected person could be documented, epidemiologic investigation did not identify any other risk factor for infection. The infected dentist, who also had a history of sustaining needlestick injuries and trauma to his hands, did not routinely wear gloves when providing dental care.

PRECAUTIONS TO PREVENT TRANSMISSION OF HIV

UNIVERSAL PRECAUTIONS

Since medical history and examination cannot reliably identify all patients infected with HIV or other blood-borne pathogens, blood and body-fluid precautions should be consistently used for all patients. This approach, previously recommended by CDC, and referred to as "universal blood and body-fluid precautions" or "universal precautions," should be used in the care of all patients, especially including those in emergency-care settings in which the risk of blood exposure is increased and the infection status of the patient is usually unknown.

1. All health-care workers should routinely use appropriate barrier precautions to prevent skin and mucous-membrane exposure when contact with blood or other body fluids of any patients is anticipated. Gloves should be worn for touching blood and body fluids, mucous membranes, or non-intact skin of all patients, for handling items or surfaces soiled with blood or body fluids, and for performing venipuncture and other vascular access procedures. Gloves should be changed after contact with each patient. Masks and protective eyewear or face shields should be worn during procedures that are likely to generate droplets of blood or other body fluids to prevent exposure of mucous membranes of the mouth, nose, and eyes. Gowns or aprons should be worn during procedures that are likely to generate splashes of blood or other body fluids.

2. Hands and other skin surfaces should be washed immediately and thoroughly if contaminated with blood or other body fluids. Hands should be washed immediately after gloves are removed.

3. All health-care workers should take precautions to prevent injuries caused by needles, scalpels, and other sharp instruments or devices during procedures; when cleaning used instruments; during disposal of used needles; and when handling sharp instruments after procedures. To prevent needlestick injuries, needles should not be recapped, purposely bent or broken by hand, removed from disposable syringes, or otherwise manipulated by hand. After they are used, disposable syringes and

needles, scalpel blades, and other sharp items should be placed in puncture-resistant containers for disposal; the puncture-resistant containers should be located as close as practical to the use area. Large-bore reusable needles should be placed in a puncture-resistant container for transport to the reprocessing area.

4. Although saliva has not been implicated in HIV transmission, to minimize the need for emergency mouth-to-mouth resuscitation, mouthpieces, resuscitation bags, or other ventilation devices should be available for use in areas in which the need for resuscitation is predictable.

5. Health-care workers who have exudative lesions or weeping dermatitis should refrain from all direct patient care and from handling patient-care equipment until the condition resolves.

6. Pregnant health-care workers are not known to be at greater risk of contracting HIV infection than health-care workers who are not pregnant; however, if a health-care worker develops HIV infection during pregnancy, the infant is at risk of infection resulting from perinatal transmission. Because of this risk, pregnant health-care workers should be especially familiar with and strictly adhere to precautions to minimize the risk of HIV transmission.

Implementation of universal blood and body-fluid precautions for all patients eliminates the need for use of the isolation category of "Blood and Body Fluid Precautions" previously recommended by CDC for patients known or suspected to be infected with blood-borne pathogens. Isolation precautions (e.g., enteric, "AFB") should be used as necessary if associated conditions, such as infectious diarrhea or tuberculosis, are diagnosed or suspected.

PRECAUTIONS FOR INVASIVE PROCEDURES

In this document, an invasive procedure is defined as surgical entry into tissues, cavities, or organs or repair of major traumatic injuries l) in an operating or delivery room, emergency department, or outpatient setting, including both physicians' and dentists' offices; 2) cardiac catheterization and angiographic procedures; 3) a vaginal or cesarean delivery or other invasive obstetric procedure during which bleeding may occur; or 4) the manipulation, cutting, or removal of any oral or perioral tissues, including tooth structure, during which bleeding occurs or the potential for bleeding exists. The universal blood and body-fluid precautions listed above, combined with the precautions listed below, should be the minimum precautions for all such invasive procedures.

1. All heath-care workers who participate in invasive procedures must routinely use appropriate barrier precautions to prevent skin and mucous-membrane contact with blood and other body fluids of all patients. Gloves and surgical masks must be worn for all invasive procedures. Protective eyewear or face shields should be worn for procedures that commonly result in the generation of droplets, splashing of blood or other body fluids, or the generation of bone chips. Gowns or aprons made of materials that provide an effective barrier should be worn during invasive procedures that are likely to result in the splashing of blood or other body fluids. All health-care workers who perform or assist in vaginal or cesarean deliveries should wear gloves and gowns when handling the placenta or the infant until blood and amniotic fluid have been removed from the infant's skin and should wear gloves during post-delivery care of the umbilical cord.

2. If a glove is torn or a needlestick or other injury occurs, the glove should be removed and a new glove used as promptly as patient safety permits; the needle or instrument involved in the incident should also be removed from the sterile field.

ENVIRONMENTAL CONSIDERATIONS FOR HIV TRANSMISSION

No environmentally mediated mode of HIV transmission has been documented. Nevertheless, the precautions described below should be taken routinely in the care of all patients.

STERILIZATION AND DISINFECTION

Standard sterilization and disinfection procedures for patient-care equipment currently recommended for use in a variety of health-care settings—including hospitals, medical and dental clinics and offices, hemodialysis centers, emergency-care facilities, and long-term nursing-care facilities—are adequate to sterilize or disinfect instruments, devices, or other items contaminated with blood or other body fluids from persons infected with blood-borne pathogens including HIV.

Instruments or devices that enter sterile tissue or the vascular system of any patient or through which blood flows should be sterilized before reuse. Devices or items that contact intact mucous membranes should be sterilized or receive high-level disinfection, a procedure that kills vegetative organisms and viruses but not necessarily large numbers of bacterial spores. Chemical germicides that are registered with the U.S. Environmental Protection Agency (EPA) as "sterilants" may be used either for sterilization or for high-level disinfection depending on contact time.

Contact lenses used in trial fittings should be disinfected after each fitting by using a hydrogen peroxide contact lens disinfecting system or, if compatible, with heat (78 C-80 C [172.4 F-176.0 F]) for 10 minutes.

Medical devices or instruments that require sterilization or disinfection should be thoroughly cleaned before being exposed to the germicide, and the manufacturer's instructions for the use of the germicide should be followed. Further, it is important that the manufacturer's specifications for compatibility of the medical device with chemical germicides be closely followed. Information on specific label claims of commercial germicides can be obtained by writing to the Disinfectants Branch, Office of Pesticides, Environmental Protection Agency, 401 M Street, SW, Washington, D.C. 20460.

Studies have shown that HIV is inactivated rapidly after being exposed to commonly used chemical germicides at concentrations that are much lower than used in practice. Embalming fluids are similar to the types of chemical germicides that have been tested and found to completely inactivate HIV. In addition to commercially available chemical germicides, a solution of sodium hypochlorite (household bleach) prepared daily is an inexpensive and effective germicide. Concentrations ranging from approximately 500 ppm (1:100 dilution of household bleach) sodium hypochlorite to 5,000 ppm (1:10 dilution of household bleach) are effective depending on the amount of organic material (e.g., blood, mucus) present on the surface to be cleaned and disinfected. Commercially available chemical germicides may be more compatible with certain medical devices that might be corroded by repeated exposure to sodium hypochlorite, especially to the 1:10 dilution.

SURVIVAL OF HIV IN THE ENVIRONMENT

The most extensive study on the survival of HIV after drying involved greatly concentrated HIV samples, i.e., 10 million tissue-culture infectious doses per milliliter. This concentration is at least 100,000 times greater than that typically found in the blood or serum of patients with HIV infection. HIV was detectable by tissue-culture techniques 1-3 days after drying, but the rate of inactivation was rapid. Studies performed at CDC have also shown that drying HIV causes a rapid (within several hours) 1-2 log (90%-99%) reduction in HIV concentration. In tissue-culture fluid, cell-free HIV could be detected up to 15 days at room temperature, up to 11 days at 37 C (98.6 F), and up to 1 day if the HIV was cell-associated.

When considered in the context of environmental conditions in healthcare facilities, these results do not require any changes in currently recommended sterilization, disinfection, or housekeeping strategies. When medical devices are contaminated with blood or other body fluids, existing recommendations include the cleaning of these instruments, followed by disinfection or sterilization, depending on the type of medical device. These protocols assume "worst-case" conditions of extreme virologic and microbiologic contamination, and whether viruses have been inactivated after drying plays no role in formulating these strategies. Consequently, no changes in published procedures for cleaning, disinfecting, or sterilizing need to be made.

CLEANING AND DECONTAMINATING SPILLS OF BLOOD OR OTHER BODY FLUIDS

Chemical germicides that are approved for use as "hospital disinfectants" and are tuberculocidal when used at recommended dilutions can be used to decontaminate spills of blood and other body fluids. Strategies for decontaminating spills of blood and other body fluids in a patient-care setting are different than for spills of cultures or other materials in clinical, public health, or research laboratories. In patient-care areas, visible material should first be removed and then the area should be decontaminated. With large spills of cultured or concentrated infectious agents in the laboratory, the contaminated area should be flooded with a liquid germicide before cleaning, then decontaminated with fresh germicidal chemical. In both settings, gloves should be worn during the cleaning and decontaminating procedures.

LAUNDRY

Although soiled linen has been identified as a source of large numbers of certain pathogenic microorganisms, the risk of actual disease transmission is negligible. Rather than rigid procedures and specifications, hygienic and common-sense storage and processing of clean and soiled linen are recommended. Soiled linen should be handled as little as possible and with minimum agitation to prevent gross microbial contamination of the air and of persons handling the linen. All soiled linen should be bagged at the location where it was used; it should not be sorted or rinsed in patient-care areas. Linen soiled with blood or body fluids should be placed and transported in bags that prevent leakage. If hot water is used, linen should be washed with detergent in water at least 71 C (160 F) for 25 minutes. If low temperature (<70 C [158 F]) laundry cycles are used, chemicals suitable for low-temperature washing at proper use concentration should be used.

IMPLEMENTATION OF RECOMMENDED PRECAUTIONS

Employers of health-care workers should ensure that policies exist for:

1. Initial orientation and continuing education and training of all health-care workers—including students and trainees—on the epidemiology, modes of transmission, and prevention of HIV and other blood-borne infections and the need for routine use of universal blood and body-fluid precautions for all patients.

2. Provision of equipment and supplies necessary to minimize the risk of infection with HIV and other blood-borne pathogens.
3. Monitoring adherence to recommend protective measures. When monitoring reveals a failure to follow recommended precautions, counseling, education, and/or re-training should be provided, and, if necessary, appropriate disciplinary action should be considered.

Professional associations and labor organizations, through continuing education efforts, should emphasize the need for health-care workers to follow recommended precautions.

MANAGEMENT OF EXPOSURES

If a health-care worker has a parenteral (e.g., needlestick or cut) or mucous membrane (e.g., splash to the eye or mouth) exposure to blood or other body fluids or has cutaneous exposure involving large amounts of blood or prolonged contact with blood — especially when the exposed skin is chapped, abraded, or afflicted with dermatitis — the source patient should be informed of the incident and tested for serologic evidence of HIV infection after consent is obtained. Policies should be developed for testing source patients in situations in which consent cannot be obtained (e.g., an unconscious patient).

If the source patient has AIDS, is positive for HIV antibody, or refuses the test, the health-care worker should be counseled regarding the risk of infection and evaluated clinically and serologically for evidence of HIV infection as soon as possible after the exposure. The health-care worker should be advised to report and seek medical evaluation for any acute febrile illness that occurs within 12 weeks after the exposure. Such an illness — particularly one characterized by fever, rash, or lymphadenopathy — may be indicative of recent HIV infection. Seronegative health-care workers should be retested 6 weeks post-exposure and on a periodic basis thereafter (e.g., 12 weeks and 6 months after exposure) to determine whether transmission has occurred. During this follow-up period — especially the first 6-12 weeks after exposure, when most infected persons are expected to seroconvert — exposed health-care workers should follow U.S. Public

Health Service (PHS) recommendations for preventing transmission of HIV.

No further follow-up of a health-care worker exposed to infection as described above is necessary if the source patient is seronegative unless the source patient is at high risk of HIV infection. In the latter case, a subsequent specimen (e.g., 12 weeks following exposure) may be obtained from the health-care worker for antibody testing. If the source patient cannot be identified, decisions regarding appropriate follow-up should be individualized. Serologic testing should be available to all health-care workers who are concerned that they may have been infected with HIV.

If a patient has a parenteral or mucous-membrane exposure to blood or other body fluid of a health-care worker, the patient should be informed of the incident, and the same procedure outlined above for management of exposures should be followed for both the source health-care worker and the exposed patient.

PERSPECTIVES IN DISEASE PREVENTION AND HEALTH PROMOTION UPDATE: UNIVERSAL PRECAUTIONS FOR PREVENTION OF *TRANSMISSION OF HUMANIMMUNODEFICIENCY VIRUS*, HEPATITISB VIRUS, AND OTHER BLOODBORNE PATHOGENS IN HEALTH-CARE SETTINGS

INTRODUCTION

The purpose of this report is to clarify and supplement the CDC publication entitled "Recommendations for Prevention of HIV Transmission in Health Care Settings."

In 1983, CDC published a document entitled "Guideline for Isolation Precautions in Hospitals" that contained a section entitled "Blood and Body Fluid Precautions." The recommendations in this section called for blood and body fluid precautions when a patient was known or suspected to be infected with bloodborne pathogens. In August 1987, CDC published a document entitled "Recommendations for Prevention of HIV Transmission in Health-Care Settings." In contrast to the 1983 document, the 1987 docucment recommended that blood and body fluid precautions be consistently used for all patients regardless of their bloodborne infection status. This extension of blood and body fluid precautions to *all* patients is referred to as "Universal Blood and Body Fluid Precautions" or "Universal Precautions." Under universal precautions, blood and certain body fluids of all patients are considered potentially infectious for human immuno-deficiency virus (HIV), hepatitis B virus (HBV), and other bloodborne pathogens.

Universal precautions are intended to prevent parenteral, mucous membrane, and nonintact skin exposures of health-care workers to bloodborne pathogens. In addition, immunization with HBV vaccine is recommended as an important adjunct to universal precautions for health-care workers who have exposures to blood.

Since the recommendations for universal precautions were published in August 1987, CDC and the Food and Drug Administration (FDA) have received requests for clarification of the following issues: 1) body fluids to which universal precautions apply, 2) use of protective barriers, 3) use of gloves for phlebotomy, 4) selection of gloves for use while observing universal precautions, and 5) need for making changes in waste management programs as a result of adopting universal precautions.

BODY FLUIDS TO WHICH UNIVERSAL PRECAUTIONS APPLY

Universal precautions apply to blood and to other body fluids containing visible blood. Occupational transmission of HIV and HBV to health-care workers by blood is documented. Blood is the single most important source of HIV, HBV, and other bloodborne pathogens in the occupational setting. Infection control efforts for HIV, HBV, and other bloodborne pathogens must focus on preventing exposures to blood as well as on delivery of HBV immunization.

Universal precautions also apply to semen and vaginal secretions. Although both of these fluids have been implicated in the sexual transmission of HIV and HBV, they have not been implicated in occupational transmission from patient to health-care worker. This observation is not unexpected, since exposure to semen in the usual health-care setting is limited, and the routine practice of wearing gloves for performing vaginal examinations protects health-care workers from exposure to potentially infectious vaginal secretions.

Universal precautions also apply to tissues and to the following fluids: cerebrospinal fluid (CSF), synovial fluid, pleural fluid, peritoneal fluid, pericardial fluid, and amniotic fluid. The risk of transmission of HIV and HBV from these fluids is unknown; epidemiologic studies in the health-care and community setting are currently inadequate to assess the potential risk to health-care workers from occupational exposures to them. However, HIV has been isolated from CSF, synovial, and amniotic fluid, and HBsAg has been detected in synovial fluid, amniotic fluid, and peritoneal fluid. One case of HIV transmission was reported after a percutaneous exposure to bloody pleural fluid obtained by needle aspiration. Whereas aseptic procedures used to obtain these fluids for diagnostic or therapeutic purposes protect health-care workers from skin exposures, they cannot prevent penetrating injuries due to contaminated needles or other sharp instruments.

BODY FLUIDS TO WHICH UNIVERSAL PRECAUTIONS DO NOT APPLY

Universal precautions do not apply to feces, nasal secretions, sputum, sweat, tears, urine, and vomitus unless they contain visible blood. The risk of transmission of HIV and HBV from these fluids and materials is extremely low or nonexistent. HIV has been isolated and HBsAg has been demonstrated in some of these fluids; however, epidemiologic studies in the health-care and community setting have not implicated these fluids or materials in the transmission of HIV and HBV infections. Some of the above fluids and excretions represent a potential source for nosocomial and community-acquired infections with other pathogens, and recommendations for preventing the transmission of nonbloodborne pathogens have been published.

PRECAUTIONS FOR OTHER BODY FLUIDS IN SPECIAL SETTINGS

Human breast milk has been implicated in perinatal transmission of HIV, and HBsAg has been found in the milk of mothers infected with HBV. However, occupational exposure to human breast milk has not been implicated in the transmission of HIV nor HBV infection to health-care workers. Moreover, the health-care worker will not have the same type of intensive exposure to breast milk as the nursing neonate. Whereas universal precautions do not apply to human breast milk, gloves may be worn by health-care workers in situations where exposures to breast milk might be frequent, for example, in breast milk banking.

Saliva of some persons infected with HBV has been shown to contain HBV-DNA at concentrations 1/1,000 to 1/10,000 of that found in the infected person's serum. HBsAg-positive saliva has been shown to be infectious when injected into experimental animals and in human bite exposures. However, HBsAg-positive saliva has not been shown to be infectious when applied to oral mucous membranes in experimental primate studies or through contamination of musical instruments or cardiopulmonary resuscitation dummies used by HBV carriers. Epidemiologic studies of non sexual household contacts of HIV-infected patients, including several small series in which HIV transmission failed to occur after bites or after percutaneous inoculation or contamination of cuts and open wounds with saliva from HIV-infected patients, suggest that the potential for salivary transmission of HIV is remote. One case report from Germany has suggested the possibility of transmission of HIV in a household setting from an infected child to a sibling through a human bite. The bite did not break the skin or result in bleeding. Since the date of seroconversion to HIV was not known for either child in this case, evidence for the role of saliva in the transmission of virus is unclear. Another case report suggested the possibility of transmission of HIV from husband to wife by contact with saliva during kissing. However, follow-up studies did not confirm HIV infection in the wife.

Universal precautions do not apply to saliva. General infection control practices already in existence—including the use of gloves for digital examination of mucous membranes and endotracheal suctioning, and handwashing after exposure to saliva—should further minimize the minute risk, if any, for salivary transmission of HIV and HBV. Gloves need not be worn when feeding patients and when wiping saliva from skin.

Special precautions, however, are recommended for dentistry. Occupationally acquired infection with HBV in dental workers has been documented, and two possible cases of occupationally acquired HIV infection involving dentists have been reported. During dental procedures contamination of saliva with blood is predictable, trauma to health-care workers' hands is common, and blood spattering may occur. Infection control precautions for dentistry minimize the potential for nonintact skin and mucous membrane contact of dental health-care workers to blood-contaminated saliva of patients. In addition, the use of gloves for oral examinations and treatment in the dental setting may also protect the patient's oral mucous membranes from exposures to blood, which may occur from breaks in the skin of dental workers' hands.

USE OF PROTECTIVE BARRIERS

Protective barriers reduce the risk of exposure of the health-care worker's skin or mucous membranes to potentially infective materials. For universal precautions protective barriers reduce the risk of exposure to blood, body fluids containing visible blood, and other fluids to which universal precautions apply. Examples of protective barriers include gloves, gowns, masks, and protective eyewear. Gloves should reduce the incidence of contamination of hands, but they cannot prevent penetrating injuries due to needles or other sharp instruments. Masks and protective eyewear or face shields should reduce the incidence of contamination of mucous membranes of the mouth, nose, and eyes.

Universal precautions are intended to supplement rather than replace recommendations for routine infection control, such as handwashing and using gloves to prevent gross microbial contamination of hands. Because specifying the types of barriers needed for every possible clinical situation is impractical, some judgment must be exercised.

The risk of nosocomial transmission of HIV, HBV, and other bloodborne pathogens can be minimized if health-care workers use the following general guidelines:

1. Take care to prevent injuries when using needles, scalpels, and other sharp instruments or devices; when handling sharp instruments after procedures; when cleaning used instruments; and when disposing of used needles. Do not recap used needles by hand; do not remove used needles from disposable syringes by hand; and do not bend, break, or otherwise manipulate used needles by hand. Place used disposable syringes and needles, scalpel blades, and other sharp items in puncture-resistant containers for disposal. Locate the puncture-resistant containers as close to the use area as is practical.

2. Use protective barriers to prevent exposure to blood, body fluids containing visible blood, and other fluids to which universal precautions apply. The type of protective barrier(s) should be appropriate for the procedure being performed and the type of exposure anticipated.

3. Immediately and thoroughly wash hands and other skin surfaces that are contaminated with blood, body fluids containing visible blood, or other body fluids to which universal precautions apply.

GLOVE USE FOR PHLEBOTOMY

Gloves should reduce the incidence of blood contamination of hands during phlebotomy (drawing blood samples), but they cannot prevent penetrating injuries caused by needles or other sharp instruments. The likelihood of hand contamination with blood containing HIV, HBV, or other bloodborne pathogens during phlebotomy depends on several factors: 1) the skill and technique of the health-care worker, 2) the frequency with which the health care worker performs the procedure (other factors being equal, the cumulative risk of blood exposure is higher for a health-care worker who performs more procedures), 3) whether the procedure occurs in a routine or emergency situation (where blood contact may be more likely), and 4) the prevalence of infection with bloodborne pathogens in the patient population. The likelihood of infection after skin exposure to blood containing HIV or HBV will depend on the concentration of virus (viral concentration is much higher for hepatitis B than for HIV), the duration of contact, the presence of skin lesions on the hands of the health-care worker, and—for HBV—the immune status of the health-care worker. Although not accurately quantified, the risk of HIV infection following intact skin contact with infective blood is certainly much less than the 0.5% risk following percutaneous needlestick exposures. In universal precautions, *all* blood is assumed to be potentially infective for bloodborne pathogens, but in certain settings (e.g., volunteer blood-donation centers) the prevalence of infection with some bloodborne pathogens (e.g., HIV, HBV) is known to be very low. Some institutions have relaxed recommendations for using gloves for phlebotomy procedures by skilled phlebotomists in settings where the prevalence of bloodborne pathogens is known to be very low. Institutions that judge that routine gloving for *all* phlebotomies is not necessary should periodically reevaluate their policy. Gloves should always be available to health-care workers who wish to use them for phlebotomy. In addition, the following general guidelines apply:

1. Use gloves for performing phlebotomy when the health-care worker has cuts, scratches, or other breaks in his/her skin.
2. Use gloves in situations where the health-care worker judges that hand contamination with blood may occur, for example, when performing phlebotomy on an uncooperative patient.

3. Use gloves for performing finger and/or heel sticks on infants and children.

4. Use gloves when persons are receiving training in phlebotomy.

SELECTION OF GLOVES

The Center for Devices and Radiological Health, FDA, has responsibility for regulating the medical glove industry. Medical gloves include those marketed as sterile surgical or nonsterile examination gloves made of vinyl or latex. General purpose utility ("rubber") gloves are also used in the health-care setting, but they are not regulated by FDA since they are not promoted for medical use. There are no reported differences in barrier effectiveness between intact latex and intact vinyl used to manufacture gloves. Thus, the type of gloves selected should be appropriate for the task being performed.

The following general guidelines are recommended:

1. Use sterile gloves for procedures involving contact with normally sterile areas of the body.

2. Use examination gloves for procedures involving contact with mucous membranes, unless otherwise indicated, and for other patient care or diagnostic procedures that do not require the use of sterile gloves .

3. Change gloves between patient contacts.

4. Do not wash or disinfect surgical or examination gloves for reuse. Washing with surfactants may cause "wicking," i.e., the enhanced penetration of liquids through undetected holes in the glove. Disinfecting agents may cause deterioration.

5. Use general-purpose utility gloves (e.g., rubber household gloves) for housekeeping chores involving potential blood contact and for instrument cleaning and decontamination procedures. Utility gloves may be decontaminated and reused but should be discarded if they are peeling, cracked, or discolored, or if they have punctures, tears, or other evidence of deterioration.

OCCUPATIONAL EXPOSURE TO BLOODBORNE PATHOGENS, 56 FED. REG. 64004, 64175 (1991) (TO BE CODIFIED AT 29 C.F.R. PT. 1910).

XI. THE STANDARD

GENERAL INDUSTRY

Part 1910 of title 29 of the Code of Federal Regulations is amended as follows:

PART 1910—[AMENDED]

SUBPART Z—[AMENDED]

1. The general authority citation for subpart Z of 29 CFR part 1910 continues to read as follows and a new citation for §1910.1030 is added:

Authority: Secs. 6 and 8. Occupational Safety and Health Act, 29 U.S.C. 655, 657, Secretary of Labor's Orders Nos. 12 71 (36 FR 8754), 8-76 (41 FR 25059), or 9-83 (48 FR 35736), as applicable; and 29 CFR part 1911.

Section 1910.1030 also issued under 29 U.S.C. 653.

2. Section 1910.1030 is added to read as follows:

§1910.1030 BLOODBORNE PATHOGENS.

(a) *Scope and Application.* This section applies to all occupational exposure to blood or other potentially infectious materials as defined by paragraph (b) of this section.

(b) *Definitions.* For purposes of this section, the following shall apply:

Assistant Secretary means the Assistant Secretary of Labor of Occupational Safety and Health, or designated representative.

Blood means human blood, human blood components, and products made from human blood.

Bloodlborne Pathogens means pathogenic microorganisms that are present in human blood and can cause disease in humans. These pathogens include, but are not limited to, hepatitis B virus (HBV) and human immunodeficiency virus (HIV).

Clinical Laboratory means a workplace where diagnostic or other screening procedures are performed on blood or other potentially infectious materials.

Contaminated means the presence or the reasonably anticipated presence of blood or other potentially infectious materials on an item or surface.

Contaminated Laundry means laundry which has been soiled with blood or other potentially infectious materials or may contain sharps.

Contaminated Sharps means any contaminated object that can penetrate the skin including, but not limited to, needles, scalpels, broken glass, broken capillary tubes, and exposed ends of dental wires.

Decontamination means the use of physical or chemical means to remove, inactivate, or destroy bloodborne pathogens on a surface or item to the point where they are no longer capable of transmitting infectious particles and the surface or item is rendered safe for handling, use, or disposal.

Director means the Director of the National Institute for Occupational Safety and Health, U.S. Department of Health and Human Services, or designated representative.

Engineering Controls means controls (e.g., sharps disposal containers, self-sheathing needles) that isolate or remove the bloodborne pathogens hazard from the workplace.

Exposure Incident means a specific eye, mouth, other mucous membrane, non-intact skin, or parenteral contact with blood or other potentially infectious materials that results from the performance of an employee's duties.

Handwashing Facilities means a facility providing an adequate supply of running potable water, soap and single use towels or hot air drying machines.

Licensed Healthcare Professional is a person whose legally permitted scope of practice allows him or her to independently perform the activities required by paragraph (f) Hepatitis B Vaccination and Post-exposure Evaluation and Follow-up.

HBV means hepatitis B virus.

HIV means human immunodeficiency virus.

Occupational Exposure means reasonably anticipated skin, eye, mucous membrane, or parenteral contact with blood or other potentially infectious materials that may result from the performance of an employee's duties.

Other Potentially Infectious Materials means

(1) The following human body fluids: semen, vaginal secretions, cerebrospinal fluid, synovial fluid, pleural fluid, pericardial fluid, peritoneal fluid, amniotic fluid, saliva in dental procedures, any body fluid that is visibly contaminated with blood, and all body fluids in situations where it is difficult or impossible to differentiate between body fluids;

(2) Any unfixed tissue or organ (other than intact skin) from a human (living or dead); and

(3) HIV-containing cell or tissue cultures, organ cultures, and HIV- or HBV-containing culture medium or other solutions; and blood, organs, or other tissues from experimental animals infected with HIV or HBV.

Parenteral means piercing mucous membranes or the skin barrier through such events as needlesticks, human bites, cuts, and abrasions.

Personal Protective Equipment is specialized clothing or equipment worn by an employee for protection against a hazard. General work clothes (e.g., uniforms, pants, shirts or blouses) not intended to function as protection against a hazard are not considered to be personal protective equipment.

Production Facility means a facility engaged in industrial-scale, large-volume or high concentration production of HIV or HBV.

Regulated Waste means liquid or semi-liquid blood or other potentially infectious materials; contaminated items that would release blood or other potentially infectious materials in a liquid or semi-liquid state if compressed; items that are caked with dried blood or other potentially infectious materials and are capable of releasing these materials during handling; contaminated sharps; and pathological and microbiological wastes containing blood or other potentially infectious materials.

Research Laboratory means a laboratory producing or using research-laboratory-scale amounts of HIV or HBV. Research laboratories may produce high concentrations of HIV or HBV but not in the volume found in production facilities.

Source Individual means any individual, living or dead, whose blood or other potentially infectious materials may be a source of occupational exposure to the employee. Examples include, but are not limited to, hospital and clinic patients; clients in institutions for the developmentally disabled; trauma victims; clients of drug and alcohol treatment facilities; residents of hospices and nursing homes; human remains; and individuals who donate or sell blood or blood components.

Sterilize means the use of a physical or chemical procedure to destroy all microbial life including highly resistant bacterial endospores.

Universal Precautions is an approach to infection control. According to the concept of Universal Precautions, all human blood and certain human body fluids are treated as if known to be infectious for HIV, HBV, and other bloodborne pathogens.

Work Practice Controls means controls that reduce the likelihood of exposure by altering the manner in which a task is performed (e.g., prohibiting recapping of needles by a two-handed technique).

(c) *Exposure control*—(1) *Exposure Control Plan.* (i) Each employer having an employee(s) with occupational exposure as defined by paragraph (b) of this section shall establish a written Exposure Control Plan designed to eliminate or minimize employee exposure.

(ii) The Exposure Control Plan shall contain at least the following elements:

(A) The exposure determination required by paragraph (c)(2),

(B) The schedule and method of implementation for paragraphs (d) Methods of Compliance, (e) HIV and HBV Research Laboratories and Production Facilities, (f) Hepatitis B Vaccination and Post-Exposure Evaluation and Follow-up, (g) Communication of Hazards to Employees, and (h) Recordkeeping of this standard, and

(C) The procedure for the evaluation of circumstances surrounding exposure incidents as required by paragraph (f)(3)(i) of this standard.

(iii) Each employer shall ensure that a copy of the Exposure Control Plan is accessible to employees in accordance with 29 CFR 1910.20(e).

(iv) The Exposure Control Plan shall be reviewed and updated at least annually and whenever necessary to reflect new or modified tasks and procedures which affect occupational exposure and to reflect new or revised employee positions with occupational exposure.

(v) The Exposure Control Plan shall be made available to the Assistant Secretary and the Director upon request for examination and copying.

(2) *Exposure determination.* (i) Each employer who has an employee(s) with occupational exposure as defined by paragraph (b) of this section shall prepare an exposure determination. This exposure determination shall contain the following:

(A) A list of all job classifications in which all employees in those job classifications have occupational exposure;

(B) A list of job classifications in which some employees have occupational exposure, and

(C) A list of all tasks and procedures or groups of closely related task and procedures in which occupational exposure occurs and that are performed by employees in job classifications listed in accordance with the provisions of paragraph (c)(2)(i)(B) of this standard.

(ii) This exposure determination shall be made without regard to the use of personal protective equipment.

(d) *Methods of compliance*—(1) *General*—Universal precautions shall be observed to prevent contact with blood or other potentially infectious materials. Under circumstances in which differentiation between body fluid types is difficult or impossible, all body fluids shall be considered potentially infectious materials.

(2) *Engineering and work practice controls.* (i) Engineering and work practice controls shall be used to eliminate or minimize employee exposure. Where occupational exposure remains after institution of these controls, personal protective equipment shall also be used.

(ii) Engineering controls shall be examined and maintained or replaced on a regular schedule to ensure their effectiveness.

(iii) Employers shall provide handwashing facilities which are readily accessible to employees.

(iv) When provision of handwashing facilities is not feasible, the employer shall provide either an appropriate antiseptic hand cleanser in conjunction with clean cloth/paper towels or antiseptic towelettes. When antiseptic hand cleaners or towelettes are used, hands shall be washed with soap and running water as soon as feasible.

(v) Employers shall ensure that employees wash their hands immediately or as soon as feasible after removal of gloves or other personal protective equipment.

(vi) Employers shall ensure that employees wash hands and any other skin with soap and water, or flush mucous membranes with water immediately or as soon as feasible following contact of such body areas with blood or other potentially infectious materials.

(vii) Contaminated needles and other contaminated sharps shall not be bent, recapped, or removed except as noted in paragraphs (d)(2)(vii)(A) and (d)(2)(vii)(B) below. Shearing or breaking of contaminated needles is prohibited.

(A) Contaminated needles and other contaminated sharps shall not be recapped or removed unless the employer can demonstrate that no alternative is feasible or that such action is required by a specific medical procedure.

(B) Such recapping or needle removal must be accomplished through the use of a mechanical device or a one-handed technique.

(viii) Immediately or as soon as possible after use, contaminated reusable sharps shall be placed in appropriate containers until properly reprocessed. These containers shall be:

(A) Puncture resistant;

(B) Labeled or color-coded in accordance with this standard;

(C) Leakproof on the sides and bottom; and

(D) In accordance with the requirements set forth in paragraph (d)(4)(ii)(E) for reusable sharps.

(ix) Eating, drinking, smoking, applying cosmetics or lip balm, and handling contact lenses are prohibited in work areas where there is a reasonable likelihood of occupational exposure.

(x) Food and drink shall not be kept in refrigerators, freezers, shelves, cabinets or on countertops or benchtops where blood or other potentially infectious materials are present.

(xi) All procedures involving blood or other potentially infectious materials shall be performed in such a manner as to minimize splashing, spraying, spattering, and generation of droplets of these substances.

(xii) Mouth pipetting/suctioning of blood or other potentially infectious materials is prohibited.

(xiii) Specimens of blood or other potentially infectious materials shall be placed in a container which prevents leakage during collection, handling, processing, storage, transport, or shipping.

(A) The container for storage, transport, or shipping shall be labeled or color-coded according to paragraph (g)(l)(i) and closed prior to being stored, transported, or shipped. When a facility utilizes Universal Precautions in the handling of all specimens, the labeling/color-coding of specimens is not necessary provided containers are recognizable as containing specimens. This exemption only applies while such specimens/containers remain within the facility. Labeling or color-coding in accordance with paragraph (g)(l)(i) is required when such specimens/containers leave the facility.

(B) If outside contamination of the primary container occurs, the primary container shall be placed within a second container which prevents leakage during handling, processing, storage, transport, or shipping and is labeled or color-coded according to the requirements of this standard.

(C) If the specimen could puncture the primary container, the primary container shall be placed within a secondary container which is puncture resistant in addition to the above characteristics.

(xiv) Equipment which may become contaminated with blood or other potentially infectious materials shall be examined prior to servicing or shipping and shall be decontaminated as necessary, unless the employer can demonstrate that decontamination of such equipment or portions of such equipment is not feasible.

(A) A readily observable label in accordance with paragraph (g)(l)(i)(H) shall be attached to the equipment stating which portions remain contaminated.

(B) The employer shall ensure that this information is conveyed to all affected employees, the servicing representative, and/or the manufacturer, as appropriate, prior to handling, servicing, or shipping so that appropriate precautions will be taken.

(3) *Personal protective equipment*—(i) Provision. When there is occupational exposure, the employer shall provide, at no cost to the employee, appropriate personal protective equipment such as, but not limited to, gloves, gowns, laboratory coats, face shields or masks and eye protection, and mouthpieces, resuscitation bags, pocket masks, or other ventilation devices. Personal protective equipment will be considered "appropriate" only if it does not permit blood or other potentially infectious materials to pass through to or reach the employee's work clothes, street clothes, undergarments, skin, eyes, mouth, or other mucous membranes under normal conditions of use and for the duration of time which the protective equipment will be used.

(ii) Use. The employer shall ensure that the employee uses appropriate personal protective equipment unless the employer shows that the employee temporarily and briefly declined to use personal protective equipment when, under rare and extraordinary circumstances, it was the employee's professional judgment that in the specific instance its use would have prevented the delivery of health care or public safety services or would have posed an increased hazard to the safety of the worker or co-worker. When the employee makes this judgment, the circumstances shall be investigated and documented in order to determine whether changes can be instituted to prevent such occurrences in the future.

(iii)Accessibility. The employer shall ensure that appropriate personal protective equipment in the appropriate sizes is readily accessible at the worksite or is issued to employees. Hypoallergenic gloves, glove liners, powderless gloves, or other similar alternatives shall be readily accessible to those employees who are allergic to the gloves normally provided.

(iv) Cleaning, Laundering, and Disposal. The employer shall clean, launder, and dispose of personal protective equipment required by paragraphs (d) and (e) of this standard, at no cost to the employee.

(v) Repair and Replacement. The employer shall repair or replace personal protective equipment as needed to maintain its effectiveness, at no cost to the employee.

(vi) If a garment(s) is penetrated by blood or other potentially infectious materials, the garment(s) shall be removed immediately or as soon as feasible.

(vii) All personal protective equipment shall be removed prior to leaving the work area.

(viii) When personal protective equipment is removed it shall be placed in an appropriately designated area or container for storage, washing, decontamination or disposal.

(ix) Gloves. Gloves shall be worn when it can be reasonably anticipated that the employee may have hand contact with blood, other potentially infectious materials, mucous membranes, and non-intact skin; when performing vascular access procedures except as specified in paragraph (d)(3)(ix)(D); and when handling or touching contaminated items or surfaces.

(A) Disposable (single use) gloves such as surgical or examination gloves, shall be replaced as soon as practical when contaminated or as soon as feasible if they are torn, punctured, or when their ability to function as a barrier is compromised.

(B) Disposable (single use) gloves shall not be washed or decontaminated for reuse.

(C) Utility gloves may be decontaminated for reuse if the integrity of the glove is not compromised. However, they must be discarded if they are cracked, peeling, torn, punctured, or exhibit other signs of deterioration or when their ability to function as a barrier is compromised.

(D) If an employer in a volunteer blood donation center judges that routine gloving for all phlebotomies is not necessary then the employer shall:

(1) Periodically reevaluate this policy;

(2) Make gloves available to all employees who wish to use them for phlebotomy;

(3) Not discourage the use of gloves for phlebotomy; and

(4) Require that gloves be used for phlebotomy in the following circumstances:

(i)   When the employee has cuts, scratches, or other breaks in his or her skin;

(ii) When the employee judges that hand contamination with blood may occur, for example, when performing phlebotomy on an uncooperative source individual; and

(iii) When the employee is receiving training in phlebotomy.

(x)   Masks, Eye Protection, and Face Shields. Masks in combination with eye protection devices, such as goggles or glasses with solid side shields, or chin-length face shields, shall be worn whenever splashes, spray, spatter, or droplets of blood or other potentially infectious materials may be generated and eye, nose, or mouth contamination can be reasonably anticipated.

(xi) Gowns, Aprons, and Other Protective Body Clothing. Appropriate protective clothing such as, but not limited to, gowns, aprons, lab coats, clinic jackets, or similar outer garments shall be worn in occupational exposure situations. The type and characteristics will depend upon the task and degree of exposure anticipated.

(xii) Surgical caps or hoods and/or shoe covers or boots shall be worn in instances when gross contamination can reasonably be anticipated (e.g., autopsies, orthopedic surgery).

(4)   *Housekeeping.* (i) General. Employers shall ensure that the worksite is maintained in a clean and sanitary condition. The employer shall determine and implement an appropriate written schedule for cleaning and method of decontamination based upon the location within the facility, type of surface to be cleaned, type of soil present, and tasks or procedures being performed in the area.

(ii) All equipment and environmental and working surfaces shall be cleaned and decontaminated after contact with blood or other potentially infectious materials.

(A)   Contaminated work surfaces shall be decontaminated with an appropriate disinfectant after completion of procedures; immediately or as soon as feasible when surfaces are overtly contaminated or after any spill of blood or other potentially infectious materials and at the end of the work shift if the surface may have become contaminated since the last cleaning.

(B) Protective coverings, such as plastic wrap, aluminum foil, or imperviously backed absorbent paper used to cover equipment and environmental surfaces, shall be removed and replaced as soon as feasible when they become overtly contaminated or at the end of the workshift if they may have become contaminated during the shift.

(C) All bins, pails, cans, and similar receptacles intended for reuse which have a reasonable likelihood for becoming contaminated with blood or other potentially infectious materials shall be inspected and decontaminated on a regularly scheduled basis and cleaned and decontaminated immediately or as soon as feasible upon visible contamination.

(D) Broken glassware which may be contaminated shall not be picked up directly with the hands. It shall be cleaned up using mechanical means, such as a brush and dust pan, tongs, or forceps.

(E) Reusable sharps that are contaminated with blood or other potentially infectious materials shall not be stored or processed in a manner that requires employees to reach by hand into the containers where these sharps have been placed.

(iii) Regulated Waste.

(A) Contaminated Sharps Discarding and Containment. (I) Contaminated sharps shall be discarded immediately or as soon as feasible in containers that are:

(i) Closable;

(ii) Puncture resistant;

(iii) Leakproof on sides and bottom; and

(iv) Labeled or color-coded in accordance with paragraph (g)(l)(i) of this standard.

(2) During use, containers for contaminated sharps shall be:

(i) Easily accessible to personnel and located as close as is feasible to the immediate area where sharps are used or can be reasonably anticipated to be found (e.g., laundries);

(ii) Maintained upright throughout use; and

(iii) Replaced routinely and not be allowed to overfill.

(3) When moving containers of contaminated sharps from the area of use, the containers shall be:

(i) Closed immediately prior to removal or replacement to prevent spillage or protrusion of contents during handling, storage, transport, or shipping;

(ii) Placed in a secondary container if leakage is possible. The second container shall be:

(A) Closable;

(B) Constructed to contain all contents and prevent leakage during handling, storage, transport, or shipping; and

(C) Labeled or color-coded according to paragraph (g)(l)(i) of this standard.

(4) Reusable containers shall not be opened, emptied, or cleaned manually or in any other manner which would expose employees to the risk of percutaneous injury.

(B) Other Regulated Waste Containment. (1) Regulated waste shall be placed in containers which are:

(i) Closable;

(ii) Constructed to contain all contents and prevent leakage of fluids during handling, storage, transport or shipping;

(iii) Labeled or color-coded in accordance with paragraph (g)(l)(i) of this standard; and

(iv) Closed prior to removal to prevent spillage or protrusion of contents during handling, storage, transport, or shipping.

(2) If outside contamination of the regulated waste container occurs, it shall be placed in a second container. The second container shall be:

(i) Closable;

(ii) Constructed to contain all contents and prevent leakage of fluids during handling, storage, transport or shipping;

(iii) Labeled or color-coded in accordance with paragraph (g)(l)(i) of this standard; and

(iv) Closed prior to removal to prevent spillage or protrusion of contents during handling, storage, transport, or shipping.

(C) Disposal of all regulated waste shall be in accordance with applicable regulations of the United States, States and Territories, and political subdivisions of States and Territories.

(iv) Laundry.

(A) Contaminated laundry shall be handled as little as possible with a minimum of agitation. (1) Contaminated laundry shall be bagged or Containerized at the location where it was used and shall not be sorted or rinsed in the location of use.

(2) Contaminated laundry shall be placed and transported in bags or containers labeled or color-coded in accordance with paragraph (g)(l)(i) of this standard. When a facility utilizes Universal Precautions in the handling of all soiled laundry, alternative labeling or color-coding is sufficient if it permits all employees to recognize the containers as requiring compliance with Universal Precautions.

(3) Whenever contaminated laundry is wet and presents a reasonable likelihood of soak-through of or leakage from the bag or container, the laundry shall be placed and transported in bags or containers which prevent soak-through and/or leakage of fluids to the exterior.

(B) The employer shall ensure that employees who have contact with contaminated laundry wear protective gloves and other appropriate personal protective equipment.

(C) When a facility ships contaminated laundry off-site to a second facility which does not utilize Universal Precautions in the handling of all laundry, the facility generating the contaminated laundry must place such laundry in bags or containers which are labeled or color-coded in accordance with paragraph (g)(l)(i).

(e) *HIV and HBV Research Laboratories and Production Facilities.* (l) This paragraph applies to research laboratories and production facilities engaged in the culture, production, concentration, experimentation, and manipulation of HIV and HBV. It does not apply to clinical or diagnostic laboratories engaged solely in the analysis of blood, tissues, or organs. These requirements apply in addition to the other requirements of the standard.

(2) Research laboratories and production facilities shall meet the following criteria:

(i) Standard microbiological practices. All regulated waste shall either be incinerated or decontaminated by a method such as autoclaving known to effectively destroy bloodborne pathogens.

(ii) Special practices.

(A) Laboratory doors shall be kept closed when work involving HIV or HBV is in progress.

(B) Contaminated materials that are to be decontaminated at a site away from the work area shall be placed in a durable, leakproof, labeled or color-coded container that is closed before being removed from the work area.

(C) Access to the work area shall be limited to authorized persons. Written policies and procedures shall be established whereby only persons who have been advised of the potential biohazard, who meet any specific entry requirements, and who comply with all entry and exit procedures shall be allowed to enter the work areas and animal rooms.

(D) When other potentially infectious materials or infected animals are present in the work area or containment module, a hazard warning sign incorporating the universal biohazard symbol shall be posted on all access doors. The hazard warning sign shall comply with paragraph (g)(l)(ii) of this standard .

(E) All activities involving other potentially infectious materials shall be conducted in biological safety cabinets or other physical-containment devices within the containment module. No work with these other potentially infectious materials shall be conducted on the open bench.

(F) Laboratory coats, gowns, smocks, uniforms, or other appropriate protective clothing shall be used in the work areas and animal rooms. Protective clothing shall not be worn outside of the work area and shall be decontaminated before being laundered.

(G) Special care shall be taken to avoid skin contact with other potentially infectious materials. Gloves shall be worn when handling infected animals and when making hand contact with other potentially infectious materials is unavoidable.

(H) Before disposal all waste from work areas and from animal rooms shall either be incinerated or decontaminated by a method such as autoclaving to effectively destroy bloodborne pathogens.

(I) Vacuum lines shall be protected with liquid disinfectant traps and high-efficiency particulate air (HEPA) filters or filters of equivalent or superior efficiency and which are checked routinely and maintained or replaced as necessary.

(J) Hypodermic needles and syringes shall be used only for parenteral injection and aspiration of fluids from laboratory animals and diaphragm bottles. Only needle-locking syringes or disposable syringe-needle units (i.e. the needle is integral to the syringe) shall be used for the injection or aspiration of other potentially infectious materials. Extreme caution shall be used when handling needles and syringes. A needle shall not be bent, sheared, replaced in the sheath or guard, or removed from the syringe following use. The needle and syringe shall be promptly placed in a puncture-resistant container and autoclaved or decontaminated before reuse or disposal.

(K) All spills shall be immediately contained and cleaned up by appropriate professional staff or others properly trained and equipped to work with potentially concentrated infectious materials.

(L) A spill or accident that results in an exposure incident shall be immediately reported to the laboratory director or other responsible person.

(M) A biosafety manual shall be prepared or adopted and periodically reviewed and updated at least annually or more often if necessary. Personnel shall be advised of potential hazards, shall be required to read instructions on practices and procedures, and shall be required to follow them.

(iii) Containment equipment. (A) Certified biological safety cabinets (Class 1,11, or 111) or other appropriate combinations of personal protection or physical containment devices, such as special protective clothing, respirators, centrifuge safety cups, sealed centrifuge rotors, and containment caging for animals, shall be used for all activities with other potentially infectious materials that pose a threat of exposure to droplets, splashes, spills, or aerosols.

(B) Biological safety cabinets shall be certified when installed, whenever they are moved and at least annually.

(3) HIV and HBV research laboratories shall meet the following criteria:

(i) Each laboratory shall contain a facility for hand washing and an eye wash facility which is readily available within the work area.

(ii) An autoclave for decontamination of regulated waste shall be available .

(4) HIV and HBV production facilities shall meet the following criteria:

(i) The work areas shall be separated from areas that are open to unrestricted traffic flow within the building. Passage through two sets of doors shall be the basic requirement for entry into the work area from access corridors or other contiguous areas. Physical separation of the highcontainment work area from access corridors or other areas or activities may also be provided by a double-doored clothes-change room (showers may be included), airlock, or other access facility that requires passing through two sets of doors before entering the work area.

(ii) The surfaces of doors, walls, floors and ceilings in the work area shall be water resistant so that they can be easily cleaned. Penetrations in these surfaces shall be sealed or capable of being sealed to facilitate decontamination.

(iii) Each work area shall contain a sink for washing hands and a readily available eye wash facility. The sink shall be foot, elbow, or automatically operated and shall be located near the exit door of the work area.

(iv) Access doors to the work area or containment module shall be self-closing.

(v) An autoclave for decontamination of regulated waste shall be available within or as near as possible to the work area.

(vi) A ducted exhaust-air ventilation system shall be provided. This system shall create directional airflow that draws air into the work area through the entry area. The exhaust air shall not be recirculated to any other area of the building, shall be discharged to the outside, and shall be dispersed away from occupied areas and air intakes. The proper direction of the airflow shall be verified (i.e., into the work area).

(5) *Training Requirements.* Additional training requirements for employees in HIV and HBV research laboratories and HIV and HBV production facilities are specified in paragraph (g)(2)(ix).

(i) *Hepatitis B vaccination and post-exposure evaluation and follow-up —* *(1) General.* (i) The employer shall make available the hepatitis B vaccine and vaccination series to all employees who have occupational exposure, and post-exposure evaluation and follow-up to all employees who have had all exposure incident.

(ii) The employer shall ensure that all medical evaluations and procedures including the hepatitis B vaccine and vaccination series and postexposure evaluation and follow-up, including prophylaxis, are:

(A) Made available at no cost to the employee;

(B) Made available to the employee at a reasonable time and place;

(C) Performed by or under the supervision of a licensed physician or by or under the supervision of another licensed health care professional, and

(D) Provided according to recommendations of the U.S. Public Health Service current at the time these evaluations and procedures take place, except as specified by this paragraph (f).

(iii) The employer shall ensure that all laboratory tests are conducted by an accredited laboratory at no cost to the employee.

(2) *Hepatitis B Vaccination.* (i) Hepatitis B vaccination shall be made available after the employee has received the training required in paragraph (g)(2)(vii)(1) and within 10 working days of initial assignment to all employees who have occupational exposure unless the employee has previously received the complete hepatitis B vaccination series, antibody testing has revealed that the employee is immune, or the vaccine is contraindicated for medical reasons.

(ii) The employer shall not make participation in a prescreening program a prerequisite for receiving hepatitis B vaccination.

(iii) If the employee initially declines hepatitis B vaccination but at a later date while still covered under the standard decides to accept the vaccination, the employer shall make available hepatitis B vaccination at that time.

(iv) The employer shall assure that employees who decline to accept hepatitis B vaccination offered by the employer sign a wavier statement.

(v) If a routine booster dose(s) of hepatitis B vaccine is recommended by the U.S. Public Health Service at a future date, such booster dose(s) shall be made available in accordance with section (f)(l)(ii).

(3) *Post-exposure Evaluation and Follow-up.* Following a report of an exposure incident, the employer shall make immediately available to the exposed employee a confidential medical evaluation and follow-up, including at least the following elements:

(i) Documentation of the route(s) of exposure, and the circumstances under which the exposure incident occurred;

(ii) Identification and documentation of the source individual, unless the employer can establish that identification is infeasible or prohibited by state or local law;

(A) The source individual's blood shall be tested as soon as feasible and after consent is obtained in order to determine HBV and HIV infectivity. If consent is not obtained, the employer shall establish that legally required consent cannot be obtained. When the source individual's consent is not required by law, the source individual's blood, if available shall be tested and the results documented.

(B) When the source individual is already known to be infected with HBV or HIV, testing for the source individual's known HBV or HIV status need not be repeated.

(C) Results of the source individual's testing shall be made available to the exposed employee, and the employee shall be informed of applicable laws and regulations concerning disclosure of the identity and infectious status of the source individual.

(iii) Collection and testing of blood for HBV and HIV serological status;

(A) The exposed employee's blood shall be collected as soon as feasible and tested after consent is obtained.

(B) If the employee consents to baseline blood collection, but does not give consent at that time for HIV serologic testing, the sample shall be preserved for at least 90 days. If, within 90 days of the exposure incident, the employee elects to have the baseline sample tested, such testing shall be done as soon as feasible.

(iv) Post-exposure prophylaxis, when medically indicated, as recommended by the U.S. Public Health Service;

(v) Counseling; and

(vi) Evaluation of reported illnesses.

(4) *Information Provided to the Health care Professional.* (i) The employer shall ensure that the healthcare professional responsible for the employee's Hepatitis B vaccination is provided a copy of this regulation.

(ii) The employer shall ensure that the health care professional evaluating an employee after an exposure incident is provided the following information:

(A) A copy of this regulation;

(B) A description of the exposed employee's duties as they relate to the exposure incident;

(C) Documentation of the route(s) of exposure and circumstances under which exposure occurred;

(D) Results of the source individual's blood testing, if available; and

(E) All medical records relevant to the appropriate treatment of the employee including vaccination status which are the employer's responsibility to maintain.

(5) *Health care Professional's Written Opinion.* The employer shall obtain and provide the employee with a copy of the evaluating health care professional's written opinion within 15 days of the completion of the evaluation.

(i) The health care professional's written opinion for Hepatitis B vaccination shall be limited to whether Hepatitis B vaccination is indicated for an employee, and if the employee has received such vaccination.

(ii) The health care professional's written opinion for post-exposure evaluation and follow-up shall be limited to the following information:

(A) That the employee has been informed of the results of the evaluation; and

(B) That the employee has been told about any medical conditions resulting from exposure to blood or other potentially infectious materials which require further evaluation or treatment.

(iii) All other findings or diagnoses shall remain confidential and shall not be included in the written report.

(6) *Medical Recordkeeping.* Medical records required by this standard shall be maintained in accordance with paragraph (h)(l) of this section.

(g) *Communication of hazards to employees*—(1) *Labels and signs.* (i) labels. (A) Warning labels shall be affixed to containers of regulated waste, refrigerators and freezers containing blood or other potentially infectious material; and other containers used to store, transport or ship blood or other potentially infectious materials, except as provided in paragraph (g)(])(i)(E), (F) and (G).

(B) Labels required by this section shall include the following legend:

BIOHAZARD

(C) These labels shall be fluorescent orange or orange-red or predominantly so, with lettering or symbols in a contrasting color.

(D) Labels required shall be affixed as close as feasible to the container by string, wire, adhesive, or other method that prevents their loss or unintentional removal.

(E) Red bags or red containers may be substituted for labels.

(F) Containers of blood, blood components, or blood products that are labeled as to their contents and have been released for transfusion or other clinical use are exempted from the labeling requirements of paragraph (g).

(G) Individual containers of blood or other potentially infectious materials that are placed in a labeled container during storage, transport, shipment or disposal are exempted from the labeling requirement.

(H) Labels required for contaminated equipment shall be in accordance with this paragraph and shall also state which portions of the equipment remain contaminated.

(I) Regulated waste that has been decontaminated need not be labeled or color-coded.

(ii) Signs. (A) The employer shall post signs at the entrance to work areas specified in paragraph (e), HIV and HBV Research Laboratory and Production Facilities, which shall bear the following legend:

BIOHAZARD

(Name of the Infectious Agent)

(Special requirements for entering the area)

(Name, telephone number of the laboratory director or other responsible person.)

(B) These signs shall be fluorescent orange-red or predominantly so, with lettering or symbols in a contrasting color.

(2) *Information and Training.* (i) Employers shall ensure that all employees with occupational exposure participate in a training program which must be provided at no cost to the employee and during working hours.

(ii) Training shall be provided as follows:

(A) At the time of initial assignment to tasks where occupational exposure may take place;

(B) Within 90 days after the effective date of the standard; and

(C) At least annually thereafter.

(iii) For employees who have received training on bloodborne pathogens m the year preceding the effective date of the standard, only training with respect to the provisions of the standard which were not included need be provided.

(iv) Annual training for all employees shall be provided within one year of their previous training.

(v) Employers shall provide additional training when changes such as modification of tasks or procedures or institution of new tasks or procedures affect the employee's occupational exposure. The additional training may be limited to addressing the new exposures created.

(vi) Material appropriate in content and vocabulary to educational level literacy, and language of employees shall be used.

(vii) The training program shall contain at a minimum the following elements:

(A) An accessible copy of the regulatory test of this standard and an explanation of its contents;

(B) A general explanation of the epidemiology and symptoms of blood-borne diseases;

(C) An explantation of the modes of transmission of bloodborne pathogens;

(D) An explanation of the employer's exposure control plan and the means by which the employee can obtain a copy of the written plan;

(E) An explanation of the appropriate methods for recognizing tasks and other activities that may involve exposure to blood and other potentially infectious materials;

(F) An explanation of the use and limitations of methods that will prevent or reduce exposure including appropriate engineering, controls, work practices, and personal protective equipment;

(G) Information the types, proper use, location, removal, handling decontamination and disposal of personal protective equipment

(H) An explanation of the basis for selection of personal protective equipment;

(I) Information on the hepatitis B vaccine, including information on its efficiency, safety, method of administration the benefits of being vaccinated and that the vaccine and vaccination will be offered free of charge;

(J) Information on the appropriate actions to take and persons to contact in all emergency involving blood or other potentially infectious materials;

(K) An explanation of the procedure to follow if an exposure incident occurs, including the method of reporting the incident and the medical follow-up that will be made available.

(L) Information on the post-exposure evaluation and follow-up that the employer is required to provide for the employee following an exposure incident;

(M) An explanation of the signs and labels and/or color coding required by paragraph (g)(]); and

(N) An opportunity for interactive questions and answers with the person conducting the training session.

(viii) The person conducting the training shall be knowledgeable in the subject matter covered by the elements contained in the training program as it relates to the workplace that the training will address.

(ix) Additional Initial Training for Employees in HIV and HBV Laboratories and Production Facilities. Employees in HIV or HBV research laboratories and HIV or HBV production facilities shall receive the following initial training in addition to the above training requirements.

(A) The employer shall assure that employees demonstrate proficiency in standard microbiological practices and techniques and in the practices and operations specific to the facility before being allowed to work with HIV or HBV.

(B) The employer shall assure that employees have prior experience in the handling of human pathogens or tissue cultures before working with HIV or HBV.

(C) The employer shall provide a training program to employees who have no prior experience in handling human pathogens. Initial work activities shall not include the handling of infectious agents. A progression of work activities shall be assigned as techniques are learned and proficiency is developed. The employer shall assure that employees participate in work activities involving infectious agents only after proficiency has been demonstrated .

(h) *Recordkeeping* — (1) *Medical Records.* (i) The employer shall establish and maintain an accurate record for each employee with occupational exposure, in accordance with 29 CFR]910.20.

(ii) This record shall include:

(A) The name and social security number of the employee;

(B) A copy of the employee's hepatitis B vaccination status including the dates of all the hepatitis B vaccinations and any medical records relative to the employee's ability to receive vaccination as required by paragraph (f)(2);

(C) A copy of all results of examinations, medical testing, and follow-up procedures as required by paragraph (f)(3);

(D) The employer's copy of the health care professional's written opinion as required by paragraph (f)(5); and

(E) A copy of the information provided to the health care professional as required by paragraphs (f)(4)(ii)(B)(C) and (D).

(iii) Confidentiality. The employer shall ensure that employee medical records required by paragraph (h)(l) are:

(A) Kept confidential; and

(B) Are not disclosed or reported without the employee's express written consent to any person within or outside the workplace except as required by this section or as may be required by law.

(iv) The employer shall maintain the records required by paragraph (h) for at least the duration of employment plus 30 years in accordance with 29 CFR 1910.20.

(2) *Training Records.* (i) *Training records shall include the following information:*

(A) The dates of the training sessions;

(B) The contents or a summary of the training sessions;

(C) The names and qualifications of persons conducting the training; and

(D) The names and job titles of all persons attending the training sessions.

(ii) Training records shall be maintained for 3 years from the date on which the training occurred.

(3) *Availability.* (i) The employer shall ensure that all records required to be maintained by this section shall be made available upon request to the Assistant Secretary and the Director for examination and copying.

(ii) Employee training records required by this paragraph shall be provided upon request for examination and copying to employees, to employee representatives, to the Director, and to the Assistant Secretary in accordance with 29 CFR 1910.20

(iii) Employee medical records required by this paragraph shall be provided upon request for examination and copying to the subject employee, to anyone having written consent of the subject employee, to the Director, and to the Assistant Secretary in accordance with 29 CFR 1910.20.

(4) *Transfer of Records.* (i) The employer shall comply with the requirements involving transfer of records set forth in 29 CFR 1910.20(h).

(ii) If the employer ceases to do business and there is no successor employer to receive and retain the records for the prescribed period, the employer shall notify the Director, at least three months prior to their disposal and transmit them to the Director, if required by the Director to do so, within that three month period.

(i) *Dates*—(l) *Effective Date.* The standard shall become effective on March 6, 1992.

(2) The Exposure Control Plan required by paragraph (c)(2) of this section shall be completed on or before May 5,1992.

(3) Paragraph (g)(2) Information and Training and (h) Recordkeeping shall take effect on or before June 4,1992.

(4) Paragraphs (d)(2) Engineering and Work Practice Controls, (d)(3) Personal Protective Equipment, (d)(4) Housekeeping, (e) HIV and HBV Research Laboratories and Production Facilities, (f) Hepatitis B Vaccination and Post-Exposure Evaluation and Follow-up and (g) (l) Labels and Signs, shall take effect July 6,1992.

ENDNOTES

1   *Kennedy v. McCarty*, 803 F.Supp. 1470 (S.D. Ind. 1992).

2   *Willis v. Freeman*, 16 National Law Journal 55 (S.D. Texas 1993).

3   *Doody v. Sinaloa Lake Owners Assn., Inc.*, 494 U.S. 1016, 110 S. Ct. 1317, 108 L. Ed. 2d 493 (1990)

4   *Poe v. Ullman*, 364 U.S. 859, 81 S. Ct. 1752 (1961)

5   *Murray v. United States*, 369 U.S. 828, 82 S. Ct. 845 (1962)

6   *Coolidge v. New Hampshire*, 403 U.S. 443, 91 S. Ct. 2022, 29 L.Ed. 2d 410 (1971).

7   *Smith v. Maryland*, 442 U.S. 735, 99 S. Ct. 2577, 61 L.Ed. 2d 220 (1979).

8   *Oliver v. United States*, 466 U.S. 170, 106 S. Ct 1735, 80 L. Ed. 2d 214 (1984).

9   *California v. Greenwood*, 486 U.S. 35, 108 S. Ct 1625, 100 L. Ed. 2d 30 (1988).

10   *O'Connor v. Ortega*, 480 U.S. 709, 107 S. Ct. 1492, 94 L. Ed. 2d 714, (1987).

11   *Marshall v. Barlow*, 436 U.S. 307, 98 S. Ct. 1816, 56 L Ed. 2d 305 (1978).

12   *Donovan v. Dewey*, 452 U.S. 385, 98 S. Ct. 2408, 57 L Ed. 2d 290 (1978).

13   *Texas v. Brown*, 460 U.S. 730, 742 (1983).

14   *Illinois v. Gates*, 462 U.S. 213, 103 S. Ct. 2317, 75 L. Ed. 2d 502(1983).

15   *Jones v. United States*, 362 U.S. 257, 80 S. Ct. 725, 4 L Ed. 2d 697 (1960).

16   *Recznik v. City of Lorain*, 393 U.S. 166, 89 S. Ct. 342, 21 L. Ed. 2d 1731 (1968).

17   *Aguilar v. Texas*, 378 U.S. 108, 84 S. Ct. 1056, 108 L. Ed 2d 222 (1990).

18   *Illinois v. Gates*, 462 U.S. 213, 103 S. Ct. 2317, 75 L. Ed 2d 502 (1983).

19   *Payton v. New York*, 445 U.S. 573, 100 S. Ct. 1371, 63 L. Ed. 2d 639 (1980).

20   *Stone v. California*, 376 U.S. 483 (1964).

21   *U.S. v. Matlock*, 415 U.S. 164 (1974).

22   *Illinois v. Rodriquez*, 110 S.Ct. 2793 (1990).

23   *Chapman v. United States*, 365 U.S. 610 (1961).

24   *Schneckloth v.Bustamonte*, 412 U.S. 218 (1973).

[25] *Florida v. Bostick*, 111 S.Ct. 2382 (1991).

[26] *Maryland v. Buie*, 494 U.S. 325, 110 S. Ct. 1093, 108 L. Ed. 2d 276 (1990).

[27] *Coolidge v. New Hampshire*, 403 U.S. 443, 91 S. Ct. 2022, 29 L. Ed. 2d 410 (1971).

[28] *U.S. v. Mendenhall*, 446 U.S. 544 (1980).

[29] *California v. Hodari*, 111 S. Ct. 1547 (1991).

[30] *U.S. v. Jacobsen*, 466 U.S. 109, 113 (1984).

[31] *Mancusi v. DeForte*, 392 U.S. 364, 368 (1969).

[32] *Schmerber v. California*, 384 U.S. 757, 86 S. Ct. 1826 (1966).

[33] *Doe v. U.S.*, 487 U.S. 201, 210 (1988).

[34] *Miranda v. Arizona*, 384 U.S. 436, 86 S.Ct. 1602, 16 L. Ed.2d 694 (1966).

[35] *Boyd v. United States*, 116 U.S. 616, 96 S. Ct 524, 29 L. Ed. 746 (1986).

[36] *Brown v. Illinois*, 422 U.S. 590, 95 S. Ct. 2254, 45 L. Ed. 2d 416 (1975).

[37] *People v. Fleming*, 560 NYS 2d 50 (1990).

[38] *State v. Nielson*, 474 P2d 725 (1970).

[39] *Michigan v. Tyler*, 436 U.S. 499 (1978).

[40] *Michigan v. Clifford* , 464 U.S. 287, 294, 104 S.Ct. 641 (1984).

[41] *Gunder v. The State*, 358 SE 2d 284 (1987). See also *Corson v. State*, 241 SE 2d 454 (1978).

[42] *Weaver v. State*, 497 So.2d 1089 (1986). A related case is *Cumbest v. State*, 456 So.2d 209 (1984).

[43] *U.S. v. Price*, 877 F2d 1174 (5th Cir., 1991).

[44a] *U.S. v. Ragusa*, 664 F2d 696 (8th Cir., 1981).

[44b] *Armstrong v. City of Dallas*, 62 FEP Cases 852 (5th Cir., 1993).

[45] *Turner v. Barr*, 811 F. Supp. 1 (D.C.D.C. 1993).

[46] *University of California v. Bakke*, 438 U.S. 265 (1979)

[47] *Higgins v. City of Vallejo*, 823 F2d 351 (1987).

[48] *Burney v. City of Columbus*, No. 93-1101.

[49] *Associated Press*, June 26, 1993.

[50] *Browning v. City of Odessa*, 990 F2d 842 (5th Cir., 1993).

[51] *Hurwitz v. Perales*, 81 N.Y.S. 2d 182, 613 N.E.2d 163 (1993).

[52] *Hafer v. Melo*, 502 U.S. 21, 112 S. Ct 358 (1991).

[53] *Young v. Sherwin Williams*, 569 A.2d 1173 (1990).

[54] *Mahoney v. Carus Chemical Company*, 102 N.J. 564; 510 A2d 4 (1986).

[55] *Kilventon v. United Missouri Bank*, 865 S.W.2d 741 (Mo.,1993).

[56] *Brown v. General Electric*, 648 F. Supp. 470 (1986).

[57] *Christensen v. Murphy*, 296 Ore. 610, 678 P.2d 1210 (1984).

58   *Flowers v. Rock Creek Terrace Limited Partnership et al*, 520 A2d 361 (1987).

59   *Hauboldt v. Union Carbide Corp.*, 467 N.W.2d 508 (1991).

60   *Washington v. Atlantic Richfield Co*, 66 Ill. 2d 103, 361 N.E. 2d 282 (1976).

61   *District of Columbia v. Air Florida*, 750 F.2d 1077 (1984).

62   *Utica Mutual Insurance v. Gaithersburg-Washington Grove Fire Department*, 455 A.2d 987 (1983).

63   *Barger v. Mayor and City Council of Baltimore*, 616 F.2d 730 (1980).

64   *Cooper v. City of Honolulu*, 1st Cir.Ct. (Hawaii 1992).

65   *Cotter v. City of New York*, 36 ATLA Law Reporter 14 (New York 1992).

66   *Archie v. City of Racine*, 826 F. 2d 480 (7th Cir., 1987); 847 F. 2d 1211 (7th Cir., 1988).

67   *City of Gary v. Odie*, 638 N.E.2d 1326 (1994).

68   *James v. Prince George's County*, 418 A.2d 1173 (Md., 1980).

69   *Prince George's County v. Fitzhugh* , 519 A.2d 1285 (Md).

70   *Malcolm v. City of East Detroit*, 437 Mich. 143 (1991).

71   *Heimberger v. Village of Chebanse*, 463 NE 2nd 1363 (1984).

72   $100,000 to Widows and Orphans of Police and Fire Personnel Killed in Line of Duty.

73   *Haavistola v. Community Fire Co.*, 3 F.3d 211 (4th Cir. 1993). But also see, *Krieger v. B.C.C. Rescue Squad*, 599 F.Supp. 770 (D.Md. 1984); summarily affirmed, 792 F.2d 139 (4th Cir. 1986).

74   *Bires v. City of Mauston*, 447 N.W. 2d 100 (1989).

75   *McCarthy v. Philadelphia Civil Service Commission*, 424 U.S.645 (1976))

76   *Hameetman v. City of Chicago*, 776 F2d 636 (7th Cir., 1985).

77   *Clinton Police Department Bargaining Unit v. City of Clinton*, 464 NW 2d 875 (1991).

78   *U. S. v. City of Warren*, 759 F. Supp. 355 (E. D. Mich., 1991).

79   *Gantz v. City of Detroit*, 392 Mich. 348, 220 N.W.2d 433 (1974).

80   *Shands v. City of Kennett*, 756 F. Supp. 420 (1991).

81   *Fagiano v. Police Board of the City of Chicago*, 456 N.E. 2d 27 (1983).

82   *Civil Service Commission of Pittsburgh v. Parks*, 471 A2d 154 (Pa.1984).

83   *Smith v. City of Newark*, 128 N.J. Supp.417. 320 A2d 212 (1974).

84   *Tye v. City of Cincinnati*, 794 F.Supp, 824 (S. D. Ohio, 1992).

85   *Harris v. City of Colorado Springs*, 9 IER Cases, (BNA) 142 (Colo. App. 12/30/93).

86   *Black Firefighters Association of Dallas v. City of Dallas*, 905 F2d 63 (1990). Also see *Wards Cove Packing Co. v. Atonio*, 109 S.Ct. 2115 (1989).

87   *Paskarnis v. Darien-Woodridge Fire Protection District*, 623 N.E.2d 383 (1993).

[88] *McArm v. Allied*, 8 IER Cases (BNA) 1317 (1993).

[89] *Quinn v. Muscare*, 425 U.S. 560, 426 U.S. 954 (1976).

[90] *City of East Providence v. Local 850, International Association of Firefighters*, 117 R.I. 329, 366 A. 2d 1151 (1976).

[91] *Sellers v. City of Shreveport et al.* (1992).

[92] *Kelly v. Johnson*, 425 U.S. 238; 96 S. Ct 1440 (1976).

[93] *Hottinger v. Pope Co. (Arkansas)*, 971 F2d 127 (8th Cir., 1992).

[94] *Stalter v. City of Montgomery*, 796 F. Supp. 489 (1992).

[95] *Weaver v. Henderson*, 8 IER Cases (BNA) 431 (1st Circuit, 1993).

[96] *Fitzpatrick v. City of Atlanta*, 62 FEP Cases (BNA) 1484, 2 AD Cases 1270 11th Cir., (1993).

[97] *Shelby Township Fire Department v. Shields*, 320 N.W. 2d 306 (Mich. 1982).

[98] *Kennedy v. District of Columbia*, 249 F. 2d 492, (D.C. Cir. 1957).

[99] *Ittig v. Huntington Manor Volunteer Fire Department, Inc.*, 95 A. 2d 829, 463 NYS 2d 870 (1983).

[100] *International Association of Firefighters v. City of Salem*, 68 Or. App. 793, 684 P. 2d 605 (1984).

[101] *Chelsea Firefighters Assn. v. City of Chelsea*, Suffolk County Superior Court, (1992).

[102] *International Association of Firefighters of City of Newburgh, Local 589 v. Helsby*, 59 A. 2d 342, 399 N.Y.S. 2d 334 (1977).

[103] *Portland Firefighters Association v. City of Portland*, 478 A. 2d 297 (1984).

[104] *City of Erie v. IAFF* (1983) and *IAFF v. City of Scranton* (1982).

[105] *City of Trenton v. Trenton FF Union* (1988)

[106] *City of Sault Ste. Marie v. Fraternal Order of Police*, 163 Mich. App. 350, 414 N.W. 2d 168 (1987).

[107] *York Firefighters Local 627 v. Pennsyslvania Labor Relations Board*, 630 A2d 527 (1993).

[108] *Police Officers Assn. of Michigan v. City of Grosse Pointe Farms*, 496 N.W. 2d 794 (1992).

[109] *EEOC v. Metzger*, 824 F. Supp. 1 (D.C.D.C. 1993).

[110] *National League of Cities v. Usery*, 426 U.S. 833 (1976).

[111] *Garcia v. San Antonio Metropolitan Transit Authority*, 469 U.S. 528, 105 S. Ct. 1005 (1985).

[112] *Nalley v. Mayor of Baltimore*, 796 F. Supp. 194 (1992).

[113] *Wouters v. Martin County*, 9 F.3d 924 (11th Cir., 1993).

[114] *Horan v. King County Division of Emergency Medical Services*, 740 F.Supp. 1471 (1990).

[115] *Abshire v. County of Kern,* 908 F.2d 483 (9th Cir., 1990).

[116] *McDonnell v. City of Omaha,* 999 F.2d 293 (1983).

[117] *Keller v. City of Columbus,* 778 F.2d 1480 (1991).

[118] *Masters v. City of Huntington,* 800 F.Supp. 355, 369 (1992).

[119] *Simmons v. City of Fort Worth,* 805 F.Supp. 420 (1992).

[120] *Johnson v. City of Columbia,* 949 F.2d 127 (4th Cir. *en banc* 1991).

[121] *Renfro v. City of Emporia,* 948 F.2d 1529 (10th Cir. 1992).

[122] *Eckles v. City of Lubbock,* 846 S.W.2d 863 (Texas 1992).

[123] *Burgess v. Catawba County,* 805 F.Supp. 341 (1992).

[124] *Carlson v. City of Minneapolis,* 925 F.2d 264 (1991).

[125] *Blanton v. City of Murfreesboro,* 658 F.Supp. 1540 (1987).

[126] *Roe v. District of Columbia,* 2 AD Cases (BNA) 1632 (1993).

[127] *Frederick, MD Post,* November 24, 1993, Page C-1 and *Washington Post,* November 24, 1993.

[128] *Frederick, MD, Post,* September 4, 1993, Page B-6.

[129] *Miller v. Sioux Gateway Fire Department,* 2 AD Cases 652 (Iowa 1993). *The Sioux City Journal,* August 9, 1991.

[130] *Blanchette v. Spokane Fire Protection District No. 1,* 67 Wash. App. 499, 836 P.2d 858 (1992).

[131] *Trautz v. Weisman,* 61 LW 2614 (SDNY, 1993).

[132] *The City of Columbus v. Liebhart,* 86 Ohio App. 3d 469, 621 NE 2d 554 (1993).

[133] *Colorado Civil Rights v. North Washington Fire Protection District,* 772 P2d 70 (Colo. 1989).

[134] *Frederick, MD, Post,* December 6, 1991.

[135] *Huber v. Howard County,* 3 A.D. Cases 262 (D. MD. 1994).

[136] *Peterson v. University of Wisconsin Board of Regents,* 818 F. Supp. 1276 (1993).

[137] *Welsh v. City of Tulsa,* 977 F.2d 1415 (10th Cir. 1992).

[138] *Burris v. Arizona,* 31 G.E.R.R. 1274 (1993).

[140] *Griswold v. State of Connecticut,* 381 U.S. 479, 85 S. Ct. 1678 (1965).

[141a] *Schmerber v. California,* 384 U.S. 757, 86 S. Ct. 1826 (1966), and *Rochon v. California,* 342 U.S. 165, 72 S. Ct. 205 (1952).

[141b] *Jones v. McKenzie,* 833 F. 2d 335 (D.C. Cir. 1987); 628 F. Supp. 1500 (D.C.D.C. 1986).

[142] *Skinner v. Railway Labor Executives Association,* 109 S. Ct. 1402 (1989).

[143] *George v. Department of Fire,* 637 So.2d 1064 (1980).

[144]   *Jackson v. Gates*, 975 F.2d 648 (9th Cir. 1992) *certiorari denied.* (1993).

[145]   *National Treasury Employees Union v. von Raab*, 109 S. Ct. 1384 (1989).

[146]   *National Treasury Employees Union v. Hallett*, 776 F. Supp 680(E. D. N.Y., 1991).

[147]   *Local 194A v. Bridge Commission*, 572 A. 2d 204 (NJ App. Div 1990).

[148]   *Lovvorn v. City of Chattanooga*, 846 F.2d 1539 (Sixth Cir., 1988).

[149]   *U.S. v. Floyd*, 418 F.Supp. 724 (1976).

[150]   *Beattie v. City of St. Petersburg Beach*, 733 F. Supp. 1455 (M.D. FL. 1990).

[151]   *Gauthier v. Police Commissioner of Boston*, 408 Mass. 335, 557 NE 1374 (1990).

[152]   *Baldini v. Ward*, 550 NY S. 2d 6456 (A.D. 1990).

[153]   *Turner v. Police*, 500 A.2d 1005 (D.C. 1985).

[154]   *Murgia v. Mass. State Police*

[155]   *Western Air Lines v. Criswell*, 105 S. Ct. 2743 (1985).

[156]   *Gately v. Commissioners of Massachusetts*, 811 F. Supp 26 (1992).

[157]   *Knight v. Georgia*, 61 LW 2788 (1993).

[158]   *EEOC v. Wyoming*, 460 U.S. 226, 103 S. Ct. 1054 (1983).

[159]   29 *Code of Federal Regulations*, Section 1604.11.

[160]   *Meritor Savings Bank v. Vinson*, 477 U.S. 57 (1986).

[161]   *McKinney v. Dole*, 765 F. 2d 1129, 1138 (D.C. Cir. 1985).

[162]   *Zabkowicz v. West Bend Company*, 789 F. 2d 540 (7th Cir. 1986).

[163]   *Jew v. University of Iowa*, 747 F. Supp. 946 (S.D. Iowa 1990).

[164]   *Waltman v. International Paper Co.*, 875 F. 2d 468 (5th Cir. 1989).

[165]   *DeCintio v. Westchester County Medical Center*, 807 F. 2d 304 (2d Cir. 1986).

[166]   *Evans v. Ford Motor Co.*, 768 F. Supp. 1318, 1322 (1991).

[167]   *Wall v. A T & T Technologies*, 754 F. Supp. 1084, 1088 (1990).

[168]   *Watts v. New York City Police Department*, 724 F. Supp 99, 108 (1989).

[169a]   *Halasi-Schmick v. City of Shawnee*, 759 F. Supp. 747 (1991). See also *Ramsey v. City and County of Denver*, 907 F2d 1004 (1990) and *Hicks v. Gates Rubber Co*, 833 F.2d 1406 (1987).

[169b]   *Wright v. Methodist Youth Services*, 511 F. Supp. 307 (N.D. Ill. 1981)

[170a]   *Dillon v. Frank*, No. 90-2290 (6th Cir., Mich. Jan 15, 1992)

[170b]   *Foster v. Township of Hillside*, 780 F. Supp 1026 (1992).

[171]   *Bowen v. Department of Human Service*, 606 A.2d 1051 (1992).

[172]   *Doe v. Heinze*, Maricopa County AZ Superior Court, CV 90-21962; 35 ATLA L. Reporter 1884 (1991).

[173]   *Burns v. McGregor*, 57 FEP Cases (BNA) 1373, (8th Cir. 1992).

174 *Washington Post*, June 13, 1994; *Fire Service Labor Monthly*, June, 1994. Volume 8, No. 6

175 *Rabidue v. Osceola Refining Co.*, 805 F.2d 611 (1986).

176 *Andrews v. City of Philadelphia*, 895 F.2d 1469 (1990).

177 *Ellison v. Brady*, 924 F.2d 872 (1991).

178 *Robinson v. Jacksonville Shipyards, Inc.*, 760 F. Supp. 1486 (M.D. Fl. 1991).

179 *Harris v. Forklift Systems, Inc.*, No. 92-1168 U.S.S. Ct. (Nov. 9, 1993).

180 *R.A.V. v. City of St. Paul*, 112 S. Ct. 2538 (1992).

181 *Adams v. Alaska*, 555 P2d 235 (1976).

182 *Camara v. San Francisco*, 387 U.S. 523, 538, 87 S. Ct. 1727, 1736, 18 L.Ed2d 930 (1967).

183 *See v. Seattle*, 387 U.S. 541 (1967).

184 *Marshall v. Barlow*, 436 U.S. 307, 98 S. Ct. 1816, 56 L. Ed. 2d 305 (1978).

185 *State v. Burke*, 582 A2d 915 (1990). See also *State v. Breton*, 562 A2d 1060 (1989) and *Camara v. San Francisco*, 387 U.S. 523, 87 S. Ct. 1727, 18 L.Ed2d 930 (1967).

186 *Van Sickle v. Boyes*, 797 P2d 1267 (1990). See also *Apple v. City and County of Denver*, 390 P2d 91 (1964); and *Penn Central Transportation Co. v. New York City*, 438 U.S. 104 (1978).

187 *State v. Andrews*, 528 A2d 361 (1987).

188 *Campbell v. Seabury Press*, 614 F2d 395 (5th Cir. 1980).

189 *Worthy v. Herter*, 270 F. 2d 905 (D.C. Cir. 1959).

190 *Pell v. Procurier* , 417 U.S. 817, 94 S. Ct. 2800 (1974).

191 *Roberson v. Rochester Folding Box Co.*, 171 N.Y. 538, 64 N.E. 442 (1902).

192 *Oklahoma v. Bernstein*, Oklahoma District Court, Rogers County, January 21, 1980, Med. L. Reptr 2313,2323-2324.

193 *Connick v. Myers.*, 461 US 138; 103 S. Ct. 1684; (1983).

194 *Thomas v. Harris County*, 784 F. 2d 648 (5th Cir. 1986).

195 *McPherson v. Rankin*, 786 F. 2d 1233 (5th Cir. 1986).

196 *Firefighters Association of D.C. v. Barry*, 742 F. Supp. 1182 (1990).

197 *Wolf v. City of Aberdeen*, 758 F. Supp 551 (1991).

198 *Holland v. Dillon*, 531 NY S 2nd 427 (1988).

199 *Burgess v. Pierce County*, 918 F.2d 104 (1990). For similar cases see: *Mitchell v. Forsysth*, 472 U.S. 526 (1985); *Manhattan Beach Police Officers Association, Inc. v. City of Manhattan Beach*, 881 F.2d 816 (1989); *Roth v. Veterans Administration*, 856 F.2d 1401 (1988); and *Pickering v. Board of Education of Township High School District*, 391 U.S. 563 (1968).

[200]   *Watts v. Alfred*, 294 F.Supp. 431 (D.C.D.C., 1992).

[201]   *Local 2106, IAFF, v. City of Rock Hill*, 660 F.2d 97 (4th Cir. 1981).

[202]   *Pesek v. City of Brunswick*, 794 F.Supp. 768 (N. D. Ohio, 1992).

[203]   *Local 1812, American Federation of Government Employees v. U.S. Department of State*, 662 F. Supp. 50, D.C.D.C., (1987).

[204]   *Glover v. Eastern Nebraska Community Office of Retardation*, 867 F.2d 461, 464.

[205]   *Anonymous Fireman v. City of Willoughby*, 779 F.Supp. 402 (N. D. Ohio 1991).

[206]   *Doe v. District of Columbia*, 796 F. Supp. 559 (D.C.D.C. 1992).

[207]   *Severino v. North Fort Myers Fire Control District*, 935 F.2d 1179 (11th Cir. 1991).

[208]   *Cavagnuolo v. Baker & McKenzie*, 1994 FEP Summary (BNA, 12 NYSHRD 1994).

[209]   *Doe v. Roe*, 599 NYS 2d 530 (1993).

[210]   *Urbaniak v. Newton*, 24 Cal. Rptr. 2d 333 (1993).

[211]   *Doe v. Borough of Barrington*, 729 F. Supp. 874, (W. D. Wisconsin 1988).

[212]   *Whalen v. Roe*, 429 U.S. 589 (1977) and *Thornburgh v. American College of Obstetricians and Gynecologists*, 476 U.S. 747 (1986).

[213]   *McGann v. H&H Music Co*, 946 F.2d 401 (5th Cir., 1991) and *Owens v. Storehouse, Inc.*, 773 F. Supp. 416 (N. D. Georgia, 1991).

[214]   *Legault v. Russo*, 64 FEP Cases (BNA) 170 (1994).

[215]   *UAW v. Johnson Controls, Inc.*, 111 S. Ct. 1196 (1991).

[216]   *Dee v. U.S.*, 111 S. Ct. 1307 (1991).